High Egg Production by Individual Hens in Pens and Flocks
A Complete Guide To Profitable Production of Eggs – How To Breed For Productivity

by Homer W. Jackson and Grant M. Curtis

with an introduction by Jackson Chambers

This work contains material that was originally published in 1922.

This publication is within the Public Domain.

This edition is reprinted for educational purposes
and in accordance with all applicable Federal Laws.

Introduction Copyright 2017 by Jackson Chambers

Self Reliance Books

Get more historic titles on animal and stock breeding, gardening and old fashioned skills by visiting us at:

http://selfreliancebooks.blogspot.com/

Introduction

I am pleased to present yet another title on Poultry.

The work is in the Public Domain and is re-printed here in accordance with Federal Laws.

As with all reprinted books of this age that are intended to perfectly reproduce the original edition, considerable pains and effort had to be undertaken to correct fading and sometimes outright damage to existing proofs of this title. At times, this task is quite monumental, requiring an almost total "rebuilding" of some pages from digital proofs of multiple copies. Despite this, imperfections still sometimes exist in the final proof and may detract from the visual appearance of the text.

I hope you enjoy reading this book as much as I enjoyed making it available to readers again.

Jackson Chambers

CONTENTS

Introduction 3
Explanation of Color-Plate Frontispiece.. 4

PART I

CHAPTER I
Possibilities in High Egg Production... 5

CHAPTER II
The Hen's Egg Organs and How They Function... 13

CHAPTER III
Egg Type and Capacity... 17

CHAPTER IV
Breeding and Management of High Egg Producers...................................... 24

CHAPTER V
Profitable Production of Market Eggs... 36

CHAPTER VI
High Egg Production by Individuals and Pens... 50

CHAPTER VII
High Egg Production by Flocks.. 54

CHAPTER VIII
The Production Possibilities of Different Breeds... 64

CHAPTER IX
Diseases of Egg Organs: Their Cause, Prevention and Treatment............. 73

PART II

CHAPTER I
A Visit to Wyckoff's Grandview Poultry Farm... 77

CHAPTER II
Where Exhibition Quality and High Egg Production Are Combined....... 84

CHAPTER III
A High-Producing Strain of Barred Rocks and How It Was Bred............. 91

CHAPTER IV
Noteworthy Achievement in Production of 300-Eggers............................... 98

CHAPTER V
High Egg Production and Superior Table Quality Combined..................107

CHAPTER VI
How Some Well-Known Breeders Have Developed High Egg-Producing Strains....113

PART III

CHAPTER I
Egg-Laying Contests and Their Lessons..126

CHAPTER II
Poultry Keeping on the Pacific Coast...135

CHAPTER III
Trap Nesting, Pedigreeing, Certification, Etc. 143

CHAPTER IV
Special Articles on Breeding for High Egg Production..............................152

CHAPTER V
Special Articles on Selection and Management of High Producers, Etc. ..160

CHAPTER VI
Opinions of Scientists on Inheritance of Fecundity....................................170

Index 176

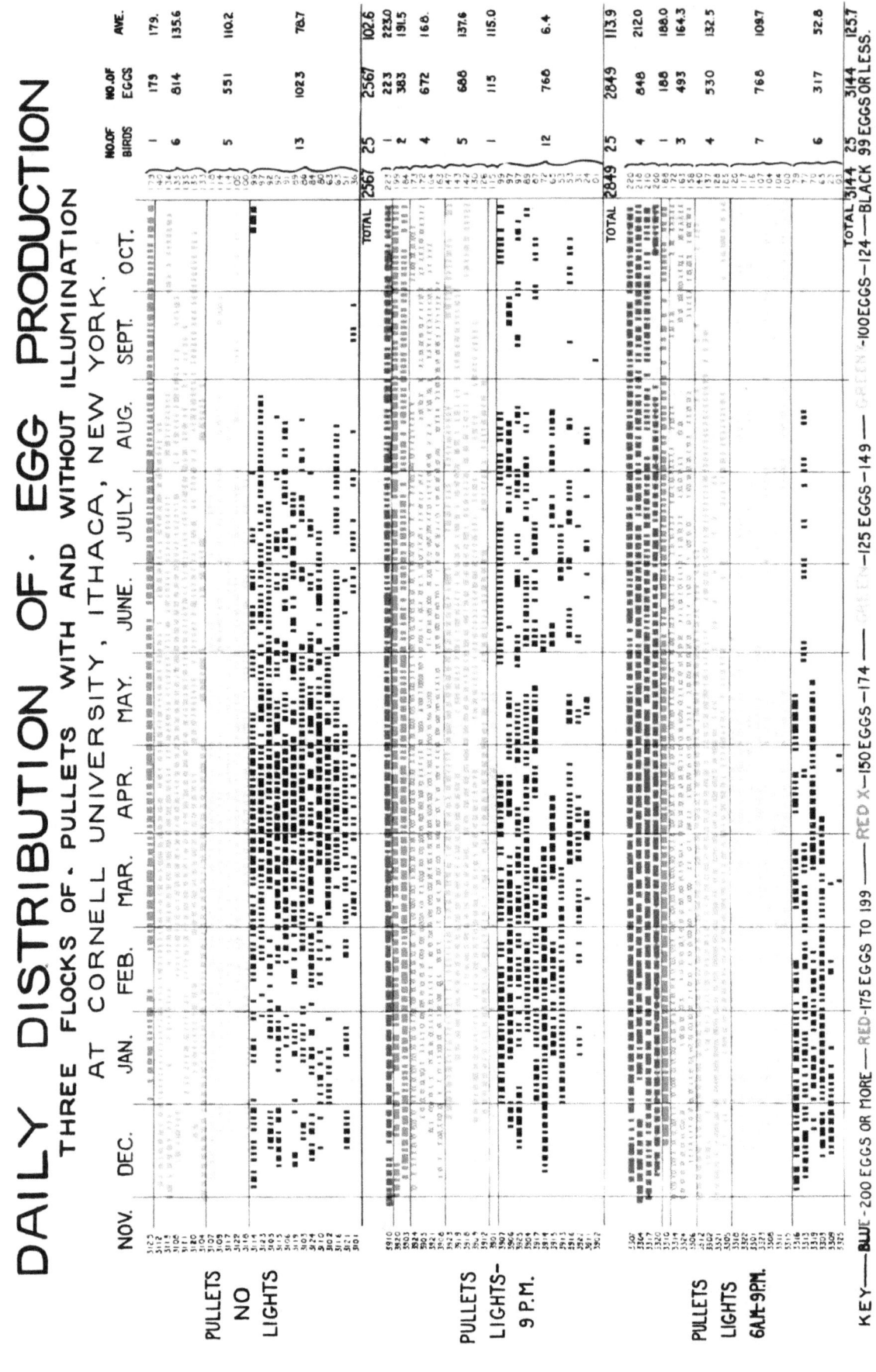

INTRODUCTION

THE Census of 1920 reveals the fact that annual egg production on farms averages only about 55.2 eggs per fowl. If ten per cent is added to this number to cover the usual (and unnecessarily high) proportion of males in farm flocks, the average still is pitifully low. Such incomplete data as are available show that average egg yields are much higher in the flocks of specialists, whether on general farms or on commercial poultry "plants," but they still fall far short of the production that may readily be secured—that MUST be secured if the poultry keeper is to realize adequate returns on his investment in capital, time and labor.

Speaking in general terms, the first 70 to 90 eggs laid by each hen in a commercial flock go to pay for feed and other expenses. It is only from the eggs laid in excess of this practical minimum that a profit can be realized; and no skill in management or economies in the purchase or use of feeds will direct a satisfactory stream of profits into the poultryman's pocket unless the production of his fowls reaches a figure considerably above that minimum.

The average farm flock doubtless returns a fair profit to its owner, in spite of its low level of productiveness, being enabled to do this through an extremely low maintenance cost. To the poultry specialist, however, with his greatly increased expenses for feed and overhead, an egg yield even double that of the average farm flock means little net income. The present widespread interest in high egg production therefore, should be considered not as a fad, but as an indication of a general awakening to its vital importance in commercial poultry keeping.

The average poultry keeper who comes to a realization of the possibilities for increased yields in his present flock through the adoption of simple, readily applied methods of management and of the further increases that can be brought about in succeeding years by systematic breeding, will quickly advance his average per hen to what a few years before he may have regarded as an impossible figure. As a matter of fact, many practical poultry keepers who could double their present net income are indirectly being held back from taking the easy and effective steps that would bring this about, by a confirmed skepticism regarding all production records that noticeably exceed their own. That some claims have been made that were unsubstantiated by facts may be conceded. There are enough high records of unquestionable authenticity, however, fully to establish the fact that averages double those secured in ordinary flocks are not only possible but readily attainable by an earnest and well-informed person.

Bulletin No. 329 of the New Jersey Experiment Station, reporting statistics secured in an unusually thorough survey of 150 poultry farms in that state, shows that the average production on these supposedly well-managed commercial farms is 109 eggs per bird—only about 20 above the self-supporting minimum in that locality and for the particular year of the census. Extension Bulletin No. 8 of the Connecticut Agricultural College, reporting a similar survey of 42 Connecticut poultry farms, gives the yearly average of eggs per hen at only 97 eggs. Contrast these figures with the averages secured in numerous large commercial flocks of superior breeding, under up-to-date methods of management, as presented in the chapters of this book, to see how far short of his possibilities the average poultry farmer is falling. It is for the purpose of bringing home to poultry keepers generally the economic importance of increased production, of showing what it is practicable to accomplish along this line, and of presenting down-to-date methods by which high and higher production may be secured, that "High Egg Production by Individual Hens, Pens and Flocks" was written.

For the convenience of the reader, the book is divided into three distinct sections. Part I is devoted to a discussion of what is possible and practical in high egg production and how it may be secured through a working knowledge of the laws of breeding and the adoption of correct methods of feeding and management, also by systematic culling. Part II contains chapters descriptive of the methods of men who have been conspicuously successful in this line of human endeavor. Through the development and distribution of high-laying stock these men have benefited the whole industry, and their successes should prove an inspiration to others. Part III contains chapters on egg-laying contests and their lessons and on the poultry industry of the Pacific Coast, where the greatest development of commercial poultry keeping that the country has ever known is even now taking place. Additional chapters reproduce in whole or in part a number of important articles that have appeared in recent issues of Reliable Poultry Journal, also experiment station bulletins of recent publication, whose practical value clearly entitles them to a permanent place in poultry literature. These corroborate and supplement the general discussion of the subject in Part I.

EXPLANATION OF COLOR-PLATE FRONTISPIECE

By PROFESSOR H. E. BOTSFORD, Poultry Department, Cornell University, Ithaca, N. Y.

ARTIFICIAL illumination as a factor in obtaining an increase in egg production is now recognized as of considerable importance when properly used on certain classes or grades of birds.

The four-color frontispiece is a reproduction of a chart prepared in the Poultry Department of Cornell University to show the result of using artificial illumination on two groups of pullets compared with a check pen on which no lights were used. One group had lights from twilight to 9 p. m., and the other from 6 a. m. until dawn and from twilight to 9 p. m. Each group is broken up into smaller groups and colored to signify the production group into which it falls. The key to this color grouping will be found at the bottom of the chart. There were twenty-five birds in each of the three pens. The bird's number, the daily distribution of production and total production of each bird is found by following each horizontal column across the chart. The number of birds in each color group is found in the first vertical column of heavy print at the right. The total number of eggs and the average of each color group are in the last two columns.

In order to see more quickly what the chart illustrates, observe that all birds laying less than 150 eggs are included in the green groups or below. The birds above the green group laid 150 eggs or over. For the purpose of this discussion, this division is made because a flock averaging 150 eggs or better is considered a good-producing flock. We shall see how the use of artificial illuminaton took birds out of the lower producing groups and placed them in the 150 group or above.

Without lights only one bird, or but four per cent of the entire flock, laid over 150 eggs. Referring to the group on which lights were used to 9 p. m., we find that 28 per cent laid over 150 eggs, the seven birds averaging 182.6 eggs, and notice further that this increase is accompanied almost entirely by a corresponding decrease in the number of birds represented by green. That is, increasing the length of day by artificial illumination has enabled all birds except the lowest producers, or those laying less than 100 eggs, to lay about 25 more eggs per bird per year. The green X birds, or those laying 100-124 eggs, in the no-light group, have practically disappeared where lights were used to 9 p. m. Apparently, lights have made it possible for these birds to pass along to the green group. Those in the green group under no lights have gone forward and swelled the number of red X and red birds, while the best bird under no lights, which laid 179 eggs, is surpassed by the best bird having lights to 9 p. m., which found conditions such that she produced 223 eggs. The black group remains the same except for one bird which moved along and laid 115 eggs, thus being the only bird in the green X group.

Birds which were uniformly poor apparently were not able to increase their production under lights to 9 p. m. only, which explains why the black group remains about the same in number.

Referring to the third table, it is obvious that even a small amount of morning light will accomplish much good in increasing egg production. Here we find 8 birds, or 32 per cent, laying over 150 eggs and averaging 191.1 eggs, and further observe that the blue has increased by three birds, these having passed up from the green and red groups. An interesting point is brought out due to the additional hour or so of morning light. Only six birds remain below 100 eggs. These six average 52.8 eggs. Apparently this is just what is needed to cause the other half dozen to move to the group above and average over twice as many eggs.

A very poor producer is not likely to make a better production as a result of artificial illumination. Morning light however, furnishes just the incentive necessary to permit many birds to draw away from their poorer associates and make a much more creditable showing than they would otherwise be capable of making. Light in the morning is beneficial in increasing production on this class of birds.

The chart points out another very important factor—the way in which illumination influences the season of production. This is especially noticeable with the poor birds represented by black. With no lights the heaviest production, as evidenced by the two months containing the greatest amount of black, is during March and April. Using lights to 9 p. m., the months of heaviest production are February and March and with lights from 6 a. m. to 9 p. m. the heaviest production falls during December and January. Lights then separate the good from the poor and allow the poor producer to lay during the winter and early spring, which is the best time for the poultryman to receive her eggs, inasmuch as she will lay but a few at the best. To a considerable extent the season of heavy production is advanced in the other groups.

There are two factors known to have an influence on total production per bird and which artificial illumination materially assists in bringing about: Precocity or early laying, and persistency or late laying. The bird which gets an early start as a layer will usually lay over a long period and hence make a higher production than if she is late in starting or ceases to lay early in the summer or fall. This does not apply to very poor birds and, as pointed out, those in the black group began early but lacked the staying power to continue, when lights were used from 6 a. m. to 9 p. m.

Referring to the chart, it will be noticed that lights to 9 p. m. did not influence the time of starting, although such birds as were laying were producing heavier through December and January. There were several more birds laying during September and October, however. Light from 6 a. m. to 9 p. m. resulted both in more birds producing in December and January and also more producing at the end of the year, except the black group. Illumination in this case at least assisted the fair to good producers by allowing them to get an early start, lay heavily through the year and hold up longer the following fall, than where birds were much later in starting, due to no lights. A poor bird is likely to cease laying earlier the following season when lights are used during fall and winter because she lays more heavily then, and comes nearer to the idea that a bird will lay only so many eggs. The fallacy of this idea as pertaining to all birds is shown by the results on the red and blue groups.

Summing up, the chart shows that artificial illumination when applied and when accompanied by the right feeding, housing and other good management factors, results in:

1st. Separating the good from the very poor producers.

2nd. Increasing the production per hen of all except the poorest layers.

3rd. Advancing the season of heavy production.

4th. Advancing the time at which birds start to lay.

5th. A more steady and longer period of laying.

(NOTE:—In this experiment, in addition to these pullet flocks, there were three flocks of yearling hens, handled in the same manner and, while the frontispiece shows only the results secured with pullet pens, those with hens were quite similar, differing only in the generally lower degree of production naturally to be expected of older birds. This experiment is most interesting and instructive but the reader should be cautioned against interpreting too literally its bearing on such details as time of day for using lights and the length of the lighted period. For complete information on these and other important factors in the successful use of lights he should secure the valuable book recently issued by Reliable Poultry Journal Publishing Company, entitled "The Use of Artificial Light To Increase Winter Egg Production." The price of this large, well-illustrated book is $1.50 and in it will be found the most complete and reliable information available on this important subject.—H. W. J.)

PART I

CHAPTER I

Possibilities in High Egg Production

High Egg Production Is Determined by Inherent Capacity and the Skill of the Caretaker—Great as Have Been the Advances Made to Date, Due to Improved Methods of Feeding and Management, There Is No Reason To Believe that the Practical Limit Has Been Reached in Either Individual, Pen or Flock Records

THE degree of success achieved by any poultry keeper is almost directly in proportion to the average production of his fowls. Under exceptional circumstances, as on farms where the flock can pick up a large part of its living, a nominal profit can be shown even on a low average egg yield, but the total income so realized will be insignificant. To make egg production worth while as a source of income or to secure adequate returns on the capital and labor invested, there must be a relatively large income per hen.

Even among many experienced poultry keepers, there often is a lack of appreciation of what high production really means in the way of increased profits. It is quite probable that this largely grows out of the general tendency to handle the laying flock as a unit and in a routine way, simply trying to give the birds good care and approved rations, and blindly letting events take their course, hoping that the balancing of accounts at the end of the year will show a reasonable profit on the enterprise. Under favorable conditions that usually is the case. But each new poultry survey taken brings out the fact that even in generally prosperous years there are many in every community who produce eggs below cost during a considerable part of the year, and some whose balances for the entire year are on the wrong side of the ledger.

Elsewhere in this book will be found data and concrete illustrations showing in a most impressive manner the vital importance in any poultry enterprise, whether conducted on a large or a small scale, of securing the maximum production of which the fowls are capable and of tolerating in the flock none that falls below the level of production where a reasonable profit is assured. This initial chapter is designed to show just what has been accomplished to date in the way of developing the productiveness of fowls, and to indicate the possibilities for further improvements that are open to earnest, industrious poultry keepers everywhere.

Productiveness of Original Fowl

If only to afford a basis for comparison, it would be of interest to know what was the productive capacity of the original fowl from which the breeds of the present day are descended. That knowledge is not available, but there are reasons for believing that the wild races of fowls that now exist, such as the jungle fowl, are capable of producing many more eggs than is commonly believed to be the case. Authentic records prove that various wild birds may and do produce eggs equal to or even in excess of the number produced by the average domestic hen. For example, Robinson in his book "Fundamentals in Poultry Breeding,"* states that "The wild Mallard Duck in captivity lays from 80 to 100 eggs. Quail under observation in scientific experiments have laid nearly 100 eggs. And such small birds as the wryneck and the house sparrow have been known to lay, in the case of the former, 48 eggs, and in the case of the latter, 51 eggs, in succession, when the eggs were removed as fast as laid. * * * * All the evidence confirms the conclusion that the original species was capable of giving immediately in domestication more eggs than the average hen lays at the present time."

It is not necessary, therefore, to assume that the ancestors of the domestic fowl were of extremely limited productiveness and that domestication has, in itself, brought about a marked increase. On the contrary, it appears more reasonable to conclude that, speaking generally, fowls have in this respect derived little if any direct benefit from domestication. In point of fact, it is difficult to see how the precarious and unbalanced rations of the average farm yard, the generally unsanitary conditions of poultry houses and the discomforts of winter in the Temperate Zone, offer any advantages over natural conditions in tropical or semitropical countries, which, it is commonly believed, were the home of the ancestors of domestic fowls.

Some Remarkable Individual Egg Records

Those who look upon the "200-egg hen" as a product of modern skill in breeding and management will learn

AUSTRALIAN RECORD LAYER
This Black Orpington has a record (semiofficial) of 339 eggs in 365 days—the highest record yet made in Australia.

---

*Published by Reliable Poultry Journal Publishing Co. Quincy, Ill.

with surprise that heavy layers are on record as far back as poultry literature carries us, though actual records are available only within a comparatively recent period. Buffon's Natural History, published in 1812, states that "a good hen lays a hundred eggs between spring and autumn," though no actual records are submitted as direct evidence. Moubray's Treatise, published in London in 1830, mentions a "half-bred Poland hen" that began to lay on December 28, 1805, and on March 1 (63 days) had laid 56 eggs. Miner's "Domestic Poultry Book," published in Rochester, N. Y., 1853, says: "Our common breeds usually lay from eighty to one hundred eggs in a year, as an average of what a flock of twenty to fifty will produce. We often see notices in the papers of instances where much larger numbers are laid, but such cases are exceptions to the general rule, as far as our native varieties are concerned. Some of the Asiatic breeds have been known to lay two hundred or more eggs in a year, and it is a settled point that some of our imported fowls do much exceed, in the number of eggs laid, any of our native tribes."

WHITE LEGHORN WITH RECORD OF 336 EGGS
The Hollywood S. C. W. Leghorn that laid 336 eggs in 365 days—the highest record made in the United States to date.

Early in the seventies, with the appearance of periodicals devoted exclusively to poultry or to poultry and pet stock, apparently authentic individual hen records became numerous. A few of these are reproduced here. Lacking trap nests however, reliable twelve-month records are extremely rare and we can only conjecture what percentage of unsuspected 200-eggers or perhaps even 300-eggers this device might have revealed in the flocks of skillful poultry keepers half a century or even a century ago.

F. H. Blanchard in "The Pet Stock, Pigeon and Poultry Bulletin" (Jan., 1873), writes that he has a Light Brahma that commenced to lay February 2, when six months and 18 days old. "For some three months she laid every day, afterward she would sometimes miss a day, never more, up to the 16th of August, in which time, say 195 days, she laid 182 eggs." In the January, 1875, issue of the same publication, W. M. Miner reports: "Last October I imported some Black Cochins from China; one hen only, lived; commenced laying early in November and laid up to April, 125 eggs." In "The Poultry World" of October, 1873, Mr. Miner states that he had "a Dark Brahma hen that laid, the past season, 111 eggs in 128 days." In the light of present knowledge it is safe to say that hens of so high-laying intensity needed only to be hatched reasonably early and properly handled to make high twelve-month records practically certain.

Three-Hundred-Egg Hens

The first reference to the 300-egg hen that has come to our notice is by L. Wright who, in the November, 1872, "Pet Stock, Pigeon and Poultry Bulletin" speaks of a friend having "a Brahma pullet which laid nearly 300 eggs in one twelve-month." It is interesting to note that there appears to have been an effort to make a systematic use of this hen in breeding for high production as the writer goes on to say that "the quality descended to several of her progeny, and I have since found other instances which prove conclusively that a vast improvement might easily be effected in nearly all our breeds."

As soon as the use of trap nests became general, reports of 200-egg hens increased rapidly in number, and again and again the latest "highest record" has been eclipsed by one still higher until now 300-egg hens are as common as 200-eggers were some years ago. The first definite claim to a 300-egg record that we have found is for "Rebecca," a Light Brahma which, in 1875-76, according to I. K. Felch, laid 313 eggs in 333 days. And it was another Felch-strain Light Brahma that, in 1917-18, made a record of 325 eggs in 365 days, her owner having made affidavit to the accuracy of this report.

The first official 300-egg record was made at the Oregon Experiment Station in 1912-13, where one of Professor Dryden's "Oregons" reached 303. In that year the same record is claimed for a French hen of unknown origin, though her production was determined by the "finger test," not by trap nesting. Another 300-egg record was made at the North American (semiofficial) Contest in 1914-15, where a White Leghorn was credited with 314 eggs. There are no more officially reported 300-eggers until 1916-17, when a White Plymouth Rock reached 304 at the National (Mo.) Contest. This year also, two Black Orpingtons and five White Leghorns were recorded in different Australian Laying Contests, whose records ranged from 300 to 315.

In 1917-18 a White Wyandotte at the Storrs Contest made a record of 308; a White Plymouth Rock at Vineland made a record of 301; and a White Leghorn at the All-Northwest Contest laid 311. The following Australian records are reported for this same year: Black Orpingtons, 335 (two), 326, 320, 312, 305; Buff Orpington, 312; White Leghorns, 332, 330 and 301. This year also, a 317-egg White Leghorn appeared in a New Zealand contest.

The year 1919-20 apparently was not so favorable to the production of high records but two Leghorns at the Western Washington Contest made records of 315 and 312, respectively, a White Leghorn at the semiofficial American (Kansas) Contest laid 306, another in Australia made the same record, and a White Orpington at the American reached 303.

At the present time the highest official record made in the United States is held by a Hollywood White Leghorn which laid 315 eggs in 365 days at the Western Washington Contest. Another Hollywood hen has a private record of 336. The world's record to date is held by a Buff Orpington which at the North American (semiofficial) Contest, 1920-21, laid 343 eggs in 365 days. For details in regard to laying contests and how records are there secured, see Chapter VI, also Chapter I, Part III.

While this book does not deal with ducks it may be stated as a matter of interest and of completing the record of what has been accomplished to date in the way of securing high production, that an Indian Runner Duck at Malvern, Pa., is credited with laying 358 eggs in 365 consecutive days, while eight ducks of this breed averaged 320 as a flock, in the same time.

In addition to the foregoing "official" performances, numerous breeders have made equally good or better private but well-authenticated individual records. Information in regard to many of these is presented in some detail in Part II.

Official, Semiofficial and Private Records

Reference has already been made to different classes of laying records. "Official" records are those secured under the supervision of heads of poultry departments in state agricultural colleges and experiment stations. Such records may be for fowls that are owned by college poultry plants or are entered in laying contests. "Semiofficial" records are those which are made at public laying contests, controlled by private individuals or business concerns. "Private" records are those supplied by poultry keepers who trap-nest their own birds or are otherwise able accurately to record daily production.

TYPICAL BRED-TO-LAY STOCK PRODUCED ON POULTRY PLANT OF UNITED STATES DEPARTMENT OF AGRICULTURE, LOCATED AT BELTSVILLE, MD.

These exhibition-quality birds—every one of high-producing ancestry—show what has been accomplished at the Government Experimental Farm by way of breeding laying strains to meet standard requirements. Starting at top of page and reading from left to right the first hen (No. 3925) laid 206 eggs in her pullet year. She is out of Hen No. 2071 (196 eggs); sire out of Hen No. 514 (213 eggs). Hen No. 408, in middle, laid 214 eggs in pullet year and 779 in 5 years. Hen at right is No. 514, with a record of 213 eggs in her pullet year, an average of 178 for 3 years. Has a daughter with a record of 206 eggs, while a son sired 16 daughters which averaged 184. Second row—Male on the left is out of Hen No. 408 (second hen in top row). His sire was out of hen with record of 202 eggs. The dam of the cockerel in the middle laid 231 eggs in pullet year and cockerel was sired by son of Hen No. 514 (third in top row) with record of 213 eggs. Cockerel on right was sired by son of hen with record of 202 eggs; his dam laid 190 eggs in her second year. Third row—First Rhode Island Red has a record of 205 eggs; the other, 206. The first Plymouth Rock (cockerel bred) laid 236 eggs in her pullet year and the second, 197.

There is a disposition in some quarters to discredit private records because of the opportunity that unquestionably exists for falsifying them. We doubt, however, whether any official or semiofficial record is so thoroughly safeguarded that it could not also be falsified if some one should set out deliberately to do it, and in passing upon the records to be used in this book it has seemed fairer to rule that, whether official or private, those are entitled to recognition that are submitted by persons of good standing, or that are vouched for by responsible persons or supported by sworn statements. Private records thus authenticated, if they approximate official records, are of value as confirming the latter, and they often are secured under conditions as regards management, size of flocks, etc., that makes them of exceptional practical value. If they exceed official records that, in itself, is not sufficient ground for questioning their accuracy, as it is conceded that private records not only may but should exceed official ones (other things being equal),

ONE OF THE HIGHEST SCORING EGG CONTEST PENS ON RECORD

This pen of S. C. W. Leghorns entered by P. B. Towne of Washington in the 1917-18 All-Northwest Contest at Pullman, Wash., made a record of 1,261 eggs, an average of 252 each.

since it is common knowledge that the conditions under which public egg-laying contests are conducted hinder rather than favor maximum production.

It is hardly to be expected that "disinterested" strangers, even though intensely loyal to the institution by which they are employed, will watch over and care for the birds with the close, sympathetic attention to their welfare that the owner would delight in giving. Moreover, the uniformity of conditions maintained between the different pens, the adherence to an arbitrarily fixed ration without regard to its adaptability to individual requirements and, finally, the constant interruptions to which contest birds are subjected by the stream of visitors passing by or through the pens all are handicaps that private operators need not meet, and with the elimination of which marked increase in production should be secured.

Some Records of Pen and Flock Production

If what has already been said in regard to the natural high productive capacity of fowls is true, it is reasonable to expect that good flock records must have been made long ago, whenever owners were willing to make an earnest effort properly to feed and care for their fowls. Very few such records, however, have been handed down to the present time. Decision as to whether this is due to the fact that they were not made, or to imperfect means of recording and preserving them, will depend largely upon individual viewpoint. It is significant, however, that with the advent of poultry papers, affording a convenient, popular means of recording results, extremely good pen and flock records immediately became common.

In the September, 1872, issue of "The Poultry World" appears an interesting communication from Samuel Seeley, of Connecticut, from which the following quotation is taken: "I have kept fowls for more than forty years, but have never kept an accurate account of the number of eggs my hens laid until within a few years. I have kept account of the number of eggs sold during each year since 1839. I have also kept an accurate account of the number of bushels of grain the fowls have consumed each year, and the market price per bushel. I see by my poultry book that I had 60 hens in 1839, and sold that year 5,501 eggs. In 1842 I had 92 hens and sold 8,756. In 1853, February 12, I counted the fowls and found 380 hens and 13 cocks. Sold 29,383 eggs." While the number of eggs sold during each year for which production is given averages less than 100 per hen, it is to be noted that eggs used and set are not included. A small flock from which 90-odd eggs per hen are sold in a year must have reached a fairly good total production.

F. H. Corbin, an early breeder of Barred Rocks, in his book "Plymouth Rocks" refers to a customer, S. F. Peck, who on the 20th of May, 1874, purchased one Plymouth Rock cockerel and eight pullets. "And from these eight pullets he obtained, during the succeeding year, one thousand, eight hundred and forty-one eggs—an average of two hundred and thirty each, or nineteen and one-sixth dozens." He also published a statement by Rev. J. M. Bates to the effect that his pullets, "in 1878, laid at the age of four and one-half months. They also averaged two hundred and twenty-five eggs each during the following year and one of them laid seven eggs in four days; that is to say, three out of the four eggs had double yolks."

In the seventies interest in breeds and their relative productiveness apparently was just as keen as now and the poultry papers of that time contain the records of a number of experiments in which the production of breeds then popular was compared. In some instances this work was conducted with painstaking thoroughness and the results secured certainly indicate that hens in those times only needed an opportunity in order to make extra-good records. It is important to bear in mind in this connection that at that time the requirements of high egg production were but imperfectly understood and, without doubt, the hens experienced much greater difficulties in making records than is now the case. As showing what could be done, not with one variety but with many, the following, by H. Langdon, in "The Pet Stock, Pigeon and Poultry Bulletin," March, 1873, is of decided interest:

"I have kept fine poultry for 21 years, and in 1869 and 1870 tried to find out which breed laid best. I send you a table of results. I kept a cock and four hens of each in yards 20 by 10 feet from May 1 to October 1, and let them run the rest of the year, after my garden was done. Fed soft food and table scraps with meat in mornings; corn and other grain in evenings. Hens all two years old; cocks, one year. The figures are all taken from my egg book." (The figures referred to give the production of each pen by months. In the following table only the totals for each year are given to save space. The columns of averages are added to emphasize the truly excellent production secured half a century ago.)

A Breed Test in 1869-70

	1869		1870	
	Eggs	Average	Eggs	Average
S. S. Hamburgs	757	189		
Spanish	687	171	529	132
Minorcas	736	184		
S. P. Hamburgs	898	224.5	828	207
Light Brahmas	470	117.5	327	81.7
Dominiques			494	123.5
Houdans			647	161.7
W. Leghorns			637	159
Black Polish			459	114.7

Another report of a similar character by Wm. A. Roosevelt, appears in the same periodical in March, 1875:

LAID 86 EGGS IN 80 DAYS

Ida-U, White Orpington pullet owned by the Poultry Department of University of Idaho, laid 86 eggs in 80 days. Is believed to have laid two eggs in one day on several occasions before trap nesting began.

"Being thrown out of employment by the panic of 1873, I set to work in December of that year with a stock of select poultry, with a view of satisfying myself as to which was the best laying breed. I purchased one cock and five pullets of each of the varieties listed in the table below; fed them with the most stimulating feed I could think of, and placed them in houses constructed by myself, unwarmed, where each flock had a run of about two acres. The runs were entirely grass fields, and the houses were each six by fourteen feet, and made of inch-grooved boards. By the 15th of January I apparently had every hen 'on the lay,' and on Sunday, February 1, 1874, commenced my monthly egg account. I have tabulated my results (from Feb. 1, 1874 to Jan. 31, 1875) as follows:

A Private Breed Test Conducted in 1874-75

Variety	Feb.	Mar.	Apr.	May	June	July	Aug.	Sept.	Oct.	Nov.	Dec.	Jan.	Total	Hen Avg.
White Leghorns	79	96	99	108	96	78	64	41	45	21	72	61	860	172
Brown Leghorns	97	103	96	94	90	80	58	37	33	34	80	73	875	175
Black Hamburgs	74	87	104	101	111	90	67	20	18	31	68	59	830	166
S. S. Hamburgs	76	95	102	104	83	77	52	45	21	49	60	70	834	167
G. P. Hamburgs	83	93	110	110	112	96	69	50	39	55	77	72	966	193
S. P. Hamburgs	92	99	122	124	128	111	75	40	27	73	74	81	1046	209
Houdans	58	84	101	99	80	72	59	49	41	39	50	48	780	156

In "Farm Poultry," under date of December, 1891, Charles B. Travis submits the following summary of the production of his fowls covering a period of eight years, for which period the average production per hen was almost 137 eggs.

Year	Number of Hens	Total Eggs	Average per Hen
1884	35	4,647	132
1885	35	4,680	134
1886	30	3,849	128
1887	20	2,994	149
1888	35	4,578	130
1889	21	2,971	141
1890	30	4,109	139
1891 to Oct. 1	25	3,554	142

In 1889, and again in 1894, the "National Stockman and Farmer" conducted egg-laying contests in which excellent egg yields were reported. In the first contest the best pen consisted of 6 Brown Leghorns with an average of 222.5 each, the second pen being White Wyandottes with an average of 200.5 eggs each. In the second contest the average production of the first pen (White Plymouth Rocks) was 286 eggs per hen, the second (8 crossbred pullets), 283. A more detailed report of this contest and of the results secured will be found in Chapter I, Part III.

Still Higher Records Possible

In the light of the records made fifty years and more ago, high egg production, as the term is commonly used, cannot by any means be claimed as a recent development. However, because of the fact that there now is a far better understanding of the principles of breeding for egg production, and that marked improvement in methods of feeding and general care have been introduced, high producers now are common instead of exceptional, and practically every earnest poultry keeper has what would have been considered a phenomenal flock years ago.

As a matter of fact, production has been pushed to so high a point among the flocks of poultry specialists that many are seriously asking whether further advances are possible. High as some of these records are, however, there is no reason to believe that the limit of what is possible has been reached, either in individual records or in flock averages. Just how much farther it is practical to go in increasing production in commercial flocks is not clear, and it would seem futile to attempt to fix a limit at the present time. When Professor Dryden (see Chapter IV) in a few years of careful breeding can produce a strain in which whole flocks of pullets, taken just as they run, develop 50 per cent or more of 200-egg birds under conditions that may readily be duplicated on any commercial plant, it must be evident that the possibilities of the situation have by no means been exhausted.

High records, whether of individuals, pens or flocks, depend upon the skill of the attendant as well as upon the hen's inherited laying capacity, and intensive feeding and care have as yet been undertaken only in a comparatively superficial manner, even in connection with the highest records so far obtained. For example, so important an aid to increased production as is represented by the use of artificial light in winter months has been adopted at only one laying contest in this country, and in all contests the number of birds which the caretaker is required to look after is so large as to prevent any chance of fully realizing the productive possibilities of individuals.

When a thoroughgoing test of the caretaker's skill and the hen's maximum capacity for production is staged, the individual birds being tested with painstaking attention to detail, such as is employed when milch cows are being fed for advanced registry, for example, there is every reason to believe that results will be secured that will exceed the highest records made to date. Such tests would be expensive and not at all practical from the standpoint of the commercial egg

WHITE ROCK THAT LAID 16 EGGS IN 12 DAYS

F. X. Bourg of Houma, La., has made affidavit to the fact that this Fishel-strain White Rock laid 16 eggs in trap nests in 12 days. Artificial lights were used after 5 a. m. and up to 7 p. m.

producer, but as a means of ascertaining the maximum possible productive capacity of fowls they have a value that unquestionably warrants the effort.

Can Hens Lay Two Eggs a Day

The ability of hens to lay two eggs in one day, as an occasional performance, is not in question, because practically every poultry keeper who uses trap nests or is otherwise in a position to note individual performance can furnish evidence to this effect from his own records. In most cases, hens that produce two eggs in one day usually skip a day either before or after the event. Such skipping however, cannot be regarded as inevitable, as the fact is already established that the hen's organs may, under some conditions, function so rapidly that two eggs are produced in one day, at least now and then, without any interruption of the regular egg-a-day schedule. To what extent this tendency to rapid egg formation may be developed by care and breeding and transmitted to offspring remains to be seen.

There does not seem to be any ground for the popular belief that a rate of production exceeding an egg a day involves the element of abnormality. Practically nothing is certainly known in regard to the actual time required for the formation of the egg. Professor Rogers of Cornell University is authority for the statement that 14 days are required for the complete development of the yolk, starting with the minute, inactive ovule. It is stated in Maine Bulletin No. 216, that from the time the ovum enters the oviduct until the egg is completely formed 16 to 20 hours are required. It seems reasonable to conclude, however, that these estimates are only relatively exact and that there may be many conditions under which the period may be either lengthened or shortened.

Even if 18 hours are accepted as the minimum of time required for the formation of the egg there are six hours to spare each day; moreover, there is no doubt that two eggs may be present in the oviduct at one time, in different stages of development. There is no apparent reason, therefore, why fowls in prime physical condition may not produce more than one egg a day where they are supplied with a sufficient quantity of easily digested and assimilated nutrients. In this connection particular attention is called to an article by John H. Robinson which will be found in Chapter V, Part III, in which this well-known writer states his belief that it is physiologically possible for a hen to lay as many as 500 eggs in 365 days.

Buffon's Natural History (1812), says: "The ordinary fecundity of hens is limited to the laying of an egg each day. There are some, it is said, in Samogitia, Malacca, and other places, that lay twice a day. Aristotle mentions certain hens of Illyria, which laid so often as thrice a day; and it is probable that these were the same with the Adrian or Adriatic hens, of which he speaks in another place, and which were noted for their prolific quality."

Aside from the foregoing "hearsay" evidence, so far as known the first published reference to the fact that hens sometimes lay two eggs in one day is given in "A Treatise on the Breeding, Rearing and Fattening of Poultry," an English book published in London bearing the date of 1819. The author of this book states: "In the number of hens there are a few whose fecundity varies; some give only one egg in three days; others lay every other day, some lay every day; and lastly, some lay two in one day, but that is very uncommon." Records of two eggs in one day are fairly common from the seventies down to date, though usually under conditions which indicate that the performance probably was the result of delay in the exclusion of the first egg rather than extra rapid development.

In recent years however, a number of instances of hens laying two eggs a day have been reported where the performance apparently was in no way due to irregularities in exclusion. Whether this results from the oviduct carrying two eggs in different stages of development, or to more rapid secretion of egg-forming material or to both, remains to be discovered. Farmers' Bulletin No. 114 (Victoria, Australia), relating to the laying contest at Hawkesbury, says: "In May, 1916, Mrs. Jobling's Black Orpington hen No. 415 laid 2 normal eggs on the 5th, 8th, 10th and 15th. In addition, she laid on preceding and succeeding days." Another unusual record is reported in a News Letter issued by the University of Idaho which states that "Ida-U"—a White Orpington pullet owned by the Poultry Department—was trap nested, beginning Feb. 17, 1920, and in the following 80 days laid 86 eggs. She laid 2 eggs on each of 11 days and laid double-yolked eggs on 3 days and skipped 5 days. She is believed to have

A LONG-DISTANCE LAYER AT CORNELL
This hen produced 1,231 eggs during her lifetime—one of the highest known records.

"MISSOURI QUEEN"—RECORD 1,280 EGGS TO DATE

A remarkable "long-distance" layer at Missouri State Poultry Experiment Station. Produced 222 eggs in her first year. Her yearly production thereafter was as follows: 187, 217, 149, 177, 147, 144. In 1921 up to June 1 she had laid 37 eggs—in all, 1,280 eggs in seven and one-half years of production.

laid two eggs on several days before trap nesting began.

In the February, 1915, issue of Reliable Poultry Journal is recorded a series of experiments conducted by Dr. E. C. Waldorf of Buffalo, N. Y., beginning in December, 1889. This article is reproduced in part in Chapter V, Part III. It is sufficient here to state that Dr. Waldorf reports that under the exceptionally favorable conditions provided by him his flock averaged to lay 10 eggs per hen per week over a period of three months, gradually falling off during the next two months and then ceasing to lay altogether.

Many records similar to the foregoing are available. Taken together they afford excellent ground for the belief that hens may not only average better than an egg a day, but that the extra eggs so produced may be normal in size and condition.

How Many Eggs May the Hen Lay in a Lifetime

There are sound practical reasons why the commercial poultry keeper aims to secure maximum production during the hen's first and second years, and takes but little interest in her possible subsequent performance. It is worth knowing, however, that productive possibilities are by no means exhausted at the age when hens generally are looked upon as having outlived their usefulness. In the last few years some remarkable records have been made by what may be called "long-distance" layers, meaning by this, hens that have persisted in laying through a relatively long term of years.

One of the first authentic records of a hen laying 1,000 eggs is furnished by the Oregon Agricultural College, where one of Professor Dryden's "Oregons" laid 240 eggs in her first, 222 in the second, 202 in the third, 155 in the fourth, 168 in her fifth year, reaching a total of 1,188 in six years, or within 12 eggs of an average of 200 per year for six consecutive years. Professor Dryden also reports that a daughter of this hen laid 1,096 eggs before she had completed her seventh year of laying. He gives four other records as follows: A-60, 1,152 in 8 years; B-42, 1,173 in 7 years and still laying; B-14, 1,238 in her eighth year and still laying; C-547, 1,104 in her sixth year.

"Missouri Queen," a White Leghorn at the Missouri State Poultry Experiment Station, has a record of 1,280 eggs in her eighth year and was still laying June 1, 1921.

For a discussion of the practical value of hens vs. pullets in the commercial flock, see Chapter V.

The possibilities of long-distance pro-

SOME HIGH-RECORD HENS AT OHIO EXPERIMENT STATION
In addition to the ten long-distance layers at the Ohio Station, whose records are given in accompanying table, there are many other high-producing birds in the station flock. A few are illustrated above. Starting at top of cut and reading from left to right, their records are as follows, the first figures being the pullet-year record and the figures in parentheses the total for two years: First row—269 (462); 245 (404); 211 (379). Second row—247 (450); 223 (400); 209 (396). Third row—240 (397); 212 (393); 228 (397).

duction have received special attention at Cornell University and Professor Rice reports the following records:

Hen No. A-3477, 1910-1919 inclusive—1,231 eggs.

Hen No. A-5727, Dec. 14, 1910 to Aug. 15, 1918—1,013 eggs.

Hen No. A-7354, Nov. 16, 1910 to Sept 20, 1917—1,000 eggs.

At the Ohio Station ten hens have the splendid record of having produced 10,000 eggs, their records being tabulated as follows:

Egg Production by Years—10 Hens at Ohio Experiment Station

No.	Hatched	1st	2nd	3rd	4th	5th	6th	7th	8th	9th	Total
8	1909	136	193	187	170	124	65	79	90	17	1,061
18	1909	64	182	150	107	123	109	64	141	52	992
C24	1910	191	189	171	163	103	113	145	104	54	1,233
C34	1910	181	160	135	129	46	86	98	75	43	910
C38	1910	202	214	181	164	104	147	83	52	27	1,171
C39	1910	181	166	145	139	48	125	116	84	13	1,017
C46	1910	174	178	130	131	60	100	147	90	59	1,069
C48	1910	178	166	152	146	36	87	129	77	62	1,028
1092	1913	245	159	183	168	161	53				969
1105	1913	211	168	153	139	115	58				834

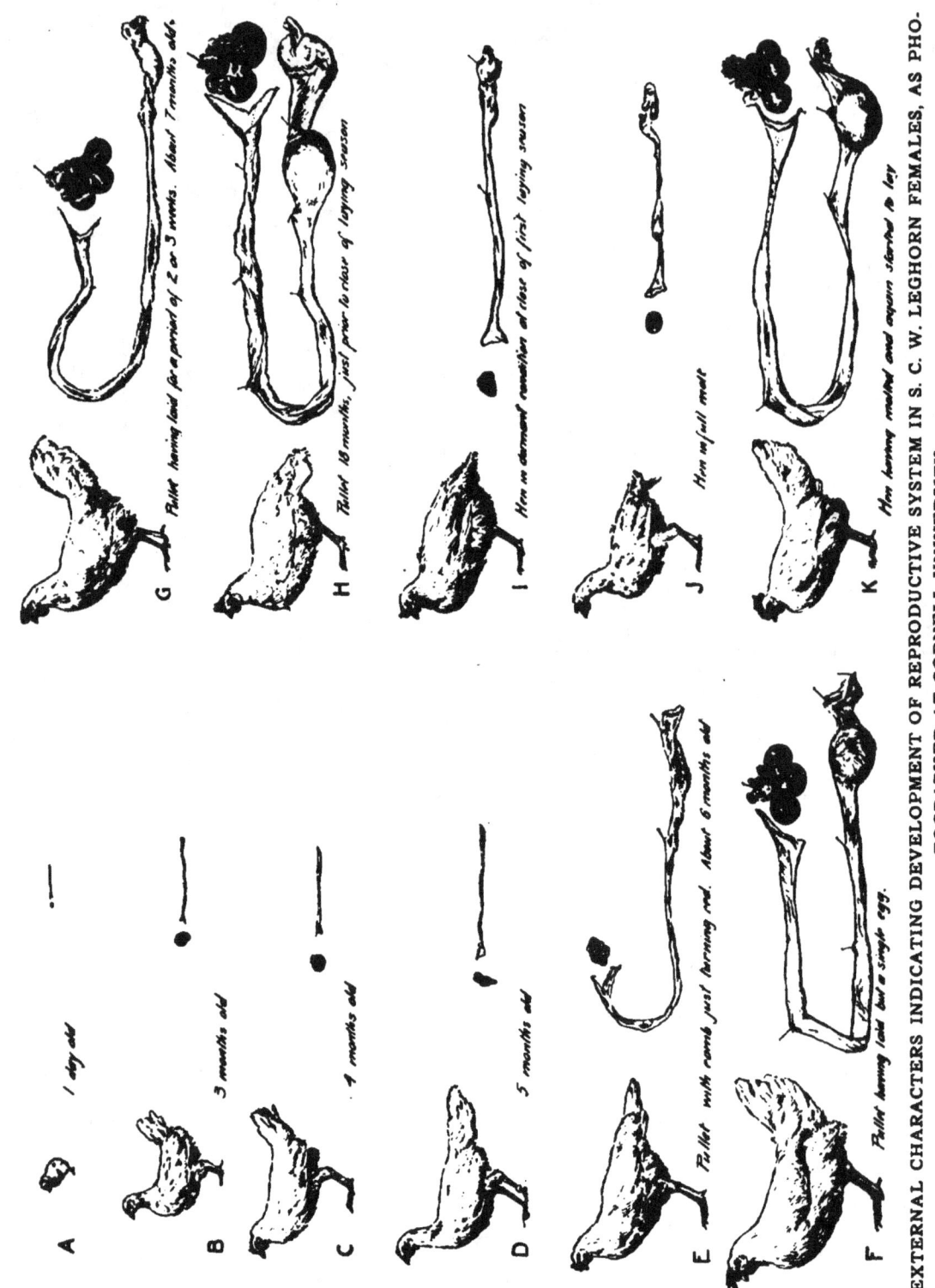

EXTERNAL CHARACTERS INDICATING DEVELOPMENT OF REPRODUCTIVE SYSTEM IN S. C. W. LEGHORN FEMALES, AS PHOTOGRAPHED AT CORNELL UNIVERSITY

CHAPTER II

The Hen's Egg Organs and How They Function

A Brief Description of the Hen's Organs, Showing How Eggs Develop, the Amount of Time Required for Their Formation and Conditions Most Favorable To Securing Good Production—Hens Are of Distinctly Different Classes as Layers, but Their Rank Quite Largely Depends Upon the Conditions Under Which They Are Kept

THE poultry keeper who aims to handle his birds to the best advantage, whether he is feeding for a record or simply for the highest production that he can secure under everyday commercial conditions, needs to know all that he can about the physiology of egg production, thus to be able to render his birds all possible assistance in the way of supplying nutritious feed, suitable environment and good general care. This knowledge should cover not only the structure of the organs but the way in which they normally function.

The egg organs consist of the ovary and the oviduct. In the former the yolk develops to its full size, after which it enters the oviduct and receives its coating of albumen, membrane and hard shell. The ovary is a small organ attached to the upper part of the abdominal cavity, immediately in front of the kidneys. It consists of vascular tissue and a large number of minute yolks known as ovules or ova. In the early stages of the growth of the female embryo two ovaries are formed, but the one on the right side never gets beyond the rudimentary stage, only the left one coming to full development. When in a dormant condition the ovary is quite small, but it gradually increases in size as the bird approaches laying condition. The variation in size of both ovary and oviduct is marked.

The interesting illustration on page 12 shows the different stages of development of ovary and oviduct in a Leghorn female. Up to the time that the pullet is fully grown these organs make but little growth, as is shown at A, B, C and D. At six months (E) the oviduct shows a marked increase in size, accompanied by the enlarging and reddening of the comb. As indicated at F, G and H, the oviduct has not reached its full size at the time laying begins, but it continues to develop slowly until toward the close of the first laying season. When the bird is in full molt (J) both ovary and oviduct shrink to extremely small proportions, but by the time she is ready to begin again both organs will have increased almost to their maximum size.

Investigations have shown clearly that so far as the number of ovules is concerned there is practically no limit to the eggs that can be produced by a hen. Counting only the ovules that can be detected by the naked eye, it has been shown that there may be as many as 4,000 in a normal ovary, and still more can be seen by the use of a magnifying glass.

Examination of the ovary of a laying hen, or one about to begin, will show quite a number of yolks in various stages of growth, though after they reach a certain size development seems to concentrate in a comparatively small number which are promptly "ripened." According to Rogers of Cornell (as stated in Chapter I) about 14 days are required for the development of the yolk from a dormant condition to full size. It is quite probable however, that the length of time required is not at all uniform but is determined by the ovarian activity of individual birds, and this in turn by season, rations, etc. That there is a marked variation in the rate or way in which ova develop is indicated by the difference in the appearance of the ovary of a heavy-laying hen as contrasted with one of lower production. The illustration presented on page 16, for example, shows the ovaries of two hens in which the contrast is quite marked.

Each ovary is enclosed in a membranous sac having connection with the ovarian artery. This sac enlarges or stretches as the yolk grows, rupturing when the yolk is fully developed, thus permitting the latter to drop into the upper end of the oviduct. There is no direct connection between the ovary and the oviduct, but when the yolk is fully formed and ready to drop, the upper funnel-shaped end of the oviduct rises and grasps or encloses the yolk. If, for any reason, the oviduct fails to function properly or if as sometimes appears to happen, it does not enclose the proper yolk, the one that is escaping from its sac may drop into the abdominal cavity instead of the oviduct, in which case it usually will be absorbed without causing any trouble.

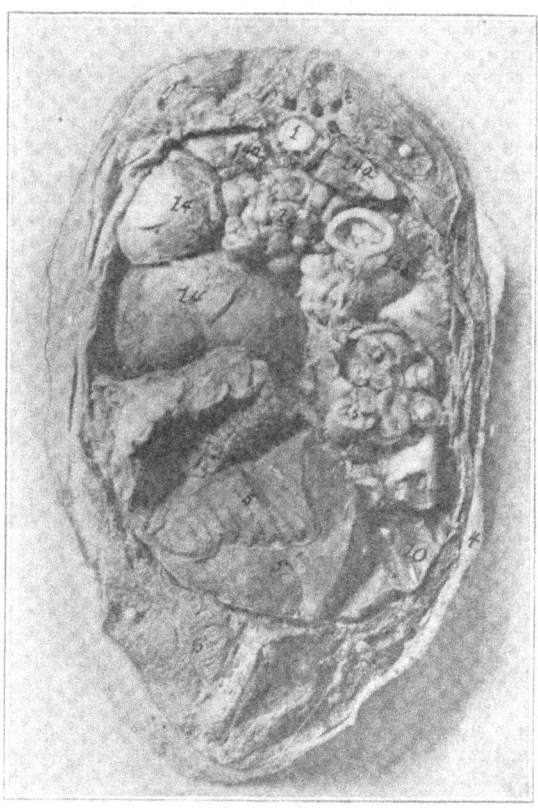

CROSS SECTION OF LAYING FOWL

This interesting photo, showing a cross section about the middle of the fowl's body, indicates clearly the position of the different organs. These are numbered as follows: 1, spinal cord; 14a, 14a, kidneys; 14, 14, 14, ovary, ova and yolks in different stages of development; 15a, section of caecum; 15, 15, small intestines; 5, 5, gizzard; 10, right lobe of liver; 4, skin; 6, breast muscles. From "Anatomy of the Domestic Fowl," by Dr. B. F. Kaupp.

The fully formed yolk consists of a white yolk center surrounded by concentric layers of yellow yolk separated from each other by thin layers of white yolk. In a hard-boiled egg it is quite easy to observe this layer

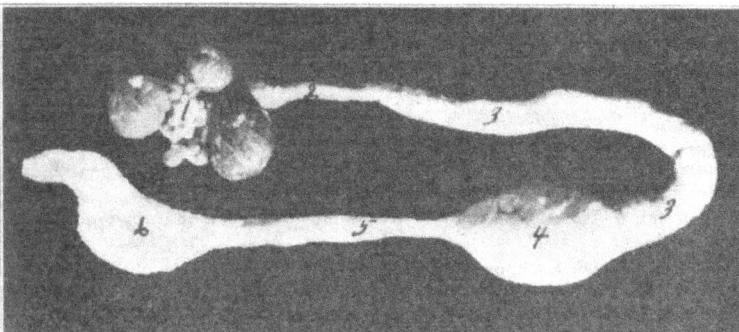

OVARY AND OVIDUCT OF LAYING HEN
Photo from C. T. Patterson, Springfield, Mo.

formation and to distinguish between the two kinds of yolk. It is commonly believed that these layers represent the deposits of yolk material on successive days.

The Oviduct

The oviduct is a whitish organ with thick walls loosely attached by means of a special membrane to the upper part of the abdominal wall. As is shown in the illustration on page 12, the dormant oviduct is extremely small, but when the hen is laying regularly it is 20 inches or more in length, and is covered with an extensive network of blood vessels which are required to carry to the various glands the materials needed to form the albumen and shell of the egg. The oviduct consists of several distinct divisions which, starting at the end next the ovary, bear the following names: the infundibulum or funnel, not indicated in the accompanying illustration; the albumen secreting portion at 2, 3 and 4; the isthmus at 5, where the shell membrane is secreted; the uterus at 6, where the hard shell is formed and the balance of the albumen added. The latter part of the oviduct, opening into the rectum, is the vagina. The albumen in the normal egg consists of three or four distinct layers of varying consistency, the secretion of which begins in the infundibulum and is completed in the uterus, a considerable part of the outer layer being deposited after the hard shell is more or less completely formed.

As regards time required for the formation of the egg, Bulletin No. 216 of the Maine Agricultural Station states that "the yolk remains in the albumen portion of the oviduct about 3 hours; about 1 hour's time is occupied in passing through the isthmus, and the addition of albumen to the egg is completed only after it has been in the uterus from 5 to 7 hours. Also for the completion of the shell and the laying of the egg from 12 to 16, or exceptionally even more, hours are required."

Assuming that the five to seven hours required for the addition of albumen in the uterus are included in the first part of the 12 to 16 hours required for the completion of the shell, this would indicate that 16 to 20 hours are required from the time the yolk leaves the ovary until the egg is laid. While nothing is certainly known in regard to the matter, it seems reasonable to expect that the time required for the secretion of the albumen and the shell must vary with individual hens and the rapidity with which laying progresses. Further observations on this point are greatly needed because of its direct bearing on the productive possibilities of the hen. If practically 24 hours are required from the time the yolk enters the oviduct until the egg is laid, then any yield in excess of an egg a day necessitates the presence of two eggs in the oviduct at one time, at different stages of development.

Laying Cycle

A clearly marked character in practically all good layers is the tendency to produce eggs in cycles, and with a fairly regular rhythm. By "cycle" is meant the number of eggs produced without skipping a day, and by "rhythm" the frequency with which the cycles are repeated. Cycles may be short or long, and usually they vary more or less with the season. The hen whose pullet-year record is given at top of page 15 had a fairly uniform cycle of three in the winter, increasing to four or five in the summer, and these cycles were maintained with remarkably uniform rhythm throughout the laying season. The hen whose record is reproduced in the cut in the center of page 15 laid almost as many eggs as the hen just mentioned but both cycle and rhythm were less uniform and as a breeder this hen would be regarded as much less desirable. The hen whose record is reproduced at foot of page 15 had much longer cycles than either of the other two but was a very inferior layer because, either through the influence of inherited characters, environment or possibly mere accident, she was later in beginning and proved to be an early quitter. With such a cycle a high second-year record is a possibility, but unless definite evidence could be produced to show that her poor winter performance was the result of accident she should have no place in either the breeding or laying pen.

Why Hens Lay

Egg laying is a reproductive function and as such cannot be "forced," strictly speaking. Rate of production however, depends on many factors and, up to the fowl's natural limit (whatever that may be), it is to a great extent controlled by environment and nutrition. The marvelous reproductive capacity of the hen has no equal

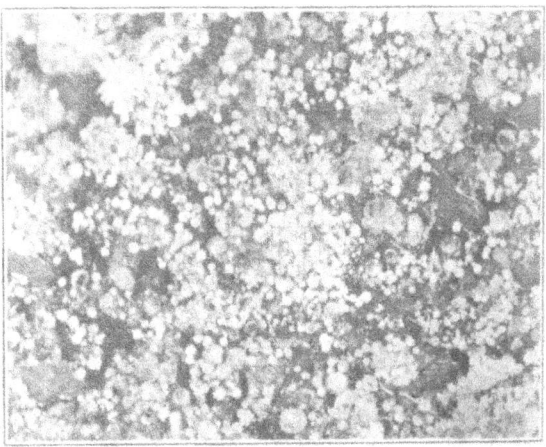

SECTION OF OVARIAN TISSUE SHOWING GREAT NUMBER OF VISIBLE OVULES
Photo from C. T. Patterson, Springfield, Mo.

among our domestic animals, and her ability to lay right along, day after day, week after week, and month after month, when properly handled, has led to a tendency in some quarters to regard her as an "egg factory." That is a practical way to look at the matter so long as it does not lead, as it often does, to handling the fowls in a mechanical manner, ignoring individuals and aiming simply at mass production.

Fowls, especially heavy producers, have highly strung nervous temperaments and must be handled accordingly, even in commercial flocks, if they are to do their best work. For example, it is a peculiar fact, recognized by all careful poultry keepers, that hens like attention. It is doubtful whether they ever enjoy being handled, but they certainly take pleasure in the company of an attendant whom they trust and like, enjoy being talked to and fed tidbits by hand, and, without question, they will respond to a marked extent in egg yield where such encourage-

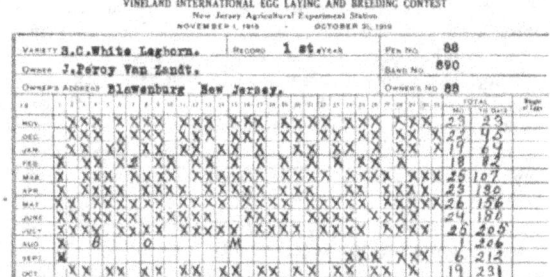

DAILY EGG RECORD OF GOOD PRODUCER BUT WITH INFERIOR CYCLE AND RHYTHM

ment is given. Frequently the presence or lack of "petting" is the only apparent explanation for differences in results secured.

On the other hand, the egg yield is certain to be unfavorably affected by rough handling and fright. In many instances flocks of layers, particularly Leghorns, not accustomed to the presence of strangers, will noticeably fall off in the egg yield as the direct result of so simple an interference as having a stranger pass through their pen. This does not necessarily lead to the conclusion that in order to secure high production, birds must be protected from strangers and seen only by their regular attendant, as there is no reason to believe that when accustomed to strangers, the egg yield will be unfavorably affected by their presence so long as the birds are not unduly alarmed. But it does point plainly to the fact that laying fowls require quiet and freedom from anything that will molest or alarm them.

Nervous and Physical Strain

The question is often raised as to the amount of strain involved in egg production. That there is some, goes without saying. Nervous strain is indicated by the fact that hens that have been laying heavily for some time often are inferior breeders—that is, their eggs are low in fertility and their chicks not strong. It is common knowledge also, that fowls that are laying heavily gradually lose flesh, and it is believed by many that discontinuance in production regularly follows loss of flesh. However, to say that loss of weight and reduced vitality are inevitable accompaniments of heavy production is to abandon accurate knowledge for conjecture.

Along with the popular belief already mentioned (that heavy-laying hens are likely to be inferior breeders) should be compared the opposite testimony of many

DAILY EGG RECORD SHOWING GOOD CYCLE AND RHYTHM

skillful poultrymen who are emphatic in their statement that their highest producers are often their best breeders and that they recognize no essential connection between heavy production and low fertility, or vitality of embryo.

In Australia, where the breeding season appears to follow closely on the completion of the laying year, contest managers pride themselves on seeing that the contesting birds finish the year in fine physical condition so that they are ready to go at once into breeding pens. For example, Hedley Jones in an article in regard to the Victoria (Bendigo) Laying Competition, says: "In connection with the feeding which has been carefully prepared and at all times fed positively fresh, two important factors have been in the mind of Manager Mitchell, viz., the getting of the maximum number of eggs and the preserving of the stamina of the bird so that if she makes a sufficiently high record she will be fit for immediate breeding. Reports from breeders show that the birds on reaching home are fit for breeding and many a man with a valuable high-scoring bird has had chickens from her by the beginning of May or just a month after her return from the competition grounds."

The question of vitality and all that goes with it is so conditioned upon the skill of the caretaker that it is not practical to attempt a too definite statement on the subject at the present time. It may safely be said however, that as fowls ordinarily are handled they certainly are weakened by long-continued production, and it is fortunate rather than otherwise that record layers need not be placed in the breeding pen until they have gone

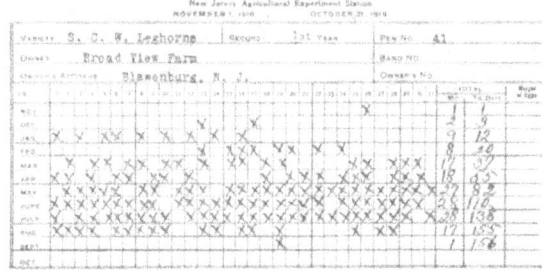

DAILY EGG RECORD SHOWING EXTRA-LONG CYCLES

through a rest period during which, if properly cared for, they may fully regain their normal weight and accumulate a store of nervous energy that will put them in good condition for breeding.

Production a Complex Character

Among many scientists there is a conviction that laying is a complex character rather than a single one, and that the various factors entering into it are subject to definite laws and capable of being brought together in various combinations just as are external characters that distinguish different breeds and varieties. Exceptional winter-laying ability, high production in the first year and relatively poor afterward, higher production in the second than in the first year, capacity for "long-distance" laying—these are only a few of the supposedly distinct factors with which the breeder has to work and which he can, by skillful breeding, assemble in such combinations as best meet his peculiar requirements.

If this theory is correct, it is of decided practical importance; but in the present stage of knowledge on the subject the commercial poultry keeper can wisely avoid dogmatizing on such points knowing, as he does, how closely production is associated with environment, feeding and care, and knowing too, to what a marked extent all three of these factors can be changed, and in their changing affect results. For example, it is easy, having trap-nest records of a number of layers, to classify them as good winter producers; good first-year or good second-year producers, etc., and that classification may be accurate for the conditions under which the birds are kept; but vary the conditions and the classifications may have to be changed. To illustrate, the colored chart which forms the frontispiece of this book, shows at the top how pullets under ordinary management classified themselves both as to total and winter production. But note how the introduction of artificial light upsets this classification, raising many 150-egg birds to the 200 class, 100-egg birds to the 150 class, and so on. Additional changes that would make still more favorable the conditions under which the birds are operating undoubtedly would necessitate new realignments.

Again, there are plenty of records to show that birds of heavy-laying ancestry and with every indication of high-laying capacity have proved inferior first-year producers but have made extremely good second-year records. It is possible that some of these birds are so constituted that they do not reach their maximum laying ability in their first year, but that is doubtful. Everything points to the conclusion that any fowl will normally produce more eggs in her pullet year than in any subsequent one, under equally favorable conditions. However, a comparatively slight handicap in some of the conditions to which the bird is exposed may place a naturally high producer at a disadvantage in her first year.

The Hen's Nesting Habits

External conditions certainly have a great deal to do with egg production and the painstaking poultry keeper can afford to give serious attention to "humoring" his birds within reasonable limits. It is worth while, therefore, to know just what are the hen's preferences as to conditions of laying. Professor George M. Turpin of the Iowa Experiment Station, a few years ago conducted some investigations in regard to the nesting habits of the hen which are of particular interest in this connection. The following paragraphs on this subject are condensed from Iowa Bulletin, No. 178:

In watching hens and in noting their choice of a nest it was found that they give a great deal more attention to this subject and take a great deal more time in making their choice than most persons realize. In starting the observations it was the purpose to keep a record of the time each nest was entered and of the time it was vacated. It was soon found however, that many of the hens often passed from one nest to another and with so little delay that the observer could not keep track of all the visits. Therefore, only a record was kept of those instances in which the hens sat down and spent at least a minute on the nests visited. Fifty Leghorns were found to make nearly as many visits to the nests in one day as 40 R. I. Reds did in two days, and they also spent nearly three-fourths as much time on the nests visited. If the actual number of times all the hens entered the nests had been counted without regard to the length of time they remained in them, a relatively larger number of visits would have been recorded for the Leghorns, as they made many more very short visits to the nests than the Reds.

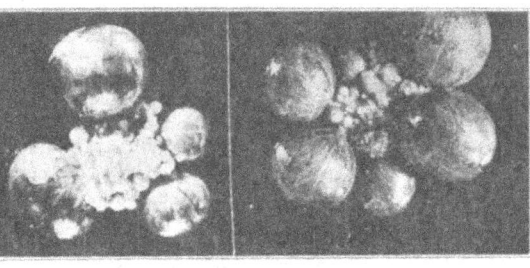

ACTIVE OVARIES WITH DIFFERENT CYCLES
The ovary on the right has more ova (yolks) under development than the one on the left, which should mean either longer cycles or less lost time between them—i. e., better rhythm. Photo from C. T. Patterson, Springfield, Mo.

Forty R. I. Reds, for example, made 8 visits of 1 to 2 minutes; 12 of 3 to 20 minutes; 9 of 21 to 40; and 3 for longer periods, a total of 32 visits occupying 8 hours and 4 minutes, these representing visits to nests where the hens did not lay. R. I. Reds spent an average of 1 hour and 45 minutes and Leghorns 1 hour and 35 minutes on the nest for each egg laid. Counting the total time spent on the nests and the number of eggs laid in each case, it was found that the R. I. Reds spent an average of 1 hour and 55 minutes and the Leghorns 1 hour and 58 minutes. In the time spent on the nest before laying and after laying considerable variation was noted, the average being: before laying, 31.7 minutes; after laying, 50 minutes.

In regard to factors influencing attractiveness of nests, it was found that those containing one or two eggs, or a nest egg, were preferred; but when the number of eggs reached four they apparently discriminated against the nests containing them. As to attractiveness of different kinds of eggs: in one experiment 108 eggs were laid in a nest with no nest egg, 214 with a china nest egg and 290 with a hen's egg. Other tests indicated that wooden or plaster-of-Paris eggs were as attractive as hens' eggs.

The influence of seclusion was also tested. The nests used were ordinary platform nests made in batteries of eight. There were two batteries in each pen, in one of which the front door was open while the other was kept closed; the hens entering in either case from the rear. Of a total of 535 eggs laid during the test 251 were laid in the exposed nests and 284 in the closed nests, or in the ratio of 100 to 113. This is rather a surprising result, as it is commonly believed that a hen particularly desires seclusion in laying. If other experiments should confirm this it would appear that seeking seclusion is more a matter of instinct than of actual preference.

CHAPTER III

Egg Type and Capacity

The Different Characters Most Commonly Associated With Heavy Production Are Here Described—Prominent College Workers, Breeders and Judges, Join in Stating Just What They Understand "Egg Type" To Be, and How It May Be Secured in Different Breeds Without Antagonizing Present Standard Requirements

"IT seems to me that type is the fundamental concept of production. Unless a fowl has the necessary physical qualifications, care and feeding cannot make her a heavy layer." Thus Professor O. B. Kent of Cornell University, summarizes his attitude toward the subject of egg type, and to this general statement we believe no one will take serious exception. It is only reasonable to expect that high egg production, or the capacity for it, should involve the presence of distinct characters which would distinguish fowls possessing them from poor layers, just as, for example, there are definite characters which differentiate the poor producer from the high producer among milch cows. And while it is to be expected that there will be exceptions to all rules, it rarely happens that a competent judge fails to discriminate between the two classes of producers from either cows or fowls.

The poultry keeper is handicapped to some extent in the observation of characters that are associated with heavy production or its absence because the fowl's contour is so greatly modified by her thick coat of feathers which to a considerable extent disguises her body shape. Only extreme variations in shape, therefore, are readily apparent to the observer. It is quite largely for this reason that it is particularly difficult to draw comparisons between fowls of different productive capacity, either before or shortly after they have begun to lay, though it is a simple matter to identify those that have been laying for a long time.

It is easy to understand that there can be no heavy production without the consumption of large quantities of feed and obviously this must, in turn, react upon abdominal development, making this one of the most certain indicators of productive ability, provided the observer is careful not to confuse large size of abdomen produced by fat with size produced by enlarged digestive organs. There are other similar, though possibly less obvious changes in type, directly resulting from high production, that the close observer can hardly fail to detect. Every poultry keeper therefore, should be willing to accept the general statement that there is such a thing as egg type, even though he may feel that the subject is not sufficiently understood to warrant any one in being dogmatic in regard to the particular characters that are assumed to indicate this type.

High and Low-Producing Types

It should be recognized that the terms "high producer" and "low producer" are only relative in their significance, and no matter where the dividing line between the two is placed there will be some birds on either side, possibly many, whose classification could readily be reversed by comparatively slight changes in environment or management. It is only when dealing with extremes that one can speak with confidence of high and low producers as distinct types. However, the study of large numbers of individuals definitely falling in one class or the other has brought out certain differences between them which utility poultry keepers generally accept as typical. These may be briefly stated as follows:

HEAD OF A HEAVY LAYER
Note texture of comb and wattles, prominent eyes and alert, intelligent expression. Photo from Cornell University.

General Health. No argument is needed to prove that birds that are high producers over long periods must be in excellent physical condition. There can be no high records made without the digestion and assimilation of relatively great quantities of feed, and capacity for keeping this up indefinitely requires that the bird be in the best of health. Some fowls that apparently have the inherent ability to produce eggs in large numbers, as indicated by high production for short periods, fail to make good yearly records because they are deficient in this particular respect. Their physical strength is not equal to the strain of transforming large quantities of feed into eggs, so they develop into spasmodic layers, producing eggs in long cycles with long rest periods or possibly, among the nonbroody races, may be fairly persistent layers through the year but in short cycles, these alternating with rest periods of varying length.

Nervous Energy. Almost without exception heavy producers are characterized by a high degree of nerve force which is evidenced quite largely by conduct. The bird that is slow or sluggish cannot be a heavy producer because, as a rule, her slowness is internal as well as external, meaning that her bodily functions are performed at a slow rate. As has already been shown, the production of an egg a day involves working the egg organs full time, or practically so, and if they function only at a slow rate they are necessarily unable to maintan a 24-hour schedule, resulting in every-other-day or every-third-day laying, or in short cycles with long rest periods, either of which, of course, makes good records out of the question.

The observer must distinguish between nervous energy however, and nervousness or timidity, which is an altogether different matter. A hen that is uneasy and "scarey," that flies to the farther end of the house or yard at every unusual sound and that cannot be approached by a reasonably quiet caretaker without showing fear, is rarely, if ever, a heavy producer. A bird that is alert, quick in her movements, not readily caught but showing no special timidity is the bird that, as a rule, will be found to have a good egg record to her credit.

Intelligence. Intelligence is one of the characters that is regularly associated with heavy production. It is one indication of nervous energy, as there rarely is a high degree of intelligence where the latter is lacking, neither is it to be associated with excess nervousness and flightiness. Along with intelligence goes docility which, however, does not mean that the bird should always be underfoot ready to be picked up and handled. On the contrary, the docile hen, though she will take an interest in her caretaker and with a little encouragement will be on familiar terms with him, is inclined to be just a little bit shy about being handled and quite apt to keep herself at a reasonable distance unless there is some good reason for coming close.

Head Parts. The head parts of the fowl are scrutinized carefully by the expert in looking for production characters. The comb, face, eyes and wattles all have their distinguishing marks and these are so distinct that the experienced handler of fowls is not easily led astray. There are some theories in regard to the significance of certain features on which authorities are not fully in accord, but in general there is substantial agreement. The comb must be of reasonably fine texture, because coarse-grained combs are indicative of coarseness in the body structure generally, with which the character of heavy production is not associated. Regardless of the size of the comb its shape is significant, particularly the serrations in a single-combed bird, which should be reasonably broad at the base, rather than narrow or what is known as pencil pointed. It is doubtful whether the size of the comb, even within the breed, can be accepted as indicating anything definitely with respect to productive capacity. Large combs, however, are not necessarily associated with high production in any breed.

The eyes of the fowl are fairly indicative of her nervous organization. Eyes that are deep-set in the head, with overhanging eyebrows, indicate the sluggish, inactive bird, while eyes that stand out beyond the head, as the bird is viewed head on, that are wide open, clear and snappy in appearance, will be found on a heavy layer. It is conceded that such eyes may occasionally be found on birds that are not heavy layers, but in that case there usually will be other characters that will guide the observer to correct classification.

The face of the laying hen is lean and clear-cut, as distinguished from the fat, stuffy face which is usually associated with low producton.

Plumage. There is a general belief that loose feathering is characteristic of low production, but this is a test that must be applied with caution, though when associated with evidence of coarseness and slow maturity, it certainly is to be avoided. To the expert observer the condition of the plumage tells much as to the fowl's probable productiveness.

Pigmentation. This is also a character which tells much regarding past performance and for that reason it more properly is treated under the head of culling in Chapter V.

The general carriage of the bird is significant in so far as it indicates the characters already mentioned, such as alertness, nervous energy, etc. There are a number of theories in regard to shape of back, point at which legs join the body, position of head, etc., that may have a practical application, but which do not appear to be sufficiently supported by evidence as yet to warrant introducing them as type characters of proved value.

Type as Indicated by Handling

Just as the dairyman recognizes certain conditions of the skin of a dairy cow to be indicative of her ability as a milk producer, so the poultryman attaches much importance to the skin in handling fowls with a view to determining their laying ability. Fineness of texture is a quality regularly sought and readily distinguished. Texture is also indicated by the comb as has already been noted. Pliability of the skin is an important factor because it determines the amount of fat underlying it, though texture is also associated with this. A bird may have a considerable amount of fat under the skin and yet be in good-laying condition, but a bird with a heavy blanket of fat and a skin that is coarse and hard is never highly productive. Pliability of the abdomen is one of the best indications of past productiveness, though it has no particular bearing upon capacity.

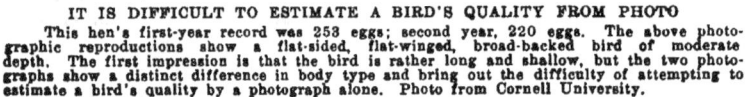

IT IS DIFFICULT TO ESTIMATE A BIRD'S QUALITY FROM PHOTO

This hen's first-year record was 253 eggs; second year, 220 eggs. The above photographic reproductions show a flat-sided, flat-winged, broad-backed bird of moderate depth. The first impression is that the bird is rather long and shallow, but the two photographs show a distinct difference in body type and bring out the difficulty of attempting to estimate a bird's quality by a photograph alone. Photo from Cornell University.

A GOOD EXAMPLE OF BODY TYPE

This hen's first-year record was 179 eggs; second year, 172 eggs; third year, 180 eggs; fourth year, 212 eggs. In her fourth year she had six daughters that averaged 210 eggs. Body is distinctly full, deep, flat and broad. The large, bright eye is especially noticeable in one of the pictures. The eggs laid by this hen averaged better than two ounces, a great many of the eggs running over two and one-quarter ounces. The two front views show her extremely deep, straight sides and well-spread legs. Photo from Cornell University.

EGG TYPE AND CAPACITY

Egg-Laying Capacity

The intestines of a fowl that is eating large quantities of feed are necessarily larger than those of one that is on a comparatively light diet. Also, the ovary and oviduct are many times larger in a laying bird than in one that is not laying (see page 12). To accommodate these enlarged organs the abdomen must be of good size. Based upon this fact, directions for selecting high layers call for full breast, widely spread processes (the flat bones that extend back along the sides from the front of the keel), well-spread pelvic arch, depth of abdomen, etc.

Depth, meaning the distance from the end of the keel bone to the pelvic arch, is one of the most important measures of capacity. In the average bird this is secured by dropping the rear end of the keel so that the lines of the backbone and the keel are more or less at an angle

HIGH PRODUCERS IN THE SAME BREED MAY VARY GREATLY IN TYPE

These reproductions of Barred Plymouth Rocks, all but two high-record birds, illustrate how radically fowls may vary in shape without any apparent effect upon their ability to produce eggs in large numbers. Starting at the top and reading from left to right their records are as follows: First row—D-18, 271 eggs; C-48, 268 eggs; D-118, 238 eggs. Second row—D-106, 225 eggs; A-77, 214 eggs; A-78, 213 eggs. Third row—C-34, 180 eggs; D-33, 85 eggs; D-7, 7 eggs. With the exception of A-77, A-78 and D-33 all the birds are of high-laying ancestry on both sides. The low producer (D-7), was bred from a hen with record of 160 and a sire whose dam's record was 203. All birds illustrated were bred by Professor James Dryden of Oregon Experiment Station, by whom photos were furnished.

with each other. The width of the pelvic arch, upon which so much stress is laid in the "Hogan test," is largely a matter of capacity, though tests have shown that width is more or less directly associated with the activity of the egg organs. A series of daily measurements taken at Cornell University throughout an entire year has shown that the pelvic arch regularly expands and contracts with the rate of production.

Capacity as determined by the position of the keel bone, depends to a considerable extent upon the breed type of the bird. For example, it is not to be expected that the Rhode Island Red and the Wyandotte will both have keel bones of the same relative length and position. So long as capacity is secured there is no reason to believe that it matters, as regards production, whether it is secured by depth or by length.

QUESTIONNAIRE ON EGG TYPE

In order to get at the views of representative breeders and college workers on the subject of egg type, some questions were addressed to a limited number and their replies are here given. A comparison of the replies indicates that there is as yet, no common understanding between breeders and college workers as to what is meant by the term, but a careful study of this questionnaire should be interesting and helpful to all.

List of Questions

1—Do you recognize the existence of a definite "egg type" in fowls?
2—What characters do you understand to be included in the term "egg type"?
3—Can a general description of egg type be formulated that will apply with equal fairness to all breeds?
4—The Purdue Utility Score Card (see Chapter V, Part III) says body must be "nearly rectangular * * *, great depth is especially desirable * * *, the general body conformation of a heavy producer conforms very closely to a rectangle with pronounced angles rather than smooth curves." Do you recognize this as a fair standard to be applied to all fowls without regard to breed type?

HEAD OF A GOOD PRODUCER
This bird is credited with 126 eggs from November to August—not a high record but good enough to present a marked contrast in outlines of head when compared with the bird on the right. Photo from Cornell University.

5—Where no allowance is made for breed type will the strict application of above description tend to a common type in all breeds?
6—If you favor the Utility Score-Card description of egg

HEAD OF A POOR PRODUCER
The estimate of production of this bird for November to August was 50 eggs. Her full, beefy face and none-too-intelligent eye identify her unmistakably as an inferior layer. Photo from Cornell University.

type in general, but would adapt it to breed characters, what changes would you make in applying it to:
  Leghorns?
  Rocks?
  Wyandottes?
  Rhode Island Reds?
7—Must body capacity be relatively large in fowls of different size? That is, must the body capacity of a 200-egg Rock, weighing 7 pounds, be approximately double the body capacity of a 200-egg Leghorn, weighing 3¼ pounds? If so, why?
8—Do you consider it desirable to modify standard weights in any of the popular breeds, with a view to improving egg production?
9—Is looseness of feathers (in standard Wyandottes, for example) a handicap in high egg production?

REPLY OF DR. O. B. KENT
Cornell University, Ithaca, N. Y.

1—I very decidedly believe that there is such a thing as egg type in fowls but do not believe that anyone can set down definite figures as to the length, breadth or depth of a bird.

2—My understanding is that those characters that are permanent in a bird are the characters that may be included under the term "egg type"—that is, such characters as length of body, breadth of body, shape of head, etc., while such characters as pigmentation, molting, size and thickness of abdomen may not be classified under the term of characters indicating egg type.

3—I think it is entirely possible to formulate a general description that will apply with equal fairness to all breeds.

4—I believe that the description given in the Utility Score Card can apply as a fairly definite body standard to all breeds. It should at once be recognized that no limit is set on length, breadth or depth, and that a rectangle may be either square or oblong.

5—I do not think that the strict application of the above description will tend to give the same external type to all breeds. It may tend to give the same flesh and bone type, but certainly not the same feather type.

Dr. O. B. Kent

6—The various breeds would differ in shape quite decidedly, depending upon the length, breadth and texture of feathers and amount of fluff in the feathers. If Wyandotte breeders desire to continue to have a short, round egg they should continue to keep the Wyandotte shape, but if a high-class market egg is desired I think that the Wyandotte should be longer in body than it is at present. I do not think that any changes are necessary in applying the score card to Leghorns, Rocks and Reds.

7—I believe that the body capacity must be relatively large in birds of different size, though it should at once be recognized that the weight of a bird increases according to the cube rather than the square and that a Plymouth Rock weighing seven pounds would not begin to be twice as large in frame as a Leghorn weighing three and one-half pounds. The increase in weight is very largely an increase in bones and flesh rather than an increase in size of organs. The organs of a Rock weighing seven pounds are only slightly larger than those of a Leghorn weighing three and one-half pounds.

8—The Leghorn standard weight should be increased or else the size of comb and adjuncts increased in order to have the Leghorn lay a first-class, two-ounce market egg. I feel at the present time that the typical exhibition Leghorn does not lay a first-class market egg. The standard weights for Rocks seem to be a little above the best producing size for the breed. Wyandottes and Reds are all right.

9—Loose feathers or long, narrow feathers are a distinct handicap in production because they tend to slow up the rate of maturity and usually cause low winter production in the various breeds. A change to a fairly short, wide feather without too much fluff would give a bird that would mature more rapidly and consequently be more economical to raise, would lay earlier and consequently would be much more profitable. The long, loose, fluffy feathers as found in the standard Wyandotte generally go with a coarse skin which is frequently associated with low production.

Note: It seems to me there is one thing that is possibly overlooked in the questionnaire. There is a distinct possibility that there is a very close relationship between the size and shape of eggs laid and the size and shape of body. A relatively long, narrow bird may lay a relatively long, thin egg and lay just as many eggs as a shorter, deeper-bodied bird, the latter laying a short, round egg.

REPLY OF JUDGE W. H. CARD, MANCHESTER, CONN.
Secretary Rhode Island Red Club of America

1—No, not one that can be detected externally without handling.

2—Wedge-shape or rectangular and deep.

3—No. It is inside, not outside.

4—Many times a deep, long body contains more intestines than egg machinery.

5—It isn't possible.

6—Should be of proportionate size in all cases.

7—Not necessarily. A Leghorn egg is generally as large as a Rock egg.

8—No. Standard weights are all right. It is the breeders who are wrong in breeding too big.

9—Not always—yet a loose-feathered Wyandotte is never as sprightly as the closer-feathered standard Wyandotte, and standard type of today tends to produce better layers in Wyandottes.

W. H. Card

REPLY OF PROFESSOR WILLIAM A. LIPPINCOTT
Kansas State Agricultural College, Manhattan

1—It depends entirely upon your definition of type. There certainly are some characteristics relative to depth and capacity which are common to high producers. At the same time one would not say that high-producing Leghorns are of the same type as high-producing Plymouth Rocks, though within the breed they certainly have characteristics in common.

2—The characteristics associated with egg type are body capacity and that general refinement which is indicative of the nervous as opposed to the phlegmatic type.

3—Within the limits indicated above my answer would be "Yes."

4—This description is very good, though a truncated pyramid instead of a rectangle would come closer to the description with the base of the pyramid toward the rear.

5—There will be the tendency indicated within a given class, such as the American class or the Mediterranean class, but it does not necessarily make the two classes approach each other very closely.

6—It is pretty early yet to be too dogmatic about definite egg types beyond the suggestions made above. These suggestions would apply to all breeds.

Prof. William A. Lippincott

7—My answer would be "No." The difference in capacity between the two breeds indicated would be in proportion to the extra amount of feed necessary for maintenance for the heavier breed.

8—The answer to your question depends upon your viewpoint. Personally I am not in favor of modifying the standard weights. At the same time I am not sure that the trimmer, smaller and more active individuals of the heavier breeds are the better. As a matter of fact, it is not necessary to modify standard weights for that will take care of itself.

9—We are not in a position to answer this question yet. The Cochins are very loosely feathered and are poor layers. The Cornish are very closely feathered and are poor layers. Not so very much can be said on this as yet either way.

REPLY OF PROFESSOR WM. F. KIRKPATRICK
Connecticut Agricultural College, Storrs

1—Yes.

2—All those characters that have to do with capacity and with vigor.

3—Yes.

4—No. Some modification must be made to allow for breeding.

5—Yes.

7—The body capacity of a Rock does not need to be any larger than the capacity of a Leghorn for the sole purpose of producing eggs. On the other hand, it must be remembered that Plymouth Rocks require more food for maintenance and therefore the capacity must be increased. In other words, the side-line job of maintaining the stock is bigger for the Plymouth Rocks than it is for the Leghorns. Thus it happens that the combined capacity for maintenance and egg production is larger in the heavier breeds.

8—Not quite sure about the Leghorns. Think probably a 4-pound bird would be better than a 3½-pound bird.

9—Yes.

Prof. Wm. F. Kirkpatrick

REPLY OF DR. H. D. GOODALE
Massachusetts Agricultural College Amherst

1—No, except in a very vague and general way, because (this being the answer to question No. 2) egg type in my mind implies definite physical characteristics or confirmation that are associated with heavy egg production. It is possible that a meat type might be defined, but there is so much variation in the types of birds that are heavy layers that I cannot see how one can speak of a definite egg type.

8—Yes, because it probably costs more to maintain a heavy bird than it does to maintain a small one, leaving out of question number of eggs laid. It appears probable that the small bird will produce more eggs per unit of food than a large one.

REPLY OF PROFESSOR HARRY R. LEWIS
New Jersey State Agricultural College, New Brunswick

1 In so far as egg production is dependent upon vitality and vigor, and vigor in turn upon a well-developed body in proportion to the size of the bird, I believe there is what we may call an "egg type." I do not, however, believe that we can lay down a rule that a bird possessed of a certain distinct type is bound to lay more eggs than a bird of another type—at least I have been unable to find it so. There is unquestionably a trend toward a certain development of body characteristics, especially the abdomen in high egg producers. I would be inclined to say that the lengthening and deepening of the body and the filling out of the breast, which is so characteristic when a hen comes into production and is generally taking place as a bird comes from the nonproducing stage into the productive, is what the most of us are thinking of when we talk about the "egg type."

2—When I use the term "egg type," I think of the contrast between the body of the heavy-producing hen and the nonproducing hen. I am thinking only of body conformation, as seen by profile and rear view. I do not consider in that understanding the condition of head parts, tail, legs, etc.

3—I do not think that a general description of "egg type" can be formulated that will apply with actual fairness to all breeds.

Prof. H. R. Lewis

4—The statement quoted from the Utility Score Card—namely, "nearly rectangular"—is probably as near a single definition as we can get in considering the appearance of the heavy-laying hen, as against the nonlaying hen. A pullet which before production may possess a rocker underbody line, will, upon coming into heavy lay, show pronounced development in breast and abdomen, which will give her more of a rectangular appearance.

5—Yes, I feel that the application of the above description will tend to a common type in all breeds, which I think would be a mistake, although I do think we will be inclined in the future to drop some of our faddisms and breed more along utilitarian lines.

6—I would make no change in the wording of the Utility Score Card when applying it to the four breeds mentioned, except to say that the greater depth and the tendency to a rectangular shape is found in the bird in heavy-laying condition and is much more pronounced and developed than it is in the same bird before she has reached production period, or while she is resting during the molt,

7—Yes, the body capacity of large hens must be relatively larger and greater, measured in every way, than the body capacity of a naturally small hen.

8—I do not consider it desirable nor necessary to modify standard weights in any of the popular breeds, with the possible exception of the Leghorn, and here it may be desirable to raise the weight one-half pound in both the pullet and hen.

9—I feel sure that any bird which has been bred for extremely loose feathers has been bred away from high egg production. We find a very close correlation, as we study all our breeds and varieties, between the tightness of the feather and production. Loose feathers go with a rather thick-meated, fleshy bird—a bird with considerable fat deposited under the skin. This is not the type of bird which produces the most eggs for us.

REPLY OF F. J. MARSHALL
Eustis, Fla.

Judge and Poultry Editor, "Southern Ruralist"

1—We believe in an egg type with modifications for different breeds.

2—Egg type as we understand it would include about the following: shape of body, condition of body obtained by handling; shape of head and adjuncts; color and condition of legs and toes—which includes whether the shanks are plump and fat, so to speak, or bony and lean, the latter being indicative of a layer.

3—The body type should be modified for the different breeds about as much as the difference between the scale of points for the different classes.

4—We would call the egg type one of rugged proportions; or rather bumpy in outline, if you will allow the expression—very much more that way in some breeds than others. For example, the Leghorn would run much more that way than the Wyandotte. The rectangular type is all right, but would vary quite a little in the different breeds. For example, the Leghorn and R. I. Red would be a long rectangle while the Wyandotte and Orpington would be a short or deep rectangle —almost a square.

F. J. Marshall

5—Were no allowance made for breed type the strict application of the above would have a tendency to unify the breeds, something that should not occur in our estimation, if we want to retain the characteristic beauties of the different breeds.

6—We would not favor a blanket description for egg type without considerable modification to suit the different breeds. In Leghorns, good length of body, somewhat deeper at rear than in front; broader at rear than in front. Rocks rather long, carried deep the full length of the body. In Wyandottes, the rectangle would be almost a square—that is, as deep as it is long. R. I. Reds would be a long rectangle—in our observation the longer the better.

7—As to body capacity: in our opinion the body capacity of a Leghorn should be relatively larger than that of a Rock, taking weights as a guide. Our reason for this is that the Rock being a real meat breed carries more flesh in proportion to the size of carcass; so much larger thighs, more meat on breast, and so on.

8—Have for years thought the weights on Rock hens and pullets too high for very best egg production. Reds are all right for they have a long-framed carcass with sufficient room to carry their weight.

9—Looseness of feathering is indicative of low egg production, as a general rule. Years ago when we bred Light Brahmas, we found the close-feathered type of a certain prominent breeder much better layers than the loose-feathered ones of another prominent breeder; Cochins the same. When we got a hen that was not much good on account of being too scantily feathered to go in the showroom, she proved to be about our best layer. We think it will hold good in Wyandottes and Orpingtons, although we have not tested it out so much as in the other breeds mentioned.

REPLY OF D. F. PALMER & SON
Yorkville, Illinois

Breeders, Barred Plymouth Rocks

1—Yes.

2—In most cases, a full plump body with plenty of capacity for food and eggs; big, broad head, etc.

3—Hardly, as different breeds have different breed characteristics of type and shape but all must have full or roomy bodies.

4—In many ways, yes.

5—Yes.

7—Yes, because we feel it takes the larger breeds a little longer to develop the egg in the body, while the Leghorn will develop an egg in a shorter number of hours.

8—No.

9—We do not think so.

D. F. Palmer

REPLY OF D. T. HEIMLICH
Poultry Judge, Jacksonville, Illinois

1—Yes.

2—"Well balanced, in that the body itself must be deep, well developed, in breast and abdomen." This from Utility Score-Card description of what can be accepted as the best "egg type."

3—Yes.

4—Yes.

5—No. Each recognized standard breed would still conform in type, size and general shape to what the standard now calls for and yet conform to what is proved the best egg type in general, as lately discovered in handling the various breeds and varieties now recognized as standard.

6—At present the tendency in Leghorns is toward larger, longer and deeper bodies, with the legs set well in the middle of the body. Leghorns of this character will prove the best lookers as well as layers. This same applies to Rocks, Wyandottes and Reds.

7—No. The Rock hen of 200-egg type, weighing seven pounds, would lay the same number of eggs as the three and one-half pound Leghorn pullet, but the eggs would weigh more than the 200 Leghorn eggs from a three-and one-half pound hen or pullet.

D. T. Heimlich

8—Yes. Rock hens at seven pounds would prove greater egg producers than the average seven and one-half pound hen.

9—Not necessarily. Good layers are found among long and short-feathered standard fowls. Having handled more than six thousand fowls, testing them out by the Hogan system this past year, and having given culling demonstrations in twelve counties in Illinois with farm advisers, to more than two thousand farmers, breeders and poultry fanciers, I want to emphasize the fact that fowls well bred and near bred to standard have proved best layers. I can recall but one cross or mongrel-bred flock where a good per cent of layers was found.

REPLY OF E. C. BRANCH
Poultry Judge, Lees Summit, Missouri

1—I do not believe that any definite type can be made to cover all breeds of fowls that can be consistently called "egg type."

2—"Egg type," to me, means capacity both in length and depth and the ability to assimilate feed.

3—I think it impossible to make a general description that will apply to all breeds, so that the average person could intelligently apply it to the numerous breeds that we have and are getting.

4—The above answer will cover this question.

5—A strict application of the Utility Score Card with no allowance for breed type would soon tend to a common type in all breeds.

6—The only changes I would make from the present "Standard of Perfection" in any of these breeds would be to be a little more definite in description of back, breast and body types, because I think after years of observation that the general standard type as now described is tending along towards capacity and therefore along lines of egg production.

E. C. Branch

7—This is a matter of proportion in the way the fowls have been bred. That is to say, the Rocks have been bred to attain nearly twice the weight of the Leghorns. Size of bone and size of frame must be taken into consideration, and while I have no accurate measurements, I doubt if the capacity of a Rock is any greater than that of a Leghorn.

8—I feel that some of the heavier breeds might well be modified in weight. Am speaking now from my judging experience, knowing how hard it is to get some of the breeds up to standard weight without fattening them to their utmost capacity.

9—I do not think looseness of feather has anything to do in the matter of egg production. That is, this one thing alone. Perhaps a loose-feathered specimen might be a poor producer or, in fact, a whole flock of loose-feathered birds might be poor producers, but I do not think this should be put down to the fact that they are loose feathered. I feel sure that the cause is not altogether in loose feathering.

REPLY OF RUSSELL F. PALMER
Poultry Judge, Kansas City, Missouri

1—I do believe in the existence of a definite egg type of fowls.

2—Egg-type characters include a well-developed, deep, full abdomen and if the female is in full flesh as denoted by an examination of the breastbone, then if good egg type she should show considerable space above the ordinary point at the rear end of the breastbone and the underedge of two pubic or pelvic bones. This particular character will apply with equal fairness to all breeds with due allowance for breed size.

3—Straight pelvic or pubic bones which are thin and pliable permit of more rapidly depositing the egg with less wear and tear on the hen's body. We have found that the length of time the average hen consumes on the nest before depositing the egg has quite a decided relationship with the straightness, thickness and thinness of the pubic bones. Fully matured females that do not develop a gristly thickness on each side and points of the pubic bones are invariably females whose characteristics lean towards manufacturing eggs. Those that do develop a hard gristly growth of considerable thickness on each side and point of the pubic bones invariably are females who utilize a large portion of the feed they consume in making fat and otherwise getting themselves into a condition which does not tend to egg production. They are usually found to be more or less glutonous eaters and invariably go broody more often, especially during warm spring and summer months, than do the others.

4—Outside of the wording "pronounced angles rather than smooth curves," the general wording is very good, but it can be modified for different breed types. One thing is positive in my opinion: that portion of the hen's carcass which houses the egg-manufacturing machinery, so to speak, must be properly proportioned so that it will not choke down when the hen is in full flesh and bound to have a certain amount of fat. Many of the very best layers of different strains that I have assisted in trap-nesting had smooth lines and curves, decidedly the opposite from any pronounced angles. This was especially true in the case of one of Thompson's Barred Rock pullets during 1917-18 in the American Egg-Laying Contest, also in the case of one of C. P. Scott's Single Comb Reds and in nearly all of Martin's White Wyandottes. It was also true in the case of two females in the Barred Rock class shown by W. D. Holterman. It was true in the case of every one of a pen of two-year-old Black Orpingtons which went through the contest two years making good records. It was also true in the case of an entry of Black Langshan pullets. It was true in the case of a White Plymouth Rock hen from Illinois and in the case of some Light Sussex from Ohio. About the only varieties where a fair to large percentage of the best layers seemed to have pronounced angles rather than smooth curves, were White Leghorns, Buff Leghorns and Mottled Anconas.

Russell F. Palmer

5—Yes, I believe if no allowance is made for breed type the strict application of the above description would tend to a common type in all breeds.

6—I would make the change of not insisting on pronounced angles in all three breeds, Rocks, Wyandottes and Reds, and I would want to be very careful in stating in the case of Wyandottes that they must conform to a rectangle. Wyandottes are deeper and broader in proportion to their length than many other breeds, which gives in most cases the same number of square inches of space within the body for use of the egg organs.

7—Yes, body capacity should be relatively large in the case of large breeds and approximately doubled that of a 200-egg Leghorn for the simple reason that the Rock is larger boned—proportioned larger all over.

8—I am not sure that I would consider it desirable to modify standard weights in any of the popular breeds with a view to improving egg production. Would have to give this matter serious thought and study in each and every individual case.

9—I am not sure that I am qualified to answer this question. My observation has been, however, that some very loose-feathered Wyandottes were splendid layers, while in the same house and yard were some decidedly hard, close-feathered individuals, also splendid layers. My opinion would be, however, that if loose feathers have a tendency to be correlated to short, cobby bodies, short shanks and short thighs, then they should be dispensed with.

REPLY OF C. T. PATTERSON, SPRINGFIELD, MO.
Breeder of Single Comb White Leghorns

C. T. Patterson

1—Yes.
2—Feminine rather than masculine.
3—Feminine and masculine characters, yes. High and low producers, no.
4—No.
5—Yes.
6—I give the word "type" a broader meaning than many do. I do not believe egg type will be a good guide in the selection of high and low producers but is fairly good in selecting producers and nonproducers.

Personally, I am convinced that hens should be selected on feminine characters and males for their masculinity. This means that in selecting males I do not use the same measurements as with females, but the opposite. For example: instead of selecting males which are wide and loose like the females they should be tight and close. While this rule for selecting the male is contrary to the rules usually given, we find that trap-nest records of dams and daughters prove it to be correct. When a high-producing hen is in a laying condition she is wide between the pelvic bones, which are thin and straight, and has a good distance from the pelvic bones to the back point of the keel bone. Using the back as one line, the keel as another, and the distance from the pelvic bones to the keel bone, as the third line, this makes a wedge with the point in front. With the male the opposite is true.

Some of the characters which should be possessed by the male are; first—size, being larger than the females in the breed; second—size and shape of head, comb and wattles; third—development of long neck, back and tail feathers; fourth—development of spurs; fifth—the habit of crowing, and sixth—the inclination to fight with males but not with females. Some of the characters which should be possessed by the females are; first—smaller in size than the males of the breed; second—smaller head, comb and wattles than the male; third—no long feathers in neck, back or tail as with the male; fourth—no spurs; fifth—singing instead of crowing, and sixth—no fighting with males.

A male may possess all the masculine and none of the feminine characters, yet those characters may be weak and not strong and pronounced, and the female may possess only female characters, yet not distinctly feminine.

7—Same sized eggs and same number require same capacity except more feed and digestive organs to support larger birds.
8—No.
9—No.

REPLY OF PROFESSOR ROY H. WAITE
University of Maryland, College Park

Prof. Roy H. Waite

1—I sometimes apply the term in a loose manner.
2—It is difficult for me to determine which part of the so-called "type" is the cause and which the effect of egg production.
3—I doubt it very much.
4—I have a feeling that past production has a tendency toward rectangular body outline. Curves seem to go with and largely make up the beauty of the ante-laying pullet stage.
5—My opinion—yes.
6—I am not certain that it is to the best interests of breeding to standardize present opinions in regard to "egg type."
7—I cannot see why it should be so large.
8—I am not as strong on standardization as some. I do not think much attention is paid to standard weights.
9—I am not competent to answer. My opinion is that in itself it is not. I can readily see how it might have become associated with low production. That is one of the objections to too much standardization.

Note: In my opinion the time to begin standardization is just about the time deterioration begins, not while progress is going on. When an individual begins to advocate standardization he unconsciously admits he has about reached his limit. I believe the movement to improve breeds in egg production is still young and flourishing.

REPLY OF JOHN S. MARTIN, PORT DOVER, CANADA
Breeder of Regal-Dorcas-Strain White Wyandottes

I certainly do not consider there is any definite egg type, because I have had 200-egg hens of so many different shapes that it is impossible to say that any one type is correct. To trap nest is the only way to estimate egg production. I do think, however, that in days gone by, we did breed our White Wyandottes too short in leg and too fluffy in feather. The present standard for White Wyandottes is a very sensible standard and I believe we are on the right track. I do not agree with some judges, however, who would do away with the fluff on the White Wyandotte and desire to make it just as tight-feathered a bird as the Leghorn. That would be all wrong. White Wyandottes have always been wonderful winter layers and the fact that they had plenty of fluff enabled them to withstand the climatic changes and thus to give a uniform egg yield the winter through. I believe in a moderate amount of fluff and a medium length of shank. In other words, I wish to see the original White Wyandotte characteristics preserved and eliminate the fad for extreme blockiness.

John S. Martin

CHAPTER IV

Breeding and Management of High Egg Producers

High Egg Production Is an Inherited Character and Readily Secured by Proper Attention to Breeding—Progeny Testing Outlined and Relative Merits of Progeny Testing, Mass Selection and Selective Flock Breeding Described—Special Suggestions on Management of Breeding Stock in and out of the Breeding Season

NO subject relating to the poultry industry has received so much attention in the past ten years as that of breeding for egg production. Starting with no more definite clue to the solution of the problem than that presented by the popular phrase "like produces like," both practical breeders and scientific investigators have established the fact not only that different degrees of productiveness are inherited and transmissible, but that the capacity for heavy production can be developed so that, year by year, successive high records may be made for which no precedents exist in the known history of the strain.

It is freely admitted that there is a good deal of uncertainty as to just what laws are invoked in bringing about the results secured. Of two persons resorting to apparently identical methods one succeeds and the other fails. But regardless of whether the exact formula for breeding is understood or not, this practical fact stands out—has been clearly demonstrated not in one but in hundreds of instances: The man who persistently breeds from heavy layers and who uses males descended from such, will certainly bring up the average production of his flock. There is, of course, a practical limit to such advances, and perhaps there may be a point at which progress must stop and the strain "break back," but nothing is certainly known in regard to this. The most reasonable position to take in the light of present knowledge is that when a limit has apparently been reached in a certain strain, it marks the range of the breeder's skill or persistence rather than of the actual possibilities for development.

In the opinion of many, one thing that probably has much to do with the disappointments and inconsistencies that have developed in efforts to breed for increased egg production is the failure to recognize the fact that "fecundity," while transmissible, is not a single character but a combination, possibly a highly complex one. It is not even clear just how much of the increase in any given strain where exceptional results have been secured, should be credited to breeding or inheritance and how much to the cumulative effect of better care, better feeding, etc., or even to more accurate records.

The Breeder and the Geneticist

The breeding of poultry has received a great deal of attention from scientists who are primarily interested in the study of inheritance and who use fowls in their experiments chiefly because results can be secured more quickly than with larger and slower-maturing animals. Along with other characters fecundity has thus been investigated from the geneticist's viewpoint and claims put forward in regard to methods of inheritance that have been the occasion of more or less controversy.

We do not feel that this controversy has any place in this book. The subject here is treated almost exclusively from the standpoint of practical breeders and experimenters who have met with definite success in the development of high producers and in the establishment of more or less permanent strains of such. These, either in written communications or in personal interviews with the editors of this book, tell in more or less detail just what they have been able to accomplish and how they did it, but it will be noted that they are extremely conservative regarding "laws of breeding," also that there is not complete uniformity among them as to methods of procedure. Apparently there are more ways than one of arriving at a similar result, under which conditions it is the part of wisdom not to be too dogmatic in regard to science or practice.

MAINE STATION'S HIGHEST PRODUCER

Breeding for egg production has been practiced at the Maine Station since 1900. This is the highest producer on record there to date—298 eggs in 12 months; two-year average, 228.

For the benefit of those who want to have, in brief form, a general statement on inheritance of fecundity from the geneticist's point of view, an article on this subject by Dr. Raymond Pearl will be found in Chapter IV, Part III. Whether Dr. Pearl's conclusions are accepted or not, the article should have the careful attention of those who are earnestly striving to get at all the facts in the case.

One of the best known illustrations of early success in building up a heavy-laying strain by selection, without the aid of the trap nest, is that of the Wyckoff line of Single Comb White Leghorns which the originator succeeded in developing to the point where a flock of 600 made an average of 194 eggs per hen in 12 consecutive months. And that this was a firmly established character is shown by the testimony of the well-known author, John H. Robinson, who states that for a period of ten or twelve years, dating from 1897, in all cases which he investigated where remarkable small-flock egg records were secured, the birds came either wholly or principally from Wyckoff stock.

On the other hand, one of the conspicuous disappointments in early attempts in breeding for increased production was at the Maine Experiment Station where trap nests were systematically used, and high-producing hens—meaning those with pullet-year trap-nest records of 150 eggs or over—were mated year after year to males from hens with records of 200 eggs or over. At the end of eight years of such breeding the average annual egg yield showed an actual decrease of 18 per cent, or from an average of 143.4 to 113.2 per hen.

A similar result was secured at Cornell University, where hens with good records were mated with males known to have come from high-producing hens with the result that production was noticeably lower at the end of a six-year period than at the start (see article by Dr. O. B. Kent in Chapter IV, Part III).

Methods of Breeding

Growing out of numerous tests made at experiment stations and in the yards of breeders three methods of breeding have developed, these being known by the terms of "progeny testing," "mass selection" and "selective flock breeding." Individual fowls have been found to vary widely in their ability to transmit capacity for heavy laying, some being remarkably "prepotent" in this respect, while others with high records have proved quite worthless as breeders. In numerous instances the daughter of a high producer may make a most inferior showing but in the second generation records equal to or better than those of the granddam may be made. The sons of high producers have been found to transmit high-laying ability with remarkable certainty and there are some who contend that transmission is solely through the male line. This belief is not generally accepted, however, and the most successful breeders insist upon having both males and females of high-producing ancestry in the breeding pen. Speaking on this subject before the Second Annual Judging School at Cornell University, Professor Dryden, whose achievements in breeding for high egg production have been truly remarkable, made the following statement relative to his methods:

Early Attempts at Breeding for Increased Egg Production

"My breeding work and experiments for a period of years have been for egg production. I have sought to discover whether, by depending on production records, we could increase production. Our first problem was to learn whether or not egg production is inherited and transmissible. In this work the trap nest is indispensable. Most of my experiments have been with Plymouth Rocks and Leghorns and with a cross of these two breeds, to which cross I have given the name 'Oregons.'

"Confining my efforts to high egg production, with but slight attention paid to exhibition points, we have

TWO BRED-TO-LAY SISTERS AT KENTUCKY EXPERIMENT STATION
Hen on left has record of 260 eggs; is daughter of 203-egg hen and full sister of bird on right, which has a record of 220 eggs. Photo from Professor J. Holmes Martin.

made some remarkable gains and in our breeding work at Corvallis are now securing a high percentage of 200-eggers in Plymouth Rocks, also a goodly number of 250-eggers, and we have been favored with two Plymouth Rocks that each have reached or passed the 300-egg mark in one year. Among the Leghorns also we have had two birds that passed the 300-egg mark, one being our top-notcher with 309 eggs, and in this breed or variety (S. C. W. Leghorns) we now find it easy to get 200-eggers.

"I am convinced that the power or ability to lay a large number of eggs is inherited and can be transmitted. Just how much influence the male has or what degree of power the female possesses in this line, I am not prepared to say. That appears to be an unanswered part of our problem.

"With us it has been a matter of good-laying vigor —meaning absolute health and good size for the breed plus selection for high egg production as recorded by the trap nest. We have worked along this line until now from our breeding pens of Leghorns we get each season over fifty per cent that lay 200 eggs or better.

"We go mainly by trap-nest records, not by type. Thus far I have not been able to get the right type in my mind, so it seems. Whenever I think I am right, the hens soon prove I am wrong. It is easy to state facts if you have them, although quite often it is hard to explain the facts. I can give you no end of facts in the form of actual production, as recorded by the trap nests, but I am still weak on explanations. Certain things are clear enough as evidence, but it is hard to tell why.

SOME SAMPLE RECORD LAYERS AT MAINE EXPERIMENT STATION
First hen's record 220 in pullet year; second hen laid 204 in first year and 210 in second; third hen laid 256 in first year; 2-year average, 204.5. Photo from Maine Experiment Station.

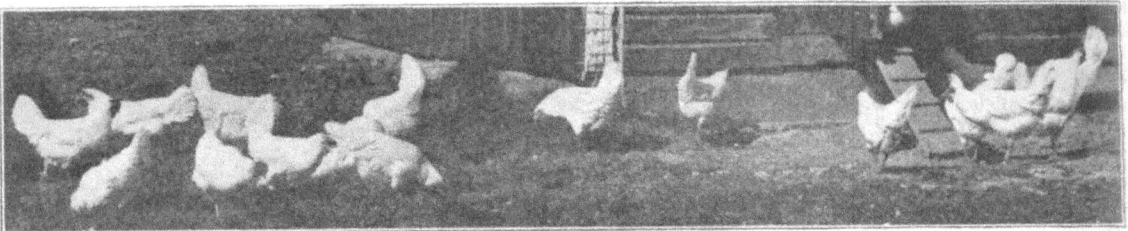

CORNELL HIGH-LINE PULLETS—COMPARE TYPE WITH THAT OF LOW-LINE BIRDS ON RIGHT
This strain has been bred from high producers for a number of years. See graph on page 27, also Dr. Kent's comments in symposium in this chapter to learn how production has increased.

"Take this question of whether the male or female really predominates in transmitting fecundity or high egg production: Our records prove it on both sides! It is possible to get a better check on the male. In one sense at least he is more important, being half of the pen. Clearly there is a remarkable difference in males as to their prepotent power, or lack of it. This has been plain to me for a period of ten years or more. We have had a few phenomenal breeders, but why? That I cannot tell you. They had the breeding power or prepotency—of this I am sure—and for the present I shall have to let it go at that, till the biologist makes plain to us what it is and how to direct it in our interests."

Experiment in Breeding at the Kentucky Station

One of the best illustrations of the fact that both male and female have a decided influence upon the productiveness of offspring, also that either may fail to transmit high production in individual instances, is given in the record of a breeding pen at the Kentucky Experiment Station. Writing in regard to the performance of this pen, Professor J. Holmes Martin, of the Poultry Department at that institution, says:

"The S. C. W. Leghorn male bird AA12 was out of a 211-egg hen, who was a granddaughter of a 260-egg hen. AA12 was sired by a son of a 314-egg hen. When mated to twelve hens with records between 53 and 258 eggs, he sired 48 pullets that averaged 152 eggs. The accompanying table gives the individual records of his daughters. Note particularly the 203-egg hen that was the mother of ten pullets which laid 1,778 eggs. This hen has proven the most wonderful breeder we have had.

"By studying the figures in this table it will be noted that the production of a hen does not give an accurate index as to her breeding power since the 203-egg hen proved to be a better breeder than the four hens with higher records than she had made. A highly interesting point in the table is the fact that a 102-egg hen when mated to AA12 had two daughters laying over 150 eggs, a 92-egg hen had two daughters laying over 170 eggs and a 53-egg hen had a daughter laying 203 eggs. Hence it is obvious that the high-producing power of these five pullets must have come to them from their sire."

Individual Records of Daughters of Hens Mated to Male AA12

Hen Record	Record of Daughters	Average
258	100	100
228	234, 163, 132	176
215	229, 188	209
205	190, 188, 181, 180, 169, 147, 134, 108	161
203	112	112
203	260, 219, 197, 194, 188, 173, 170, 138, 137, 102	178
196	195, 176, 124, 120, 115, 84, 80, 79	122
145	171, 165, 147, 117, 109	142
131	150, 60, 40	83
102	188, 153	171
92	193, 171, 89	151
53	203	203

20 Pullets sired by AA12 averaged 197½ eggs each.
48 Pullets sired by AA12 averaged 152 eggs each.
26 Pullets laid over 152 eggs each.

Progeny Testing

In order to be certain of breeding only from individuals that are capable of transmitting these characters, progeny testing is in high favor. This involves the breeding of high-producing females to males from high-producing females and the systematic use of trap nests to determine which individuals are capable of transmitting this character, with a view to using their offspring alone in the breeding pen. Progeny testing amounts to judging the hen by the performance of her offspring, not by her own record. There are few instances in which this method of breeding has not proved successful, though in some cases the first generation has fallen considerably short of the production of the original dam. Where this is the case the second generation usually recovers lost ground and sometimes establishes new high records.

This method is open to the objection that it involves the use of trap nests and the keeping of elaborate records, both of which are expensive operations and for that reason not well adapted to the use of the average poultry keeper. The earnest breeder, however, who is seeking to establish a distinct high-producing strain cannot afford to neglect progeny testing, because in the long run it is the best known means of securing the desired result. For practical suggestions in breeding by progeny test, see article entitled "Trap Nest Pedigree Breeding for Egg Production" by Dr. Kent, in Chapter III, Part III.

Mass Selection

The method of breeding followed at Maine up to 1908, also the first six years of breeding at Cornell, to which reference has just been made, is what is known as mass selection, in which

HOUSES IN WHICH CORNELL HIGH AND LOW-LINE STRAINS ARE KEPT

CORNELL LOW-LINE PULLETS—COMPARE TYPE WITH THAT OF HIGH-LINE BIRDS ON LEFT
This strain has been bred from low producers for comparison with high-line strain. Note graph on this page showing how high and low-line strains compare in productiveness.

females known to be high producers are mated to males descended from similar or better females, but without any attempt at pedigree breeding or to check up on individual capacity for transmitting fecundity. This method of breeding is by no means certain in its results, as has been demonstrated at Maine and Cornell and in many private breeding yards. It has, however, proved successful too many times to be dismissed as ineffectual. As a matter of fact, the conditions of scientific experiments in breeding are such that the failures of mass selection in the two notable instances just referred to may possibly have to be attributed to conditions in no way associated with the method of breeding adopted.

The following instructions in practical breeding given by Professor Dryden at the conclusion of the address referred to on page 25, while they include progeny tests, by no means discredit mass selection:

"First: Use the trap nest; then from the records in the pullet year select the breeders for the next year. Select those that have made high records and those that may not have made a high record but have laid heavily in the fall months at the beginning of the year, or those that have laid heavily late in the following fall, or those that lay heavily in the two months of heaviest production.

"Second: Use males from highest record hens and of good vigor.

"Third: Don't expect too much in the first generation. Mate up, if possible, a number of pens of good-laying hens. Trap nest the pullet progeny and before the breeding season comes on, you will get a line on the breeding quality of the sires of the pullets. If any one male produces daughters of higher fecundity than the others, use that male again on the same hens if possible. You may not find an exceptional male in the first year, but don't be discouraged. Keep at it. BY USING HIGH-RECORD MALES EACH YEAR WITH HIGH-RECORD HENS THERE IS BOUND TO BE AN INCREASE. Other things being equal, the males or the hens with the longest line of heavy producers behind them will be the best.

"Fourth: Don't offset high-record pedigree with type, though the temptation may be great sometimes to sacrifice egg-laying pedigree for type.

"Fifth: It is not necessary to inbreed to get good layers. Avoid it as much as possible. Line breeding is as near as you need to come to inbreeding."

Selective Flock Breeding

This method differs from mass selection in this respect: While in the latter trap-nest records are required so that only females of known high production are used, choice of individuals in selective flock breeding is determined by external characters recognized as identifying heavy layers. This method of selecting breeders is much simpler than trap nesting and, obviously, less accurate as regards the exclusion of low or medium producers. For this reason it is not to be expected that it will result in as rapid improvement as either progeny testing or mass selection, but there is ground for expecting good results, particularly if progeny-tested males are procured for mating with the selected females, as is earnestly recommended in all cases. For instructions in selection of layers by external characters see Chapter V.

Breeding Methods and Results at Experiment Stations

Space is not available in this chapter for recording the details of breeding for high egg production as practiced at numerous state experiment stations, but in Part III will be found several articles showing how fowls have been successfully bred for high egg production at these institutions, and to these the reader is referred for details.

Line Breeding, Inbreeding and Outbreeding

The scope of this work does not include a complete discussion of the subject of mating and practical management of breeding fowls. For this information the reader is referred to "Fundamentals in Poultry Breeding," a comprehensive and authoritative book published by Reliable Poultry Journal Publishing Company. There are, however, a few special points to which, because of their practical importance, attention should here be called.

Probably no definition of the terms line breeding, inbreeding, etc., could be formulated at the present time

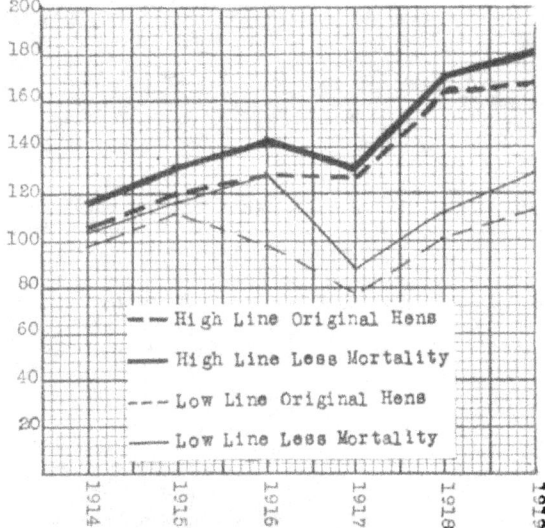

GRAPH SHOWING PRODUCTION OF HIGH AND LOW-LINE STRAINS AT CORNELL UNIVERSITY

that would pass everywhere unchallenged. But in this book these terms are used in the following sense:

Inbreeding is the mating of **closely** related individuals.

Line breeding is the systematic mating of males and females originating within the same strain, and therefore related. It usually starts with inbreeding but once the "line" is established, close-relationship matings are avoided.

Outbreeding is the mating of unrelated individuals of the same breed and variety.

The last named is the method practiced by most poultry keepers. It is the "safe" way—and the poorest as regards improvement in egg production or any other desired character.

Line breeding is generally recognized as the quickest and, in fact, the only method of "fixing" desired characters—of establishing a strain that can be depended upon to reproduce itself with reasonable certainty, at the same time avoiding the dangers that admittedly accompany too close inbreeding. The popular prejudice against all forms of inbreeding is based on many known instances of degeneration that have resulted from relationship matings where proper precautions were not observed. In point of fact, however, there are few well-established producing strains that are not more or less inbred, nor are such strains necessarily at any disadvantage as regards health, vigor, fertility of eggs or any of the characters the absence of which denotes physical degeneracy. It may or may not be advisable for the individual poultry keeper to practice inbreeding, but it should be clearly understood that there is no "curse" on relationship matings, close or distant; nor will there be any degeneration, provided proper attention is given to health and vigor. Vigorous fowls will produce vigorous fowls, regardless of whether they are related or not.

CORNELL HIGH-LINE HENS HAVE GOOD-LAYING TYPE

One method of inbreeding, which may be called indiscriminate inbreeding for lack of a better name, is entitled to more serious attention than it often receives from the beginner. This method, which has the approval of some well-posted breeders, consists simply in confining breeding operations within a given strain or "line," but entirely ignoring the degree of relationship—that is, mating the fowls to the best apparent advantage regardless of whether they are closely akin or not. In a flock of fair size such a method may result in a limited amount of close inbreeding, but there is no reason to believe that, in the long run, fowls so produced will be inferior in vigor to the most elaborately line-bred stock.

It is not to be expected, of course, that any other method will, or can, produce fowls equal to those secured in painstaking line breeding. But not every one will care to secure high quality at such a price, and there are thousands of beginners who, when confronted with the complications of that method, will throw up their hands in despair. It should be encouraging to these, therefore, to realize that they can keep their strains pure, and can secure many of the advantages of inbreeding without endangering the health of their fowls, and at the same time may greatly simplify the details of care and management.

Perhaps as fair a statement of the relative values of the two systems as can be given is to say that for the highest development and improvement in fowls—for definite strain building—line breeding is indispensable. But for those whose ambitions are more modest, indiscriminate inbreeding will answer their purpose. In fact it may be doubted whether the average beginner will have anything more to show for the labor and involved record keeping of systematic line breeding than he could achieve with much less trouble and expense by simpler methods. At the risk of repetition, however, it should be clearly stated that, in either method, slackness in selection or carelessness in using fowls in anyway inferior in vigor will soon bring disaster.

Special Points in Breeding and Management

Breed Only Standard Fowls. It is not necessary that the flock of the commercial poultry keeper should be of exhibition quality, but it is unwise to concentrate on egg yields to the exclusion of everything else. Increased productiveness, it is true, can be obtained in this way, but it is believed to be much less certain in transmission, and the stock will be of so inferior quality that it will have no value for breeding, and in the long run the owner will discover that for the sake of a little more rapid improvement in average egg yield he has been sacrificing his opportunity for producing good-quality breeding stock eggs for hatching, etc., which, if from an established high-producing line, are always salable at good prices.

Hens and Pullets as Breeders. In choosing breeding stock the preference usually should be given to hens, rather than pullets, for various good reasons. Only those who use birds that have gone through at least one laying year can be reasonably certain that they have in their breeding pens nothing but high-record individuals. In addition, one or two-year-old hens, if they have been properly handled, will be found to be in much better breeding condition than pullets and therefore more likely to be able to transmit such characters to their offspring. Heavy-laying pullets, especially if they have matured early and have been laying continually since fall, are apt to be more or less exhausted and unfit for breeding. It is possible to use pullets successfully, especially where the birds are trap nested, as suggested by Professor Dryden in his article in Chapter V, Part III, but the practice calls for extreme care and it is not a method to be recommended to the inexperienced. In this connection it is worth while to keep in mind the belief of many good authorities that the characteristic exhaustion of pullets after months of laying is not inevitable, but is the result of improper handling and feeding. Where the birds are properly fed and kept up in weight and in condition, it is believed that not only can good hatches be secured from their eggs but that they are capable of transmitting high vitality to their offspring.

Breeding for Early Maturity. It is common experience among those who use trap nests or otherwise are able accurately to observe individual production to find that the first pullets to mature and begin laying in any given brood are the highest producers for the year. Early maturity is clearly an inherited character and it is desirable to encourage it so far as it can be secured without actual forcing or resulting in underdevelopment. Particularly in the larger breeds early maturity is desirable, as it not only reduces the length of time that the pullets must be fed before they become profitably productive but, by bringing them into early laying, the opportunity for making good records before the next summer's molt is greatly increased. Pullets that are late in beginning to lay are rarely able to overcome the handicap.

That early maturity is inherited and may be made a definite character in a strain is proved by the experience of Professor W. R. Graham of Ontario Agricultural College who, through years of selection has developed a strain of Barred Plymouth Rocks regularly maturing at less than six months of age—the same time required by his strain of White Leghorns. (See interesting article on the way in which this strain was developed, in Chapter VI, Part II.) Leghorns naturally develop at a much earlier age than fowls of the larger breeds. Illustrations of two exceptionally early-laying pullets are shown on pages 33 and 34. Strains in any breed however, can be brought to mature regularly in much less time than the average for the breed, when selection is based on that character.

Breeding for Quality. A point that should receive particular attention is quality in eggs, meaning by this the characters that make them most acceptable in high-class markets, such as size, shape, color of shell, etc. In many instances the eggs laid by record-making hens are noticeably inferior in size. It requires some strength of character to exclude from the breeding pen a hen that may have the best record in the flock, doing this because her eggs are inferior in one or more of the foregoing characters. Each of these, however, is directly transmissible, and indifference to this fact will, in a short time, bring about marked deterioration in the strain. No matter what method of breeding is followed, all eggs set should be carefully selected, using none that have characters that would interfere with their ready sale in first-class markets (see interesting article on this subject in Chapter IV, Part III).

Selecting Males. In the selection of males, tests for egg type and egg capacity are applied just as in the case of females. It is not expected, of course, that males will show the same development of abdominal capacity, but there is a marked difference in the physical proportions of a male from a high-line ancestry as compared with one from a low line, and it is not a difficult matter to distinguish between them, particularly the extremes. Select males in all cases, whether cocks or cockerels, that are masculine in appearance but not coarse, giving careful consideration to intelligence, fineness of skin and constitutional vigor. Full breasts, good width of the pelvic arch and reasonable depth between arch and keel are indispensable characters where a good degree of productiveness in the offspring is desired. The combination of characters such as are here described, along with heavy-laying ancestry, should be the breeder's constant ideal.

A QUESTIONNAIRE ON BREEDING FOR IMPROVEMENT IN EGG PRODUCTION

To secure an expression of opinion regarding various practical problems relating to this subject a list of 15 questions was prepared and sent to a few experiment station workers who have specialized in this line. Their replies which follow should prove extra interesting and suggestive as to the present trend in theory and practice.

List of Questions

1—When did you begin breeding your present high egg production strain?

2—Was the strain founded on a single individual, a pair, or a number of high producers?

3—What was the source of the foundation stock used in breeding?

4—What were the records of the birds used in the foundation pen?

5—Have you at any time after the first year introduced unrelated blood in your strain?

6—If so, have you found that new blood has any influence on production, favorable or otherwise?

7—Has there been a steady improvement in average production since your line was started, or have there been "recessive cycles," alternating with further advances?

8—Give the yearly flock average, if available.

9—What has been the record of your highest hen, year by year, from the start?

10—Comparing present generation with foundation birds, what changes have occurred in size and general appearance or type?

11—Have you observed that the highest producing birds are poorer breeders as regards fertility or livability of chicks than average of flock?

12—Have you found medium-good layers better for bringing up the average than high-record layers?

INFERIOR TYPE OF CORNELL LOW-LINE HENS IS READILY NOTED

13—Is there anything in your records to warrant the statement sometimes made that daughters of extra-high producers are comparatively poor layers, but exceptionally good breeders of high producers?

14—Is there anything in your experience to indicate that a "break-back" is inevitable in strains that reach a very high level of production, or can the production of such families be maintained at a high level and possibly further improved by constant selection and breeding?

15—Do you regard it as possible or practical to maintain production at a high level by using unrelated blood, year after year, even when it comes from high-producing strains?

REPLY OF DR. O. B. KENT
Cornell University, New York College of Agriculture

1—Our general breeding began in 1906 and pedigree breeding really began in 1913. Prior to 1913 the female line was the only one that was studied or pedigreed.

2—The strain that we have goes back in a large measure to ten hens that we had in 1906. Some of the hens were good and others were fairly poor.

3—I believe that the stock goes back ultimately to the Wyckoff stock.

4—I do not know just what were the records of the original ten hens. There is a confusion in the records

A NORTH CAROLINA PEDIGREED HIGH PRODUCER
Record, 263 eggs in 12 months. Her dam was from a dam with record of 223 eggs; great-granddam laid 174. Sire was from hen that laid 234 eggs; granddam, 242. Photo from North Carolina Experiment Station.

of 1906 so that I have not felt confident as to the records of the birds. Unfortunately the same numbers were used on hens in several different pens. I am sure that the records were pedigreed from only one pen but our records do not definitely establish which hens were the ones that were used.

5—We have a number of times introduced unrelated blood in the flock.

6—We have generally found that the introduction of new blood has improved the earliness of maturity and consequently has increased to a considerable extent the winter production of birds, but has not in all cases improved the annual production. In fact, I will say that in most cases while the winter production has been higher the annual production has been lower.

7—As given above, when all birds were bred from and little attention paid to selection the egg production decreased year by year. Since then it has been practically a continuous improvement.

8—Beginning with 1914 our high lines—that is, birds bred for egg production—have averaged as follows:

```
1914—106        1917—127
1915—120        1918—164
1916—128        1919—169
```

These figures are based on the number of hens originally put in the pens, and not on the number of hens alive at the end of the year. The records for the hens that lived throughout the year beginning with 1914 are:

```
1914—117        1917—132
1915—132        1918—171
1916—144        1919—182
```

It will be noticed that the record for 1917 is a drop off from 1916. I believe that was due to rather late hatching and possibly to poor handling at the end of the summer. The same birds laid an average of one hundred and thirty-six based on the original number and one hundred and fifty on the basis of those that lived in their second year, which is not normal production. During the same period the low lines averaged as follows for the original number of hens:

```
1914— 98        1917— 78
1915—112        1918—102
1916— 98        1919—118
```

The hens that lived throughout the year:

```
1914—105        1917— 89
1915—118        1918—111
1916—128        1919—128
```

By comparing the two sets of figures it will be noticed that in the last three years the range between the high and low-producing birds has been about the same, and that the range in the last three years is very much greater than that obtained in the first three years. As to how long it will take us markedly to increase the difference I do not know. My present feeling is that it will be extremely difficult to breed a hen so that she will lay less than one hundred eggs when well fed and taken care of. I rather suspect that the low line will tend to remain constant rather than decrease, and that whatever difference is obtained will come through the raising of the high line rather than the lowering of the other.

The graph on page 27 will show the difference possibly better than the figures will do. I feel that by proper selection and handling it is going to be possible for us to get flocks of birds that will do well over two hundred eggs, within a comparatively short space of time.

9—Beginning with 1908 our highest year's record has been as follows:

```
1908—216        1914—205
1909—257        1915—210
1910—243        1916—209
1911—212        1917—258
1912—223        1918—251
1913—260        1919—257
```

10—Since I was not here when the foundation birds were started I cannot tell as to just what changes have taken place. However, we still have some of the second generation birds on the farm at the present time. We have been selecting for practically the same ideal or type of birds during the fourteen years. Our present birds are larger and neater than the birds of 1914-15 and are probably better finished than the foundation birds.

11—In our observation the highest producing birds, as a general rule, have been poor breeders in the early part of the hatching season before the birds have gotten outdoors on the grass. In the latter part of the hatching season—that is, May— there seems to be comparatively little difference between the highest producing birds and the remainder of the flock.

12—I do not believe that the average of the flock can be raised by breeding from the medium hens. Whatever increase comes must come from breeding from the best in order to use any chance variation or improvement that may have come. Of course, the chances of obtaining a high-producing hen from the highest producing hens in the flock are much less than of obtaining a high-producing hen from the medium hens in the flock, but that is not because of the inferiority of the high hen as a breeder but merely because of our inferiority in numbers. I believe that the chances of obtaining a high hen from a high hen are better when an equal number of hens are used.

It should at once be recognized that there are high hens and high hens. While eggs may be eggs hens are not hens by any manner of means. Whenever we have a small hen that makes a high record by laying a small egg and is practically always on the ragged edge of condition we do not consider that such a hen is worth breeding from except as a curiosity to see what she will do, but a good, well-built hen that lays a good-sized egg and makes a high record is more likely to produce daughters that are good than a hen that makes a lower record.

If however, the high hen is forced to stop laying in the fall so as to get in condition for the next breeding season her chicks will undoubtedly be much better. Probably the birds that come from a hen that

A PEDIGREED MALE AT WISCONSIN STATION
The dam of this bird is 254-egg hen shown on opposite page; sire is from hen with record of 239 eggs; grandsire from 248-egg hen.

has only laid two hundred eggs because of being stopped would be much better than birds that would come from the same hen if she were allowed to go on and lay two hundred and fifty eggs and not go into the breeding season in good condition.

13—It occasionally happens that there is a tendency to skip a generation but am becoming more confident that it is due to the failure on our part to properly handle a hen and properly prepare her for the breeding season. A dairyman does not expect a good calf and good production unless he gives a cow a good rest prior to calving, and yet there is a tendency for poultrymen to force hens the year around.

14—I do not believe that a break-back is necessarily inevitable. It seems to me that a break-back is probably due to climatic conditions or to the human factor—such as if the birds do not happen to be handled equally well. Every year the seasons are not equally favorable and uniform production could not be expected each year.

15—Am not at all settled in my own mind as to whether it is possible to maintain high production by using unrelated blood or whether it can be better done by using related blood. I rather suspect that the results obtained in breeding corn would also give the same results in breeding for egg production. That by developing, say, highly inbred lines and crossing these lots, that it would be possible to produce a large flock of high producers. East and Jones have been very successful in doing this with corn. By intensively inbreeding the various defects and weaknesses appear and are eliminated so that when the inbred lines are combined much greater strength results. It should be possible by improving inbred lines to increase the general flock production.

REPLY OF PROF. J. G. HALPIN
College of Agriculture, University of Wisconsin

1—We began breeding our high-producing strain of Single Comb White Leghorns in 1912—that is, we began actually to breed that year for high production.

2—Our strain was founded largely on four individuals. Two of these have proven to be exceptionally good breeders and have given us a lot of our best stock.

3—The foundation stock used in our breeding was an old strain of Single Comb White Leghorns that the writer has been breeding since 1905.

4—At the start we had one hen with a record of 220, another with a record of 202, and another with a record of 198 eggs. From the first two hens we raised sons that gave us splendid producing daughters. From the third hen we raised some of our very best producing hens.

5—We have introduced new blood, especially from the Kansas Experiment Station and in one instance from the Oregon Agricultural College. Some of our best producing pullets have been the result of a cross of our strain with the Kansas birds.

6—We have had all sorts of answers to this question. In some instances the new blood has been a decided help. In other instances we have not had any favorable production after the introduction of new blood, and we have discarded all of the birds raised from such a cross.

7—Our progress has been steady. Our average production at the beginning was 120 eggs for the entire University flock of W. Leghorns and last year's average was 185 eggs with a steady increase each year. Our highest record has been as follows: First our original hen with a record of 220 eggs who gave us a granddaughter by a son with a record of 248 eggs. Then followed the record of 252, then 262 and 264, and last year 284.

WISCONSIN STATION HEN NO. 242
RECORD 264 EGGS
Laid 264 eggs in first year, 200 in second and 99 in third.

10—I am not sure, but I think our birds have been getting a little bit larger than they were originally. The difference is slight, if there is any, and does not show in the weights that we have taken. I mean by that, that although we have weighed a good many pullets each year, yet by November 1 weighings there is not very much of a change. There has been some difference in weight each year but that has depended upon our feeds and feeder more than upon the birds, in my estimation.

11—We have observed that the highest producing birds are excellent breeders in regard to fertility or livability of chicks in some instances and in other instances we have high producers that are very low. Our average has been worked out several times and I am glad to say that we are extremely fortunate in the taking in any year. Our ten highest producing hens and our ten lowest producing hens have been pedigreed and our ten highest have always given much the better results. It has also happened that our ten highest hens have done better than the general flock average. However, as I said above, we have some cases of high-producing hens that are extremely poor as breeders.

ANOTHER WISCONSIN PRODUCER
RECORD 254 EGGS
Hen No. 337 laid 254 eggs in her first year; 212 in her second and 157 in her third. Is dam of male shown on opposite page.

12—This depends entirely upon the individual, in my estimation. If you can get high-producing hens that come from a high-producing family, then you are very likely to have good breeding results from them. On the other hand, it is very unusual in my experience to have a high-producing hen that comes from a poor line and have her amount to anything as a breeder. In case of a pullet of poor-producing ancestry that comes high, the only thing to do is to try her out as a breeder. She isn't worth anywhere near as much as a high-producing hen with several generations of high production behind her.

13—We have had all sorts of results. Our first high hen gave us ordinary daughters (average 150 to 160 eggs) till her last year, when she was mated to a Kansas cockerel, and her daughters averaged around 220 eggs apiece for the year.

14—I think there is a limit. After you reach that I think you might as well be satisfied to stay there.

15—Yes. If you can get new blood from the right place. Keep bringing in unrelated blood from a high-producing strain.

REPLY OF PROF. GEORGE R. SHOUP
Western Washington Experiment Station

Professor Shoup's reply to the list of questions addressed to him was not made item by item, but he gave detail information on certain points of his breeding practice which is here repeated, in substance, as he wrote it. Some of the results secured by him are presented in the accompanying table:

With birds handled in far greater units than larger animals, the problem of carrying accurate individual blood-line charts is too complicated for most, so, to get the work on a practical basis, outbreeding is practiced here. Outbreeding is crossing birds of like breed, but of different family. There are in this territory quite a number of breeders with high-grade trap-nested birds so that it is a fairly simple matter to get unrelated stock of like grade. By using unrelated males the pitfall of accentuating defects by line breeding is avoided, and eventually enough families are carried on one plant to permit of what amounts to outcrossing within the strain. By raising the grade of breeders as much as possible each year through selection and mating them to "get" high producers, the curve of production gradually rises until, as has happened at this station, whole flocks of breeders up to 400 or more are of the 200-egg grade or better.

Record of Pullet Flock of S. C. W. Leghorns at West. Wash. Experiment Station Showing How the Stock is Gradually Making Higher Averages Along With Larger Flock Units

Pullet year	1915-1916	1916-1917	1917-1918	1918-1919
Average eggs from whole flock record	95.3	149.5	184.4	178.4
Average of all birds marketed and kept over	98	158	194	196
Average of all birds selected to keep longer than one year	172.3	197.1	219.1	218.8
Total number of birds in group	204	330	558	767

At the end of the first laying period the birds physically fit and having records from 200 to 250 eggs are placed in the breeding pen. No bird, however, no matter how high her record, is selected as a breeder unless her eggs are chalk white and average 24 ounces to the dozen and her physical conformation shows her to be qualified to be a potential mother hen.

Vigor, we claim, must always be the first requisite in selection of the male; high egg production of the immediate dam, second; high production of other ancestors, third; and standard qualification, fourth. The breeding pen illustrated on page 35 consists of some of our best birds, selected by trap nest and physical conformation from more than 1,000 layers of previous years. The male happens to be entirely unrelated to any of the females as he was secured through an exchange with one of our best known trap-nest breeders. His dam's record is only 280 eggs, but he had special vigor and desirable standard qualifications. This pen will produce the males which, as cockerels, will be used in the general breeding pens next year. The hens in these breeding pens will be one year older than the cockerels and will be distantly related to the males, but there will be no mating of these cockerels from this selected pen which will show more than one-eighth relationship. We believe one-eighth relationship is sufficient distance to insure continued vigor and stamina in the offspring. We never breed one-fourth related birds to say nothing of one-half, but we do favor crossing unrelated strains of known egg production.

SOME CHOICE LINE-BRED "BELLE OF JERSEY" COCKERELS AT NEW JERSEY EXPERIMENT STATION

REPLY OF PROF. H. R. LEWIS
State Experiment Station, New Brunswick, N. J.

1, 2 and 3—Belle of Jersey was hatched in 1911, being a direct descendant, two generations removed, from an exceptional, heavy-producing Leghorn hen, 70-C, on the College Farm. 70-C was inbred to her son and I have tried to get as much of her blood into both the male and female lines as was consistent with high vitality and good constitution.

4—Belle of Jersey was mated in 1913 to her father and produced seven pullets and four cockerels. All of these pullets laid over 200 eggs each during their first laying year. In 1914 she was mated to one of her sons, No. 1286, and produced nine pullets and five cockerels. All but two of these pullets succeeded in laying over 200 eggs during their pullet year. This male, No. 1286, has been used each succeeding year since 1914 and has been bred back to his daughters of the previous year, using each year only a very few of the earliest hatched, quickest maturing daughters for this line-breeding work.

5—Yes, we have in two previous instances made distinct outcrosses on the Belle of Jersey line, with the primary object to increase size. We have succeeded in doing this just as we have planned and we have maintained our average egg production, but we get a greater range between the production from the individual families from such a mating than we do from the line birds themselves. The line birds themselves laid somewhere between 110 and 240 eggs, the fact being that most of the birds laid around 200 eggs.

6—From my experience in handling this strain, I do not feel that new blood has influenced the productive quality to any extent. We were careful, of course, in using line-bred cockerels of known production.

7—There seems to have been no pronounced improvement in the Belle of Jersey stock from the beginning to the present time—that is, marked improvement. We have gradually increased our average a little, which I think has been done more by eliminating low-producing individuals from more close inbreeding.

8—The flock average is 190. The average for the line-bred Belle of Jersey pullets raised here, some being trap nested on the farm, some at my house and some at Vineland, was 208 eggs.

9—I do not know that I can give you the record of the highest hen each year, year by year, but I will say that Belle herself, laid 246 eggs. Two years from that time one of her granddaughters laid 267 eggs. In 1915, one of the line-bred daughters laid 264, with eight out of ten in one trap-nest pen averaging over 200 and the pen averaging 211. Last year at Vineland, Belle of Jersey birds did not come through quite as well as they did at New Brunswick, although the full-line pens averaged 200 eggs with no great range in production either way—more than twenty—the highest record lay here being 259 eggs.

No. 10—No marked changes have appeared. In fact, the tendency has been for the line-bred type to come to a more definite type and to possess more definite characteristics, namely: body shape, comb, neck, shape of eye, etc. It has been possible for a number of our poultrymen to go through our pens at Vineland and here and there pick out our line-bred Belle of Jersey birds, owing to their uniformity in appearance.

11—No, we have not observed that our high-producing birds are poor producers, as regards fertility and livability of chicks. I do not think there is any correlation either way.

12—Given the opportunity to choose, I would always breed from the heavy producers, provided they had stood up under production and shown their ability to produce and at the same time maintained themselves in good-producing condition.

13—No, there is nothing in our records that would warrant the "skip generation" theory. I think that when problems of this kind have appeared, the one analyzing the data has not taken into account the weakened condition of the heavy producer; that they are not in a condition to produce good chicks, hence they did not develop the real inherited traits that were back of them.

14—So far as we have seen there is no "break-back" in our line. We are going to carry it on indefinitely.

15—I do not consider that it is nearly as easy nor as sure a way of maintaining high production by using unrelated lines, as it is to use line-bred birds. In fact, the ideal way to found a high-producing strain is to start with one superior hen, mating her to one of her sons and closely line-breeding her descendants. A strain thus established should be much more uniform with respect to heavy-laying ability than one produced by more or less outcrossing.

PRACTICAL MANAGEMENT OF BREEDING STOCK

Speaking generally, the treatment of breeding fowls need not differ materially from that given the laying flock, aside from the fact that it is desirable to let the birds have more room, both floor space and range, and to feed more conservatively—that is, with a view to keeping the fowls in fine physical condition at all times and avoiding any approach to forced production. Starting at the end of the first or pullet laying year, the treatment of the birds that are to be carried over for the breeding flock should be planned with a view to insuring their complete recovery from the exhausting effects of a year of heavy laying so that at the beginning of the breeding season they will be able to impart the highest degree of vitality and vigor to the embryos. This applies with equal force, of course, to all birds carried over for the next breeding season, regardless of whether they are pullets or one or two-year-old hens; also to all males.

BELLE OF JERSEY—A REMARKABLE BREEDER
Record, 246 eggs in 365 days; descended from an exceptionally good winter-laying hen and used as foundation of a strain that has made a wonderful record. See Professor Lewis' article herewith.

Summer Feeding and Management of Breeding Stock

Breeding fowls often suffer serious injury through neglect during the nonproductive period. When the hens are laying heavily and their eggs are selling at good prices, enthusiasm runs high, and it is a pleasure rather than a task to give them the care and attention necessary to keep them in first-class condition. But when the breeding season is over and production falls off or entirely stops, too often the birds are left to shift for themselves under highly unfavorable conditions, so that instead of building up health and vigor they become still further weakened, losses are heavy, and results realized the next season from the survivors are disappointing and unprofitable. It is not necessary to make the care of breeding stock a burden, but fowls that for several months have been supplying valuable hatching eggs and are expected to do so again the following year, certainly are entitled to something more than neglect during the rest period.

Fowls of good quality have a value much beyond the service they render the first season. Both males and females of known breeding ability are good for several seasons' use if properly handled, and rightly are regarded by the experienced breeder as a definite asset in his business—valuable property to be cared for and conserved to the greatest practical extent. In both male and female breeders, age is an advantage rather than otherwise. The breeder who uses his fowls a single season and then sells them is throwing away opportunity.

No matter how well bred the fowls may be, their first breeding season must be mainly a tryout. But knowing the results of the first year's matings, the observant breeder is in position to mate with greater certainty thereafter, because of the opportunity he has had to eliminate poor layers and those of low vigor and inferior breeding power. Careful culling each succeeding year leaves old fowls that are the cream of the breeder's flock.

All this is conditioned however—very positively conditioned—upon the fowls receiving proper care, particularly during the nonproductive period. Where this requirement is not met it is common experience to find old fowls giving even poorer results than young stock early in the breeding season. In fact, there are many who frankly admit that they do not expect good fertility or strong chicks from their old fowls until spring has come and they can get out on range and gradually put themselves into fair breeding condition.

Such a result is partially due to the way the fowls are handled after they are placed in winter quarters, but in a number of instances the foundation for the trouble is laid in the inexcusable neglect to which the fowls are subjected during summer and fall. We are all apt to think of winter as the chief season of discomfort for fowls while, as a matter of fact, they probably suffer a great deal more in hot weather than in cold. This is true not because such suffering is unavoidable, but because the comparatively simple precautions needed to protect them are completely neglected.

How Breeding Stock Should Be Handled in the Off-Season

After the breeding season is over the fowls should be given free range if it is possible to do so. Exercise is, beyond question, the best conditioner, and breeding birds cannot have too much. If they can be transferred to colony houses and these moved to outlying fields and located in or near a grove, or at least in a comfortable shaded spot, this should by all means be done. If the fowls must remain in their regular quarters and be confined to yards, make them as comfortable as possible.

The most serious handicaps that breeding fowls have to encounter in warm weather are: confinement to small bare yards, hot roosting quarters, lack of shade in daytime, lack of water, irregular or improper feeding, and lice. None of these is especially difficult to overcome with the exception of close confinement, which sometimes is unavoidable. Where this condition must be met, its injurious effects can be quite largely avoided by providing compulsory exercise.

Most houses can be made reasonably comfortable in summer by providing free circulation of air, doing so by having rear windows under the droppings platform, to be left open at this season, or by installing ventilator openings immediately under the eaves of the rear wall. Make these 6 to 10 inches wide, extending the en-

BUFF LEGHORN PULLET LAID AT 8 MONTHS AND 20 DAYS OLD
Was hatched February 18 and laid her first egg June 7. Bred by R. E. Sims, Little Rock, Ark.

tire length of the house. Do not fail to provide tight-fitting doors for these openings so that there will be no drafts from them in the winter.

Males that are not to be kept over for the next season's use should, of course, be disposed of immediately at the close of the breeding season. The desirability of isolating the males that are to be carried over depends upon the conditions under which they are to be kept. Where there are a number of these and they can be properly cared for, it will be well to separate them from the hens. But if isolation means that they are to be closely confined to uncomfortable, unsanitary houses with small, bare yards, and neglected in their feeding and general care, as is far too often the case, then it will be a great deal better to let them run with the flock.

The practice of putting males into small coops, such as exhibition coops, and keeping them confined there for months, is one that cannot be excused on any practical ground. If that is the best that can be done for them, either leave them in the flock or sell them. Breeders who have valuable males frequently find it practicable to confine them singly with small pens of nonproductive females. If large comfortable yards can be supplied this probably is the best method where open range is out of the question. Where a number of males are to be confined together there is great danger of injury from fighting. This may be prevented to some extent by a little attention on the part of the caretaker, while the fowls are getting acquainted. In the case of particularly "scrappy" individuals, trimming the beak so as to make it blunt and a little sensitive will help to prevent fighting.

Do Not Feed Now for Egg Production

It is not advisable to feed for egg production after the breeding season, though if the hens persist in laying on ordinary rations do not try to stop them by starving or making violent changes in the feeding. During the summer various digestive disorders, particularly liver troubles, are apt to develop in adult fowls as a result of heavy feeding, too little exercise, or too high a percentage of fat-forming feeds. The ration should be nonstimulating, and the amount of feed that should be given will depend quite largely upon the supplies available on

AN EXCEPTIONALLY EARLY-MATURING PULLET

This S. C. W. Leghorn pullet began laying at 116 days old and produced six eggs in the first eight days. Another pullet of the same hatch laid at 125 days; most of the pullets in this hatch were laying by the end of the fourth month. Bred by Professor E. P. Clayton, Mississippi Agricultural College.

the range, but should be regulated so as to keep the fowls in good condition, but not overfat or inactive.

It is advisable to feed dry mash in hoppers, always. If the consumption of this part of the ration is found to be running too high, it may be reduced by suitable changes in the composition of the mixture. The consumption of mash can be quite generally controlled by varying the proportions of ingredients so as to increase or decrease its palatability. If there are plenty of bugs and other insects available on the range it will not be necessary to give any meat scrap at all, but under ordinary conditions it will be found desirable to provide ten per cent of this in the mash.

Oats may be fed liberally. If the grains are plump and heavy they may be fed dry, otherwise they should be soaked or boiled.

A PREPOTENT SIRE

This S. C. W. Leghorn male, bred and owned by Storrs Experiment Station, has the unusual record of having sired 128 daughters, not one of which has proved to be a poor producer. Photo from Professor W. F. Kirkpatrick.

Only a limited amount of corn should be supplied. Use wheat or barley instead, if available. Large fowls, even on free range, are apt to take too little exercise when they are well fed, and this should be guarded against. If they are found to be spending too much time in idleness, their feed must be reduced or the method of supplying it changed. If there is a reasonable amount of feed obtainable on the range, one grain feed a day will be sufficient, giving this in the evening. Grit, green feed and water should be provided as for laying flocks.

Methods of Mating

Those who keep the heavier breeds like Plymouth Rocks, Reds, etc., try to have their breeding pens mated up by about the first of the year as a general average, and by about February 1, in the case of Leghorns. This insures the birds getting acquainted and well settled in their new quarters by the time eggs are to be saved for hatching.

Three general methods of mating are practiced, known as stud, pen and flock mating. In stud mating the males are kept in separate coops or pens and mated with the females as the latter lay and are released from trap nests. This method is followed where extra-careful pedigree breeding is practiced.

In pen mating a single male is mated with a selected flock of females. In connection with the use of trap nests and practical records it is possible to know the pedigree of each chick hatched. The method involves a separate pen for each small flock, and, where many fowls are to be mated, largely increases the labor cost of caring for them. It also demands comparatively expensive housing. For this reason, where only medium quality is aimed at, as in commercial poultry breeding generally, it is customary to mate up regular laying flocks, numbering from fifty to several hundred females, providing males in about the proportions indicated in the table given elsewhere and allowing all to run together. This is flock mating. Obviously, it must be inferior to pen mating in quality of the stock produced, but it is practical and desirable where low-cost production is important.

Proportion in Which to Mate

The number of females that can be mated with a male, with reasonable assurance of securing good fertility, will depend on the breed, the season and the individuality of the male. In the natural breeding season and with the fowls on open range, excellent fertility has been secured in a flock of forty Barred Plymouth Rock hens and pullets mated to a single cockerel. On the other hand, with fowls in close confinement and using old males, it sometimes is necessary to reduce the number to 5 or 6. In general, the proportions given in the accompanying table may be accepted as fair averages. It may be necessary to reduce these numbers under some conditions, particularly with fowls in close confinement in cold weather, while they may be increased in exceptional cases where it is desirable to do so.

NUMBER OF FEMALES TO ONE MALE

General-purpose breeds in confinement, mate 8 females with cock.
General-purpose breeds in confinement, mate 10 females with cockerel.
General-purpose breeds on range, mate 10-12 females with cock.
General-purpose breeds on range, mate 12-15 females with cockerel.
Leghorns in confinement, mate 12-15 females with cock.
Leghorns in confinement, mate 15-20 females with cockerel.
Leghorns on range, mate 15-20 females with cock.
Leghorns on range, mate, 20-25 females with cockerel.

How to Feed Breeding Stock

As soon as the birds get accustomed to what are to be their permanent quarters and pen associates throughout the breeding season, they may be placed on a good but not extreme laying ration. About the same grain and mash mixtures may be used as are supplied the laying pen, only reducing the proportion of meat scrap or limiting the amount of mash to not more than one-third of the day's total feed consumption. This is recommended in order to avoid a ration that will stimulate production beyond the point where the birds will keep in full vigor, which is of the first importance in maintaining high fertility through a long breeding season. There is a general impression that meat scrap cannot be used to good advantage in the breeding ration and some experiments have been reported in which its use has been held responsible for poor fertility. Other experimenters have reported exactly opposite results, however, so it does not appear probable that there is anything in meat scrap that will necessarily produce unfavorable results when properly fed and too heavy production avoided.

Of greater importance than any particular ingredient of the ration is an abundance of exercise. Birds that are kept in comparatively close confinement, as is usually the case with breeding stock, must receive special attention in this respect. It will pay to go to any reasonable length in providing compulsory exercise, giving the grain part of the ration in small installments during the day, scattering it in deep litter and burying it if necessary, so that the birds will have to scratch for all they get. There will be few complaints as to vitality of embryos and vigor of chicks if this detail is properly looked after.

"Vitamines" are regarded as of special importance in the breeding ration, as their presence is thought to have a strong influence on the vitality of the chicks. These nutrients, while not thoroughly understood, are known to be present in milk, also different forms of green feed. For that reason milk is regarded as especially invaluable in feeding breeders and where an abundance of skim milk, buttermilk, semisolid buttermilk or powdered milk is available, no meat will be needed. Such green feeds as cabbage, kale, grass, clover, etc., are also believed to be high in vitamines, while mangels, turnips, etc., are extremely low.

It does not matter greatly what the housing conditions are so long as the pens are well lighted and ventilated, and the birds have plenty of floor space. It will pay to allow considerably more than is assigned to the laying flock—at least 5 or 6 square feet per hen, thus to avoid crowding, insure better ventilation and generally to promote individual comfort. Breeding stock should not be any more closely confined than is imperative, and should have access to yard or range even in comparatively bad weather. If it is possible to give the birds open range this should by all means be done.

SPECIAL BREEDING FLOCK ON THE POULTRY PLANT OF THE WEST. WASH. EXPERIMENT STATION
The male here shown is of good standard quality and his dam has a record of 280 eggs. His mates were selected from 1,000 layers with first-year records running up to and over 300 eggs.

CHAPTER V

Profitable Production of Market Eggs

Monthly Minimum Standards of Production Will Enable the Producer Definitely to Maintain His Flock on a Profitable Basis at All Times—Special Points in the Management of Laying Flocks that Help to Increase Average Production—Influence of Culling and Use of Artificial Light on Production Records

THE 1920 Census statistics for poultry production were not available at the time this book went to press, but the estimate in the 1919 Yearbook of the Department of Agriculture for total egg production in that year is 1,957,000 dozens of eggs, and it is also estimated that 600,000,000 fowls were raised. The production of backyard flocks is not included in either estimate. As the market value of these products, on the farm, largely exceeded $1,000,000,000, it is evident that poultry production is one of the Nation's great agricultural industries, and this is true notwithstanding the fact that the average laying flock is operated at a comparatively low level of efficiency as regards its production possibilities.

Such statistics as are available indicate that the average egg yield per hen, taking the country over, does not exceed 75 eggs a year, and probably is considerably less. As contrasted with the egg yields regularly secured in well-managed flocks everywhere, such an average appears inexcusably low and in every case the first step in striving for increased production should be to make certain that full advantage has been taken of the opportunity afforded in every flock to bring about a marked increase in the average per hen by eliminating poor producers and giving the birds that are left good everyday care and feeding.

It will be helpful to anyone who keeps fowls for profit to understand clearly just how little "average" production means in the way of net income, how readily and markedly this average can be increased by better care and breeding, and what such increase will mean financially. While poultry keeping is by no means an exact science it still is possible for any earnest individual to work out standards of production for his flock that will enable him to know just what his cash balance will be at a given level of production. And it is safe to say that when this is done there will be not only a keener appreciation of the importance of getting increased egg yield, but means will be found for promptly bringing it about.

To show how such a scale or standard of production can be worked out, the accompanying table has been prepared from data published in Storrs (Conn.) Bulletin No. 100:

The number of eggs required to meet the cost of maintaining hens naturally varies with the market prices of eggs and feed respectively. Speaking on this subject before the 1918 National Poultry Conference, in Chicago, Professor Dryden of Oregon Agricultural College stated that at 1916 feed prices, when it cost $1.80 each to feed hens averaging to lay 180 eggs each, 77 eggs were required to pay for the feed (average price of eggs 28c per dozen), while in 1917 when feed cost $2.97 per hen, it took 97 eggs at an average of 36.74c per dozen to meet each hen's feed bill.

In working out standards of production for personal use, the farmer, the back-yard poultry keeper and the commercial egg farmer, each approaches the subject from a somewhat different angle, but it is not a difficult matter for any poultry keeper to adapt the foregoing table to his own scale of prices for feed and eggs, thus to know at any day of the year just what his balance is. And if that balance is on the wrong side of the ledger—outside of the three low-production months of November, December and January when even a flock averaging 15 eggs daily per 100 hens may show a small deficit—it is because the caretaker has fallen short either in the management or in the selection of his birds.

Putting Returns on a Labor-Income Basis

Up to this point consideration has been given to feed cost only, in its relation to egg production. The poultry keeper however, has many other expenses to meet and where the work is to be placed on a thoroughly businesslike footing, all cash outlay and fixed charges, such as interest, repairs, depreciation, etc., must be taken into account. When this is done and the returns figured on a "labor-income" basis, better average production becomes increasingly important. For illustration: The Connecticut Agricultural College, after a careful survey of 42 farms on which poultry keeping is the most important industry, has published, in Extension Bulletin No. 8, the following table to show the relation of different averages of production to profits or labor income. The bulletin does not give complete details in regard to feed and egg prices but the figures in the table are based on an average expenditure of $1.92 per hen for feed, which probably is a little low for the specialist but a fair average on the farm.

Table Showing Eggs Per 100 Hens Required to Pay for Feed, Month by Month, and Profit Realized at Different Levels of Production

Month	Price per doz. eggs	Eggs per day to pay for feed	Eggs per day hens laying 61-100 per yr.	Gain or Loss	Eggs per day hens laying 101-140 per Year	Gain or Loss	Eggs per day hens laying 141-180 per year	Gain or Loss
Nov.	$ .645	15	7	$—.43	9	$—.32	16	$ .054
Dec.	.645	16	6	—.58	8	—.36	17	.053
Jan.	.525	19	6	—.57	7	—.46	16	—.13
Feb.	.47	21	16	—.20	28	.27	36	.59
March	.375	26	41	.47	53	.84	59	1.03
April	.37	26	52	.80	61	1.05	65	1.20
May	.38	26	52	.82	65	1.23	70	1.39
June	.38	26	43	.54	60	1.08	72	1.46
July	.425	23	30	.24	53	1.06	66	1.51
Aug.	.515	18	18	.00	43	1.07	61	1.80
Sept.	.56	17	8	—.28	16	—.05	38	.98
Oct.	.66	15	2	—.71	3	—.66	11	—.225

The first column in above table gives month of year; second column, the average price of eggs received for month; third column, number of eggs per 100 hens that must be laid daily to pay for feed when the ration costs $3.50 per 100 pounds; fourth column, average number of eggs that will be laid in any month by hens whose total production for the year ranges from 61 to 100; fifth column, monthly profit or loss on eggs laid by such hens; columns 6 to 9, the same data for hens at higher levels of production.

Relation of Egg Production per Hen to Labor Income

Egg Production per Hen	Average Labor Income
Over 130	$1,364
100 to 130	911
70 to 100	327
Less than 70	—553

Plainly, under Connecticut conditions, the poultry specialist who secures an average of only about 100 eggs per hen has no hope of realizing a comfortable living, and when the average falls much below that point his flock becomes a source of loss instead of profit. During 1915-16 the Department of Farm Management of the New Jersey Agricultural College made an exhaustive survey of 150 egg farms in that state, most of these being devoted almost exclusively to commercial egg production. The relation between egg production and labor income under New Jersey conditions is shown in the following table:

Egg Yield and Labor Income on 150 New Jersey Egg Farms

Eggs per Hen	Egg Receipts per Hen	Average Labor Income
46	$1.30	$—176
68	1.90	— 67
91	2.30	312
108	2.90	775
126	3.40	1,173
155	4.20	1,823

During the year represented in above table the cost of feed per bird was $1.92, the cost per 100 pounds was $2.07, while the average price of eggs was 34c per dozen. In both the New Jersey and Connecticut surveys "feed cost per hen" is secured by dividing total feed bill by average number of fowls, hence there are included the rations consumed by whatever chicks were raised. The cost of feed consumed by the hens alone would be considerably less than here given, but data are not presented in sufficient detail to make it possible to arrive at the exact amount. It is important to remember, in considering the last table, that New Jersey poultry keepers are especially favored in the prices they are able to secure for eggs, and when they must get 90 eggs per hen, or thereabouts, to have any labor income at all, it is safe to say that those who are compelled to accept a much lower price for eggs will have to do even better as to average yield unless they can correspondingly reduce production costs.

Illustrations similar to the foregoing could be multiplied but these should be sufficient to show the practical importance, in any poultry enterprise, whether on a large or a small scale, of securing the maximum production of which the fowls are capable, and of tolerating in the flock no fowls that are conspicuously below the level where a reasonable profit is assured.

Shifting the Peak of Production

One of the poultry keeper's constant and important problems is that of shifting the peak of production more nearly to coincide with the price peak. This can only be approximated, of course, but every successful step in this direction is so substantially rewarded that it is well worth the earnest attention of every person interested in producing market eggs at a profit. The chart presented on page 38, illustrates graphically the practical importance of this. It is here shown that the peak in market prices is reached in December, as a general average, whereas the peak in production is in April (for the average flock). This chart illustrates the fact that almost the entire egg yield of a flock averaging eighty-seven eggs per hen must be sold at a price below the average for the year, while with high-production flocks by far the greater number of eggs are laid when prices are above the yearly average. The commercial poultry keeper's net profit is largely determined by the extent to which he is able to go in making his production line correspond with Curve No. 3 rather than Curve No. 2. There are a number of ways of bringing this about.

Choice of breed has a direct bearing upon winter production, particularly in sections where severe cold is to be expected. Leghorns are not well adapted to extremely low temperatures or to sudden changes, and at laying contests in cold climates it is almost invariably the case that fowls of the heavier breeds prove better winter producers, though this depends somewhat upon the season. There may be little contrast in a mild winter or one of fairly uniform cold. Comparing the production of the three popular medium-weight breeds, Plymouth Rocks, Wyandottes and Rhode Island Reds with Leghorns at the Storrs Laying Contest in 1916-17, the Leghorn production for the months of December, January and February was 31.3 per cent, while the average for the other three breeds was 34.5. In the winter of 1917-18 which, as the reader may recall, was exceptionally severe, Leghorn production for the three winter months was 26.9 per cent as compared with 33.6 per cent in the case of the other breeds mentioned.

As showing the influence of a milder climate, the production of Leghorns at the Vineland Contest in 1916-17 was 36.4 per cent and 36.8 per cent for Reds, Wyandottes and Rocks. Farther north, or where the winters are much colder than at Storrs, still wider differences in winter production are recorded. In comparing Leghorns with fowls of the larger breeds, however, it should be remembered that cost of production with Leghorns is so much lower that the handicap due to reduced winter yield may be largely overcome or entirely wiped out so far as net returns are concerned. See Chapter VIII for more on this subject. While there are some who hold that the capacity for winter production is in itself a distinct character that can be increased by selective breeding, the general feeling among poultry keepers is that it is almost entirely conditioned upon early maturity which,

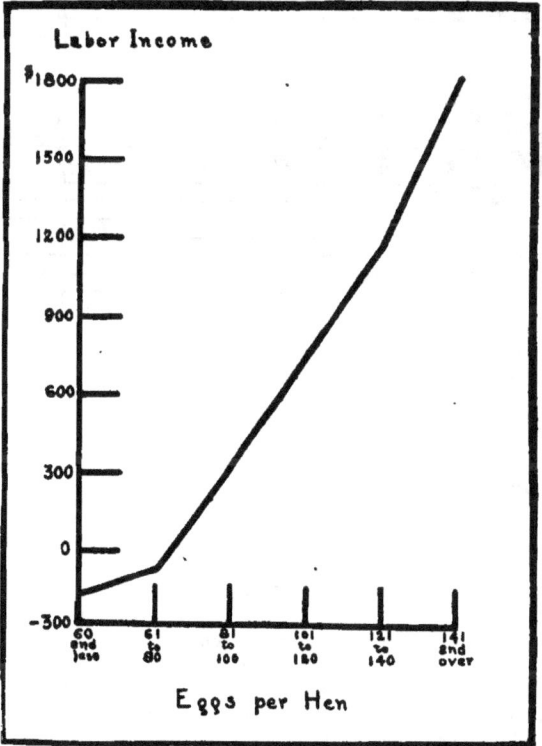

RELATION OF AVERAGE EGG YIELD TO LABOR INCOME
This graph shows how labor income is affected by average production of flock. An average of 61 to 80 eggs means no income at all, but an actual loss, while an average of 141 and over gave a labor income of $1,800. Data for this graph were secured by a careful survey of 150 poultry farms in New Jersey. Reproduced from N. J. Bulletin No. 329.

in turn, depends not only on breeding but on good care all along the line from the day the chick is hatched.

Special Treatment of Pullets

The poultry keeper who has an early maturing strain has a distinct advantage in the fact that he not only can bring off his hatches later in the season, thus largely reducing the difficulties and the expense of brooding, but the shorter growing period naturally reduces the feed cost of bringing the birds to adult size. Good winter production, however, is not limited to quick-growing birds, as the poultry keeper who knows the length of time required for maturing his pullets, whether they are of an early or late-maturing strain, has only to arrange his hatching accordingly. Making due allowance for differences in strains, Leghorn pullets should begin to lay at 5 to 5½ months of age; Wyandottes and Reds at 6 to 6½, and Plymouth Rocks at 7 to 7½ months.

Throughout the North generally the best time to have pullets begin laying is during October. Later than that much of the advantage of the high-price season is lost. On the other hand, pullets (particularly Leghorns) that begin laying in September require careful handling to prevent them from going into a fall molt after they have laid a few eggs, in which case they will be of little value as winter layers. It is not a difficult matter for the experienced poultry keeper to determine, along in September, just what progress his pullets have made toward development, and whether or not he should try to hold them back or push them forward.

Early laying should, of course, be avoided in pullets that are to be used as breeding stock the next season. If they lay heavily throughout the fall and winter they will be in no condition to produce the highly fertile eggs and vigorous germs that will be expected of them. If there is danger of their doing this it is possible to delay maturity by suitable feeding, supplying a ration that will promote growth but not sexual development.

Whole oats are one of the best grains that can be given where delayed maturity is sought and, if boiled or soaked for several hours before feeding, may be supplied freely. Barley and wheat also are excellent, but only a limited amount of corn should be fed. Meat scrap or other animal feed should be reduced, or entirely omitted if there are many insects on the range. Do not make the mistake of trying to retard development by limiting the rations however, or by increasing too greatly the bulk. To do this may result in permanent injury to the fowls. As the pullets approach maturity, or by October 1 as a general rule, they should be placed in permanent quarters and left undisturbed thereafter. Moving fowls from one place to another is almost certain to delay egg production or to stop it after it has begun.

Something can be done to promote laying maturity in late-hatched or underdeveloped pullets, but extreme measures are unwise. Pullets never begin laying until they are fairly well matured and rounded out, and until this condition is reached they should be fed a ration designed to hasten maturity rather than one intended to promote egg production. On the ordinary laying ration with its high percentage of protein young pullets are apt to keep on growing instead of maturing, and so long as rapid growth continues there is no hope of egg production. What is needed at this time, to hasten production, is not a laying ration, but a fattening one which will check growth and cause the pullets to take on flesh instead.

To promote early laying, therefore, the ration should contain liberal proportions of corn and meat, and, if possible, a daily feed of a fairly rich, moist mash. Continue this ration until the pullets get plump and begin to sing and otherwise indicate that the egg organs are developed, after which it may gradually be changed to the regular laying formula. It is not necessary nor desirable to resort to extreme measures to provide exercise for young pullets. Plenty of compulsory exercise is helpful in preventing adult fowls from becoming fat, which is a sufficient reason for avoiding it with the immature pullets that are being deliberately fattened. Stimulants and condiments are of limited value in hastening egg production. There is nothing that will "force" either hens or pullets to lay when they are not in condition to do so, and after they are in proper condition good wholesome feed and plenty of it is all that is needed to start them.

Uniform Production

In order to take full advantage of the high-price season the market egg producer tries not only to get early winter production but to carry summer production over as late as possible into the fall. Some adopt the plan of giving their birds a modified rest in midsummer with a view to bringing them back into laying again during August and September. Whether it is practicable to do this or to try to keep the birds up to their maximum straight through the season will depend to some extent upon breeds and individual skill. Where

THE NEARER PRICE AND PRODUCTION PEAKS CAN BE BROUGHT TOGETHER THE MORE PROFITABLE THE FLOCK CAN BE MADE

The high point in prices comes in December (Curve 1), while the high point in production in the average flock comes in April (Curve 2), or in May with a heavy-laying flock (Curve 3). Average price for the year is indicated by straight line 4. Note that practically all eggs produced by the average flock must be sold in the low-price season, while in the case of the pens whose production line is shown in Curve 3 (average for Second Vineland Contest), the bulk of the eggs were laid when prices were above the average.

Plymouth Rocks, Wyandottes, Reds, etc., are kept, the birds are apt to take care of the rest period themselves by going broody, and it is well known that hens that have had a long broody period and perhaps have hatched and raised a flock of chicks in early summer can be brought back into excellent production later on. At laying contests however, or anywhere that the birds can have proper attention, it is possible to keep them productive straight through to October, as is shown by Curve No. 3 in the chart on page 38.

There is another method of securing increased fall production which, in the last few years, has come into prominence as a practical measure, particularly in Leghorn flocks: This is the plan of bringing out extra-early pullets which will begin laying in late summer. Such pullets usually molt in the fall (except possibly where artificial lights are used), but if they are hatched early enough so that good production can be secured before they go into the molt they may be utilized to excellent purpose in keeping up the volume of production at a time when, otherwise, there is certain to be a big drop. The advantages of having a flock of February-

PRIZE-WINNING MARKET EGGS SHOWN AT PURDUE EGG SHOW

Eggs illustrated above were exhibited at the Thirteenth Annual Purdue Egg Show. Starting at top, plate on left shows a dozen Light Brahma eggs; plate on right, Black Langshan. Second row—plate on left, Barred Plymouth Rock eggs; plate on right, White Wyandotte. Third row—plate on left, White Leghorn eggs; plate on right, Black Minorca. Photo by Franklane L. Sewell.

hatched Leghorn pullets are clearly stated by E. H. Wene in an issue of "Hints to Poultrymen," a monthly publication of the New Jersey Experiment Station. Extracts from this article follow, including a table showing the production of a flock of 190 February-hatched Leghorns at the New Jersey Experiment Station and the profits realized from them:

"The March and April-hatched birds lay an excellent production during the year. There is, however, a period when they must molt. This usually occurs during the late summer or the months of September and October of the following year. If the poultryman is going to keep these birds the following year for layers, they must be carried through the winter at a loss, which must be covered in some way if a suitable profit is to be secured. Perhaps it has not occurred to the poultryman that a small flock of February-hatched pullets will come into maturity about the last of July and will lay a fair production during these months of usually low production thus insuring a more continuous supply of eggs. Futhermore, the eggs would be sold at a time of the year when prices are highest, making a good profit from the number of eggs sold.

"February hatching has not been carried on to a great extent in the past, possibly because of the fact that market eggs sell for such high price at that time that the poultrymen have not considered early hatching profitable. The fertility is apt to run lower during the winter months, a fact which also would tend to discourage the practice. Although there are one or two drawbacks to early hatching, there are a number of very important ways by which it can be made of great help to the poultryman. It is our firm belief that the farmer can plan to have from one-quarter to one-third of his chicks hatched during February. This would give him plenty of time to care for them before the rush of spring work. Also, by early hatching it is possible for the poul-

A SIMPLE, EASILY CONSTRUCTED SELF-FEEDER
Bottom should be 10 inches wide, straight sides 6 inches high, sloping sides 4 inches. Space at top is about 6 inches wide. May be made any length desired, allowing one lineal foot for 5 to 7 hens. In use at N. J. Experiment Station.

tryman to get a greater supply of hatching eggs during the following spring from his February-hatched birds, for such birds make the best possible breeders by the next April, being yearlings in fact, in that they are over one year old and have molted."*

Market Quality in Eggs

The commercial poultry keeper gives unusual attention to the matter of quality in eggs and it is chiefly due to this fact that he is able to secure a marked increase in price over ordinary market grades. As a matter of fact, with higher costs all along the line, he could not maintain his flock on a profitable basis if he had to sell his

Production of 190 February-hatched Pullets at New Jersey Experiment Station—
One Year Record

Month	Number of eggs	Per cent production	Total cost of feed	Value of eggs	Profit or loss
August, 1914	1245	21.1	$17.43	$39.42	$21.99 profit
September, 1914	1756	29.6	17.23	57.07	39.84 profit
October, 1914	1934	32.9	14.51	79.78	65.27 profit
November, 1914	593	1.1	15.58	29.11	18.53 profit
December, 1914	132	0.2	16.75	6.38	10.37 loss
January, 1915	847	1.4	24.51	32.47	7.96 profit
February, 1915	2155	40.0	20.99	62.85	41.85 profit
March, 1915	4014	68.0	27.50	83.62	56.15 profit
April, 1915	3984	68.0	27.84	79.68	51.83 profit
May, 1915	3619	61.1	28.50	72.48	43.98 profit
June, 1915	2927	51.4	28.75	64.48	39.78 profit
July, 1915	3029	52.0	21.94	71.96	50.02 profit
	26280	426.8	$257.53	$679.80	$421.88

Note in this table the excellent egg yield secured during the months of August, September and October. Beginning apparently about the first of November, these pullets went into a general molt and during November, December and January, production was at a very low ebb. But with February they came back strong, and for the year ending July 31, showed an average production of 138.3 eggs each, and an average net profit over cost of feed of $2.22.

products at ordinary prices. Every first-class city market, however, discriminates sharply in favor of eggs of best quality, meaning not only those that are strictly fresh but that meet special requirements as to size, color, shape, etc. Some of the requirements of city markets may appear to be superficial, but if the buyer is willing to pay a substantial premium to get just what he wants it certainly will pay to cater to his preferences. Eastern producers, who for years have had a practical monopoly on the New York City fancy egg trade, recently had an unpleasant illustration of this when they learned for the first time that it not only was possible for Pacific Coast producers to ship their eggs across the continent and sell them in the New York market at the same price as that received by near-by producers, but that this market actually preferred the California eggs because of the greater uniformity with which they are graded and packed.

The preference of the New York market for white-shelled eggs is well known, but too many producers do not realize that the best prices are paid only for eggs that are chalk white, and that those with shells that are tinted even slightly cannot be sold to this exacting trade. The color of shell is an inherited character and those who make it a practice to reject all eggs in hatching that are not of the true chalk-white color will in a short time have a high degree of uniformity in this respect. The same is true with regard to shape of egg. The market demands eggs of uniform oval shape and it is a short-sighted policy to incubate those that are noticeably longer or shorter than the standard. As long as such eggs are incubated they will be produced, and must either be disposed of as culls or, if they are included in the regular shipments, the price for all will be discounted. Size also is an inherited character and those who persist in incubating small eggs are simply perpetuating this character in their flock to their continued financial disadvantage.

The standard size usually is fixed at 24 ounces to the dozen (in some instances 25 ounces), and those that are

---
*Note: This statement applies when the pullets have molted in the fall and have taken a rest from production, but not where, through the use of artificial light or otherwise, they are kept steadily productive right through the fall and winter.—Ed.

undersize will bring a lower price in any discriminating market. No producer who wishes to sell eggs to the best advantage can afford to ignore the requirements of his market, whatever these may be, and to do so in regard to the characters just mentioned is inexcusable since they can so readily be eliminated from the flock simply by refusing to incubate eggs that have these defects. See special article on breeding for market quality in Chapter IV, Part III.

Selling the Product

To a considerable extent, success in commercial egg production is a matter of salesmanship. Those who have the good fortune to be located where they have the advantage of community marketing organizations are largely spared the necessity for considering this phase of the business, but those who must market their own products should give close attention to it. The ideal method of selling eggs is to dispose of them at retail, thus getting the full retail price and establishing the personal touch that has so much to do with marketing a premium-price product. It is not always practicable to do this, however, and particularly where large numbers of eggs are produced it is necessary to find some method of disposing of them at wholesale. Frequently it is possible to make arrangements with city retailers, thus eliminating all other middlemen and keeping in closer touch with a market which, if it is more exacting, is also more profitable. Dealing with retailers in this way however, sometimes is difficult when poultry keepers are not able to make fairly uniform shipments most of the year, which, as a rule, they cannot do.

Those who are at considerable distance from market, particularly if they must ship to cities, usually find it necessary to sell through commission merchants. Often it is possible to secure through these a better price than the individual would obtain by dealing direct with retailers, so that while the practice involves an added middleman and some delay in marketing, selling through commission merchants is the most convenient and most practical method for many to adopt.

No matter how sold, freshness and careful sorting to eliminate eggs distinctly below grade as to size, color, shape, cleanliness, etc., are of the first importance. Where eggs are shipped by express, careful packing is imperative. In spite of whatever efforts the express company may make to secure more careful handling, eggs shipped in this way have a good many chances of going through in bad order unless the producer is extra careful in packing. Many go so far as to refuse to pack in used cases under any condition. Others employ used cases after carefully nailing them up, but discard all used fillers. The quality of material now employed in ordinary fillers is such that they usually break down so much the first time they are used as to make it unsafe to ship in them a second time. With new or thoroughly overhauled cases, new fillers, double excelsior pads for top and bottom (some careful packers put an extra pad in the middle of the case or below the top layer), the eggs should go forward with only moderate breakage.

Under the head of selling the product comes the disposal of surplus fowls. This can be made a source of considerable revenue if the poultry keeper has room enough to handle the birds properly so as to insure being able to send them to market in the best condition. The producer who keeps Leghorns usually feels that his surplus fowls are not worth any special attention, however, and he aims to get rid of them in the easiest possible way. The cockerels are sold as soon as someone can be found to take them off his hands, while the cull hens are sold at the end of the laying season almost regardless of price.

As a result, comparatively little revenue is received from this source. For example, in New Jersey Bulletin No. 329, giving data from a survey of 100 poultry farms (chiefly Leghorn flocks), the egg sales are shown to amount to 63 per cent of the total revenue, while sales of pullets, hens and cockerels, including value of birds used at home, amount to but 13 per cent. Wherever the larger breeds are kept, however, and on farms generally, the sale of surplus fowls is a much more important source of income.

OKLAHOMA HAS SOME HIGH PRODUCERS ALSO

This hen, owned by the Poultry Department of the State Experiment Station, has a record of 243 eggs in first 12 months of laying. She also laid 147 eggs in the second year and 108 in her third. She weighed 5½ pounds at completion of first year, and lays an extremely large egg. Photo from Professor Harry Embleton.

Taking the United States Department of Agriculture's estimate of fowls raised and eggs produced on farms in 1919 as quoted at the beginning of this chapter, a fair estimate of the relative market values of the two classes of products would be around 65 per cent for eggs and 35 per cent for fowls.

SPECIAL PROBLEMS IN COMMERCIAL EGG PRODUCTION

Pullets Versus Hens

The maximum production of fowls usually is reached in the pullet year, though it frequently happens that, owing to lateness of hatching or to other unfavorable conditions, some individuals make better records in the second year than in the first. Occasionally, whole flock averages will be found to be higher in the second year. For example, Utah Bulletin No. 135 gives the average production of a particular flock as 117 eggs in the first year, 146 in the second year and 117 in the third. As a rule, however, a considerable drop occurs in the second year. Tabulating first, second and third year production reported from different sources, Utah Bulletin No. 148 gives the average for selected flocks of Leghorns at 180 eggs per hen in the first year, 146 in the second and 119 in the third. In the case of general-purpose breeds the first year average for selected flocks was 177 eggs per hen; second year, 116; third year, 93. The drop from first to second and from second to third year usually is greater where comparatively high first-year production is secured. Compare the drop of 34 eggs in the second year and 23 in the third in the case of the selected Leghorn flocks just mentioned with the record of an unselected flock of Leghorns of the Utah Station plant whose average production from the first to the third year is given as follows: first, 124; second, 119; third, 106.

An especially good comparison of the relative merits of pullets and hens is obtained by taking the figures secured in the first and second years of the Vineland, N. J., Laying Contest. Pullets were entered the first year and all living birds were carried through to the end of the second year, under which condition an average of 161.8 eggs per bird was secured in the pullet year as compared with 129.5 in the second year. During the pullet year 3.9 lbs. of feed were required for each pound of eggs produced, while 4.9 lbs. were required the second year. The mortality in the pullet year was 10 per cent and 14.7 in the second.

In order to show the relative value of pullets and hens as winter layers, the monthly production (in percentages) for the first two years of the Vineland Contest is presented in the following table. Production of Plymouth Rocks and Leghorns is given separately thus affording a comparison of the relative merits of the two breeds as winter producers under New Jersey conditions.

Monthly Per Cent Production in First Two Years of Vineland Contest

Month	First Pullet Year		Second Year	
	Plymouth Rocks	Leghorns	Plymouth Rocks	Leghorns
November	17.	34.	18.1	7.9
December	24.8	33.8	13.	6.4
January	34.5	32.2	17.1	16.5
February	51.7	43.2	30.3	33.9
March	67.1	64.5	59.4	58.4
April	66.8	69.6	59.	70.1
May	61.2	67.4	44.2	66.1
June	53.6	66.3	43.	60.9
July	43.8	60.	35.7	54.4
August	32.3	48.	30.5	41.8
September	27.6	22.3	27.6	26.3
October	20.8	14.4	14.	9.3

A study of this table will show that even under skillful contest management the production of the hens did not approximate their pullet records until March. Wyandottes and R. I. Red hens (data omitted to save space) did a little better than the Plymouth Rocks. Under the mild winter conditions of New Jersey, Leghorn pullets proved to be slightly better in winter production (November to February) than either Rocks, Wyandottes or Reds. In the second year, however, it will be noted that Plymouth Rock hens laid much better than Leghorns, and the same was true of both Wyandottes and R. I. Reds.

At Storrs a series of averages for 5 years shows the following percentage of winter production for Wyandottes and Leghorns, these being estimated from a graph in Storrs Bulletin 100. They are based on averages for 4-week periods, instead of being calendar-month records as in the Vineland table.

	Nov.	Dec.	Jan.	Feb.	Mar.
Wyandottes	15	26	34	42	52
Leghorns	19½	22½	21	29	47

"Long Distance" Production

Just what can be done in the way of developing strains of "long-distance" layers is a matter of decided interest (see Chapter I for numerous illustrations), though its practical value is questionable. Among commercial egg producers comparatively little importance is attached to the production of hens after the second year because reduced winter yield and increased feed cost per dozen eggs both operate to discourage keeping any more hens than are required for breeding purposes or to make up a given number.

In the commercial egg flock the percentage of pullets to hens hinges chiefly on the number of early-hatched pullets that can be raised, and their cost as compared with the cost of carrying hens through the nonproductive molting season. In a poultry survey reported in New Jersey Bulletin No. 329 the percentage of pullets to hens was carefully considered.

By dividing the flocks according to the proportion of pullets and yearlings, the advantage or disadvantage of keeping a large proportion of yearlings in the laying flock should be shown. Table given on next page shows this relation as it existed on these farms.

Other tables in the bulletin referred to show that poultrymen whose flocks have 50 to 70 per cent of their number in pullets obtain greater total receipts per flock and per hen than any other class. They also receive the highest number of eggs per hen, except the class having 80 to 90 per cent of their flock in pullets. Hence it seems clear that a well-balanced flock must carry 30 to 50 per cent of yearlings.

As the nonproductive period of good hens will not exceed three months and may be considerably less, while the cost of producing a pullet is variously estimated at from 75 cents to double that amount, with the minimum cost rarely attained on commercial plants, it is apparent that it is considerably cheaper to carry hens over than to replace them with pullets. However, the winter production of pullets is so much greater than that of hens that the better returns thus secured may be more than sufficient to wipe out the extra cost. For that reason the argument for carrying hens over, even to their second year, on the commercial egg farm, must rest chiefly on their value as breeders.

Keeping Up the Laying Flock

The average operator looks upon everything not directly related to egg production as merely an added com-

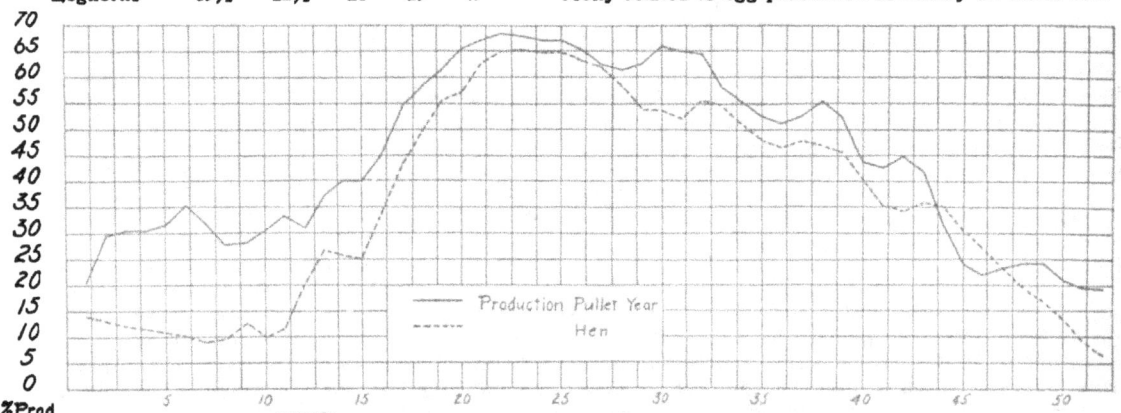

PRODUCTION OF PULLETS COMPARED WITH HENS
This graph shows the production of the pullets entered in the First Vineland Contest and their production the following year; also indicates plainly their relative productiveness and the superiority of pullets as fall layers. Reproduced from N. J. Bulletin No. 338.

plication, and the tendency, especially on intensive plants, is to avoid such as far as possible. The poultry keeper must face the fact, however, that his permanent success depends not only upon getting good egg production during the present year, but in keeping up his flock both with reference to the number and quality of the birds

Relation of Proportion of Pullets per Flock to Profits on 150 Poultry Farms in New Jersey

Per cent of pullets laying in flock	No. of farms	Pullets per farm	Yearlings per farm	Cockerels per farm	Per cent of pullets	Birds per farm	Labor income
50 or less	19	276	475	19	37	771	$ 511
50.1 to 60	37	450	405	19	52	874	1034
60.1 to 70	36	556	309	23	64	888	1062
70.1 to 80	32	470	172	16	73	659	445
80.1 to 90	12	412	83	8	83	503	430
90.1 to 100	14	312	00	4	100	316	272
	150	442	278	17	61	787	$ 780

—also improving quality, as regards both average productiveness and breeding, so far as practicable. The first step toward failure is taken by many poultry keepers when, often unwittingly, they allow their flocks to start on the down grade by letting the percentage of early-hatched pullets fall off, by keeping too many old hens over, or by allowing the flock to deteriorate in breeding.

There are three general ways of keeping up the flock. For a quick start, also as a method of securing the annual supply of pullets when it is not practical to have one's own breeding pens, the purchase of baby chicks, partly grown or ready-to-lay pullets is adopted by many. This plan enables the operator to concentrate his attention almost exclusively on egg production (though this is not always a practical advantage) and to reduce investment in equipment. It also requires much less expert knowledge, as the hatching and brooding of chicks is generally considered to be the most difficult part of the work. It has however, the disadvantage of decidedly increasing the cost of renewing the flock each fall, and there sometimes is uncertainty as to quality of stock thus secured. Also there is an indirect but important disadvantage in the fact that the buyer, in order to reduce expenses, is constantly tempted to skimp in numbers or in quality—mistaken economies that are almost certain to result in reduced profits.

Another method, and the one most commonly practiced, is to depend upon the regular laying flock for hatching eggs. This is the cheap and easy way. It is fairly satisfactory if eggs are saved only from hens (not pullets) and if these are carefully culled so that only the best layers are retained. By penning the hens separate from the pullets, and following the methods outlined under the head of "Selective Flock Breeding" in Chapter IV, good results should be secured, though improvement in breeding will be comparatively slow. It should be clearly understood however, that if this method is accompanied by carelessness in selection, use of hatching eggs from pullets that have been pushed for winter production, or neglect to provide males from high-laying ancestry, deterioration is almost certain to follow.

The best method of keeping up the laying flock and of improving it is for the operator to qualify himself as a breeder as well as an egg producer, and to give to his breeding stock a full measure of attention. This calls for greater skill, but it gives the maximum of independence and of economy in replacement, and it makes the poultry work not only more profitable but more interesting. The last consideration is by no means unimportant, for a large percentage of failures is directly traceable to the depressing influence of the monotonous daily routine on too highly specialized plants. Moreover, everyone who is reasonably successful in breeding good fowls with better-than-average production records finds, almost without seeking, a constant demand for eggs for hatching, day-old chicks and breeding stock, and at prices much in advance of what these products would bring in an ordinary market, thus largely adding to the net income.

Influence of Environment on Production

Egg laying to a great extent is affected by physical comfort. A well-nourished hen or pullet that is contented and comfortable is bound to be productive up to her normal limit—can scarcely be otherwise. So the wise caretaker makes it his constant study to contribute to the contentment of his fowls, knowing well that if he

BREED FOR LARGER SIZE AS WELL AS GREATER NUMBERS
This White Plymouth Rock was bred for size and for large eggs. She produced 178 eggs in 229 days, which averaged to weigh about 29 pounds, and when 10 months old she weighed 8¾ pounds. The six eggs shown in carton weighed 16 ounces and were the last six to be laid consecutively before her photograph was taken.

succeeds in doing this the egg yield will largely take care of itself. Warmth in winter, coolness in summer, sanitary housing conditions, provision of healthful activity, gentle treatment—all have a direct influence on production. The influence of severe cold has already been considered but it is worth noting here that skillful operators, particularly those who keep fowls of the larger breeds and who have warmly built houses, often find that the egg yield goes up rather than down when winter sets in, provided that the fowls are not exposed to temperatures low enough to cause frozen combs or severe physical discomfort.

It is quite probable that artificial heat can be utilized to contribute to maximum production, but whether this is practical for commercial laying flocks is an open question. Recent experiments in the use of artificial heat have generally been with the object of keeping the house dry and protecting the fowls from extremes of cold, but without attempting to make the house warm. Some favorable reports have been received, though these are not sufficient as yet, to warrant generally recommending the practice.

There is a very general tendency on the part of those who seek to keep fowls comfortable in winter to go to injurious extremes in cutting off ventilation. Fowls are distinctly fresh-air animals and their respiratory systems are so developed that they are at a great disadvantage when compelled to breathe moisture-ladened air. Fowls in stuffy, damp quarters are not comfortable, even if warm, and all efforts to protect them from cold must stop before that condition is reached. Cold dry air is much to be preferred to damp warm air reeking with ammonia fumes.

Without doubt fowls suffer greatly in extremely hot weather, often dying in considerable numbers from heat prostration. Plenty of shade, loose, moist earth to wallow in, cold drinking water and a free use of water sprinkled about the house and yards, when available, will do much to prevent deaths and a falling off in production at such times.

Soil and general environment have much to do with the health of the flock and therefore with production. Intensive poultry keeping is rarely advisable on soil that is not quite sandy, as soils that are not thoroughly well drained soon become contaminated when heavily stocked, and losses from disease are almost unavoidable. Low-lying sections where air drainage is not good are also to be avoided as they are damp and unhealthful. Extremely exposed locations are objectionable because they increase extremes of both winter cold and summer heat.

So far as egg production is concerned it does not appear to matter whether the fowls are confined, yarded or on open range, provided methods of care are adopted to conditions. However, maximum yields are rarely secured on free range, particularly with Leghorns, as the fowls appear to use up their force in muscular activity rather than in production. Confinement in warm weather is liable to have an unfavorable influence because of practical difficulties in the way of making the fowls comfortable indoors. The usual practice in the North, therefore, when high production is desired, is to keep the fowls in confinement from November to March or April and to give them access to yards of moderate size the rest of the time.

The house to which the birds are confined should be designed for the comfort and convenience of the caretaker as well as for the fowls. If it is reasonably well built and adequately ventilated without exposure to drafts or direct air currents, details of construction are not highly important to the fowls, though in many small ways their comfort can be ministered to by careful planning. Reasonable floor space (not less than four square feet per fowl when constantly confined), plenty of sunlight in winter, with windows high enough above the floor so that the birds will not be tantalized by outdoor views when they cannot be allowed to run out, feed hoppers and water vessels large enough so that there need be no crowding, and located in a good light so that the birds can see to eat and drink early in the morning and late at night, perches at a reasonable height so that heavy hens can readily reach the floor, good-sized exit doors so that one or two birds cannot obstruct the passage— these and many other similar details have a bearing on production because they directly affect comfort.

Fowls in confinement require especial attention to exercise. They are naturally highly active and where closely confined and heavily fed are quite apt to turn surplus feed into fat rather than eggs when conditions are not entirely favorable to production. It is not necessary nor advisable to go to extremes in this respect, and it is rather a fine point to determine just where to draw the line, but with fowls of any breed and at any age, enough compulsory exercise should be provided to promote digestion, to prevent the accumulation of too much fat and to contribute to the general contentment of the fowls. The amount of exercise that must be provided to bring about these conditions depends upon various factors, and a hard-and-fast rule is apt to be misunderstood and misapplied. However, it is safe to say that where one flock is overexercised many more have far too little to do.

Importance of Correct Feeding

The essentials of correct feeding of the laying flock can be stated in a comparatively few sentences, but the details of their practical application are capable of being so widely varied that the range of choice often proves confusing. The marked differences in practice among skillful producers, which sometimes appear to involve even the accepted essentials, are proof that methods and rations—the best of them—depend for their efficiency largely upon skillful management. A good feeder can get better results with a comparatively poor ration than will be secured by a careless, indifferent person with the best theoretical combination of feeds than can be made up. So while feeding for eggs is somewhat of a science it is much more of an art, and best results will always be secured by those who most carefully observe their fowls and who are quick to make changes as required.

Speaking generally there is no such thing as overfeeding a laying hen. On the contrary, egg production is almost directly in proportion to feed consumption, up to the limit of what the hen can digest and assimilate. Fowls can be improperly fed, their rations can be unwisely selected and balanced and thus not well adapted to the requirements of egg production, but the only correct answer to the question "how much shall I feed" is given by the appetites and health of the hens.

A good general outline for the laying ration may be stated as follows: the ration should be around 1.4 to 4.5 in warm weather and 1.5 to 5.5 in winter; it should be composed of about equal proportions of grain and mash; at least one-third of the total protein should be in the form of meat or milk; the percentage of crude fiber should be around 4 per cent; and there should be a fair amount of fat or oil in the ration, though its correct percentage does not appear to have been exactly determined.

Starting with the foregoing general statement of requirements, the operator will introduce variety, will discriminate in the selection of grains and meals to meet market prices, the needs of the fowls or the availability of supplies, and in general will study to adapt himself to circumstances. Other things being favorable, his success—his percentage of production—will be measured exactly by the skill and judgment with which he does this. Various practical rations and methods of feeding are given in Parts II and III and need not be repeated here. For detail information on all phases of practical feeding the reader is referred to the new book, "How to Feed Poultry For Any Purpose With Profit," published by Reliable Poultry Journal Pub. Co.

PRACTICAL SUGGESTIONS ON CULLING THE LAYING FLOCK

No Poultry Keeper Can Afford to Neglect Culling His Flock, Whether Large or Small—Suggestions Are Here Given as to When to Cull, Methods That Should Be Adopted, Etc.

TRAP-NEST records, both public and private, have shown that in almost every flock there is the widest range in individual production, varying all the way from hens that lay no eggs at all to those whose records exceed 300 eggs in twelve months—and this is true almost regardless of breed or breeding. Hens that have no organic defect but that lay only a limited number of eggs may do so because of an inherited incapacity for heavy egg production or because they lack the vigorous digestive organs essential to such production, or are deficient in other important respects. Inferior production in the pullet year may also result from late hatching or from some special and often unknown cause.

Practically all good-sized, unculled flocks contain some hens that are nonproductive or practically so, and at least a few that are capable of laying 200 eggs or more in twelve months. In this respect the difference between ordinary and bred-to-lay flocks is simply one of degree. That is, there are apt to be 200-egg layers in each, but there will be a much greater percentage of them in flocks that have been carefully bred for high production. And in even the best of flocks enough inferior birds will be found to repay the owner well for the trouble of culling. In flocks of poor breeding the percentage of unprofitable layers may amount to as much as 50 per cent of the entire number.

In the last few years several methods of culling have been developed, which make it possible to estimate the productiveness of hens with a high degree of accuracy by observing certain external characters. Those most commonly considered in culling are:

General appearance and conduct
Head points
Plumage and molt
Egg type and abdominal capacity
Condition
Pigmentation

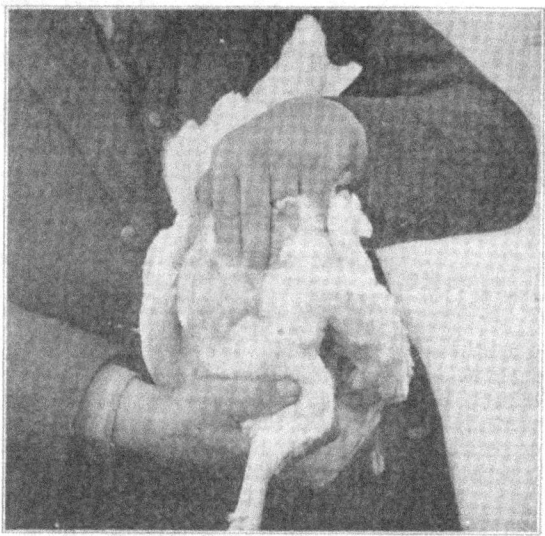

MEASURING SPREAD OF PUBIC BONES
The distance between the pubic bones (the so-called "lay bones") varies more or less, and is greatest in high producers. The spread of the pubic bones is generally stated as being so many "fingers." The fowl here shown is a "three-finger" fowl. Reproduced from "Profitable Culling and Selective Flock Breeding."

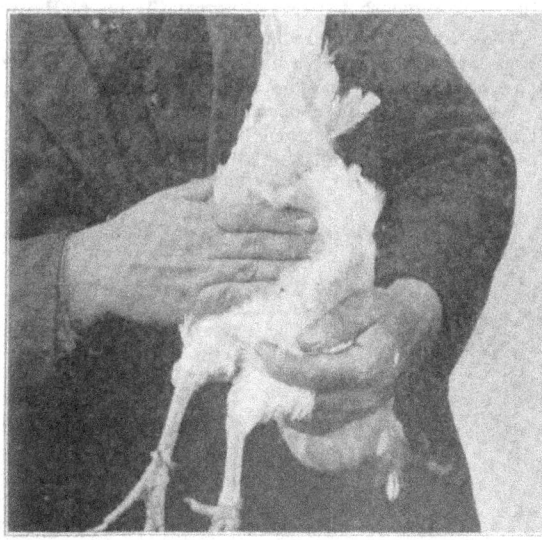

MEASURING CAPACITY OF ABDOMEN
Heavy-laying ability depends quite largely upon abdominal capacity. This is conveniently measured by the distance between the rear end of the keel and the pubic bones. When these are far enough apart to admit four fingers, as in the case of the above pullet, high production is to be expected. Reproduced from "Profitable Culling and Selective Flock Breeding."

General Appearance and Conduct

These afford an excellent index to health, which obviously bears an important relation to productiveness. The hens that are active and alert in bearing, not timid and wild nor, on the other hand, so indolent that they will hardly keep from underfoot, are usually the best layers. The same is true of those that are last on the perch at night and first off in the morning, and busy all day long.

Comb

At any season in the year the layer's comb affords a good clue to her productiveness. The size of the comb does not seem to bear any direct relation to egg yield though, as a rule, fowls with large combs for the breed are apt to be the better layers, particularly if the combs are fine in texture and (in the case of single-combed fowls) with serrations broad at the base instead of being pencil pointed. The color and size of the comb vary greatly in different birds and with different stages of laying, and are not a reliable test for the inexperienced, though to those who have made a careful study of the comb, its condition tells many interesting facts regarding production. The eyes of the good layer are prominent and more or less parallel in the head, as viewed from the front, while the eyes of the poor layer "toe in" toward the beak. It is stated by some observers that the eyes of the good layer are inclined to be oval rather than round, with the pupils set rather back of the center, whereas in the poor layer they are apt to be in front of the center.

Plumage

In late summer and early fall the condition of the plumage is quite helpful in determining whether or not

EVERY NEST OCCUPIED AND A GOOD WAITING LIST
The owner of this commercial flock was dissatisfied with his summer production, and culled his hens. Above photo was taken in the house containing the selected layers showing every nest occupied. Compare with cut at foot of page. Photo by Prof. Roy E. Waite, Md.

the bird is still laying. Fowls that have stopped laying at this time and have begun to molt will rarely become profitably productive until well along in winter, if ever. Practically everybody understands, of course, that late molters are the best layers in the flock, as a rule, provided the delay is not due to ill health.

Egg Type

The argument for "egg type" is based largely upon its indication of capacity—that is, heavy production demands large capacity to eat and digest food and ample room for the egg organs. Hence, any characters indicating such capacity may be assumed to be an indication of good-laying ability. Egg type in fowls is modified by breed type, though the latter does not necessarily have much influence upon laying ability except in extreme cases. A conservative general statement on this subject, therefore, is that while there is no particular bodily conformation, so far as now known, that certainly identifies a hen as a good layer or a poor layer, heavy production ordinarily is accompanied by or brings about certain changes in shape and condition which, taken together, may fairly be called "egg type." See Chapter III for full treatment of this subject.

Condition

The amount of fat carried by the bird is an important "point" in culling. The abdomen of the laying hen has a distinctive "feel," due chiefly to the fact that there is but little fat under the skin, as a result of which the abdomen is soft and flabby. On the other hand, the well-fed hen that has not laid for some time will have a heavy layer of fat under the skin, making the abdomen feel hard and stiff by comparison. Both pullets and hens take on a good deal of fat as they approach laying and if they are properly nourished will keep in moderately good condition throughout the laying period; but any hen with a thick layer of fat under the skin of the abdomen should be viewed with suspicion.

Pigmentation Tests

In fowls having yellow skin, pigmentation tests are particularly helpful in estimating length of laying period up to about three months. It has been found that production usually results in the prompt fading of the yellow, starting with the skin around the vent and rapidly extending to the eye ring (and ear lobe in the case of Leghorns), beak and, last of all, the shanks.

For example, a hen that has been laying in the neighborhood of three months will be bleached out in practically all pigmented (yellow) sections. If there still is a little pigment left on the back of the shank near the hock joint, the hen has been laying only about six or eight weeks. If the yellow color of the beak is entirely gone and has just begun to bleach out of the shank, the productive period probably has not exceeded four to six weeks. If the bleaching process has hardly begun to show on the shanks, and the beak still has a little pigment at the tip, the hen has not been laying more than a month. Two weeks of laying will affect the pigmentation at the base of the beak only. One week to ten days should remove the pigment from the eye ring and the ear lobe, while only two or three days of laying will remove it from about the vent. In some instances this may have disappeared entirely by the time the bird has laid her first egg. In attempting to estimate closely the length of laying period it must be kept in mind that the rapidity with which the pigment fades out or reappears after laying has ceased, depends quite largely upon the breed, the conditions under which fowls are kept, the rations fed, etc. Leghorns fade out much more rapidly in close confinement than where they have open range, particularly if they have access to succulent green feed. Under any given condition Leghorns will lose their pigment quicker than fowls of the larger breeds.

How to Distinguish Between Different Classes of Hens

The Nonproducer. In many instances nonproducers, if laying internally, will show many of the points of a good layer but can be distinguished by handling, as there usually is an abnormal accumulation or a tumorous growth in the abdomen. If the nonproducer's ovary is not functioning at all she will be easily detected as

NOT A NEST OCCUPIED IN THIS PEN OF CULLS
In this pen the culls from a commercial flock were placed. Photo was taken same day as the one reproduced at top of page. Not a hen on a nest, and not an egg in the nests. Getting rid of the poor layers is the quickest and surest way to increase profits. Photo by Professor Roy E. Waite, Md.

she will have all pigmented parts normal in color, her pubic bones will be close together, and her abdomen will be relatively small and hard.

The Poor Layer. Many hens that seem to have no organic defects nevertheless produce so few eggs that they are a source of loss in any flock. Assuming that the fowls are well fed and cared for, low productiveness may be due to age, poor health, limited powers of digestion, natural (inherited) inferiority, or to other causes not clearly identified. The poor layer at any season will have her pubic bones close together, her keel often will be short and "tucked up" behind, and her abdomen usually will be small. However, in detecting low producers, the "feel" of the abdomen is a more reliable guide than its size, as such hens are overloaded with fat and the abdominal walls will lack the pliability that is regularly associated with laying. Except in the early summer when they may be fairly productive, poor layers are apt to have normally yellow beaks and shanks, and their head points (eyes, comb, etc.) are permanent indications of natural inferiority. Such hens usually stop laying and begin molting in July. At this time pinfeathers will be found in the neck, and an occasional feather will fly out of the back when the fingers are drawn rapidly through from tail to neck. One or more wing feathers also may be dropped.

Medium Layers. It is chiefly in this class that the operator is most liable to make mistakes as there is no test that will determine certainly whether or not a hen has been laying for three months or for six. If the three-month layer is handled near the end of her laying period she may be as completely faded in beak and shank as the hen that has laid twice as long and possibly had produced three times as many eggs. Speaking in averages, however, the short-time layer will have less abdominal capacity, her pubic bones will be closer together, and her head points will be unfavorable. If the culling is done in August or September, however, the short-time layer will have stopped to molt and yellow pigment will be coming back or will be entirely restored. With fowls of such breeds as Rocks, Wyandottes, etc., it is necessary to discriminate between the short-time layer that has stopped for the molt and the good layer that has stopped for a broody spell and will soon resume laying if kept in the flock.

The Best Layers. The best layers in any flock can readily be detected at almost any time. It is only necessary to select those that are faded out in all pigmented sections, that have the desired head points, large abdominal capacity, well-spread pubic bones, a soft, pliable abdomen, late molt, etc. In early-summer culling some medium layers may be included by the inexperienced, but a second culling in early fall should eliminate practically all of these.

An interesting article on culling, By Dr. O. B. Kent of Cornell University, will be found in Chapter V, Part III. For complete information in regard to all methods of culling the reader is referred to "Profitable Culling and Selective Flock Breeding," a new and splendidly illustrated book published by Reliable Poultry Journal Publishing Company.

USE OF ARTIFICIAL LIGHT FOR WINTER LAYERS

Artificial Light Properly Used, Will Enable Even the Inexperienced to Double the Number of Eggs That Would Be Secured From the Same Hens Without Lights

ONE of the most important developments of recent years with reference to winter production is the discovery of the close relationship that exists between the number of hours of activity and egg yield. There are good physiological reasons why the eight or nine-hour winter day places the hen at a great disadvantage. The use of artificial light is not a "forcing" method at all but one that simply gives the hen a normal day for eating, digesting and assimilating her food and for healthful activity, instead of compelling her to spend 15 or 16 hours on the perch with an inevitable slowing down of bodily functions, which is directly opposed to egg production. The increased egg yields that, with scarcely a single exception, are reported wherever lights are properly used, are practical proof that this reasoning is cor-

HOUSES AND PENS USED IN TESTING VALUE OF ARTIFICIAL LIGHT AT UTAH EXPERIMENT STATION

rect, and few energetic commercial egg farmers now neglect this important aid to better winter production.

While increased feed consumption usually, though not always, accompanies the use of light, this is incidental rather than the primary object of the practice. According to Professor Lewis of New Jersey Experiment Station, an increase of four or five pounds of grain per bird (covering 12 months) is necessary where lights are employed throughout the short-day period. Experiments conducted at Cornell University also show an increase though not a uniform one. The following table condensed from data secured at Cornell University (published in complete form and illustrated with a number of 3-color charts in the new book, "Use of Artificial Light to Increase Winter Egg Production," issued by Reliable Poultry Journal Pub. Co.) shows the value of feed consumed and number of eggs produced by hens and pullets with and without light:

Influence of Lights on Feed Consumption
November 28-April 16

	Value of feed	No. of eggs laid
25 hens without lights	$26.92	765
25 hens with lights (average of 3 pens)	31.07	1,229
25 pullets without lights	32.05	1,331
25 pullets with lights (av. of 3 pens)	33.60	1,629

The check pens without lights consisted of 25 birds each. There were three pens of hens and three of pullets under lights for varying periods, each pen containing 25 birds. In this table the data for the lighted pens have been averaged for pens of 25, so that the feed cost and production of fowls with and without lights may be directly compared. It will be seen that the increase in feed consumption was slight. The number of pounds of feed is not given in the table, but if we assume an average cost of $3.50 per 100 pounds, increased consumption amounted to 4.8 pounds per hen and 1.8 pound per pullet.

If the object were simply to get the hens to eat more it would be an easy matter to accomplish this by making slight changes in the ordinary ration, and without the use of lights at all. Probably, therefore, the chief advantage derived from lights is the increased activity of the fowls, particularly internal activity. The importance of this will be better understood when it is realized that the fowl is naturally quite active, as is clearly indicated by a high body temperature—106 degrees or over. It ought not to require argument to show that compelling such animals to remain on perches for fifteen or sixteen hours out of every twenty-four is bound to produce unfavorable effects, particularly when associated with heavy feeding on highly concentrated rations. Under lights however, the birds are active for several more hours daily, with the result that, while they keep in good flesh, they carry less surplus fat and are less liable to become logy. Their bowels are in a more laxative condition also and their egg organs appear to function at a more rapid rate or, perhaps more accurately, for a longer period during the day.

Having decided to meet the hen's requirements for a longer working day by supplying artificial light, the beginner needs to understand the dangers as well as the benefits of the method, thus to guard against the former and insure maximum results not only in the winter but throughout the entire laying period. The following represent the more common difficulties that he is liable to encounter:

What Causes the Premature Molt

The average poultry keeper looks upon the winter molt as the direct cause of a slump in production, whereas the slump always precedes the molt. It may originate in a variety of conditions which unfavorably affect the fowls, their presence often being unsuspected until the egg record calls attention to them. A falling off in production is not always followed by a molt, but it is quite apt to be, and those who are feeding for maximum production must constantly be on guard against it. Fowls may go through a light (neck) molt without dropping off much in production, but a heavy molt will result in their remaining unproductive for many weeks during the very season in the year when eggs are most in demand and sell at highest prices.

Those who are feeding for maximum production often find that their birds, even though they may be getting a good ration and plenty of it, fall off in weight after a time. This is something that must be prevented, and any steps necessary in order to do so should be taken, as loss of weight is a certain forerunner of reduced production. The higher the egg yield secured the more careful must the poultryman be to see that his birds are well nourished. The most persistent layer may continue to produce eggs for a time while losing weight, meantime drawing upon her bodily reserves; but complete cessation in production is inevitable unless this condition is soon

SCENE ON EGG FARM OF E. H. WENE AT VINELAND, N. J., WHERE "LIGHTS" ARE REGULARLY USED
All houses on left were lighted one winter and Mr. Wene states that as a direct result the egg yield from November to March inclusive, WAS DOUBLED, the lighted pens averaging 79 eggs per hen during these months as compared with an average of only 41 from unlighted pens in the row of houses on the right.

corrected. No advice is more important to the winter egg producer with or without lights, than this: KEEP THE WEIGHT OF THE BIRDS UP TO NORMAL.

Why Forced Production Is Unwise

Experience has shown that where lights are used for comparatively long periods so that the birds have a 15 or 16-hour day, production can be forced up to 70 per cent or above. Practically without exception however, those who do this find that the hens soon break down and that, taking the season through, they may even produce fewer eggs than those in pens not lighted. It may not be a physical impossibility for fowls to eat and assimilate the quantities of feed required to meet their bodily requirements and lay eggs in such numbers in winter, but clearly there are few persons who are sufficiently skillful feeders to maintain production at that high level for any length of time. Around 50 to 60 per cent is about as high production as can be maintained indefinitely, and no matter how easy it may be to get the birds above this, the poultry keeper can depend upon it that in the long run he will lose out by so doing.

The unfavorable results that sometimes are secured where lights are turned on too early in the fall or traceable, for the most part, to the fact that pullets are not fully matured or the hens are not through the molt, the birds thus being brought into production while they are physically immature or below par. As they usually are handled and fed they are not able to produce eggs and at the same time build up reserves so, sooner or later, they drop below the point in condition where production can be maintained.

Pullets and Hens Should Be "Ripened"

The use of lights is especially effective when applied to late-hatched pullets as it brings them along rapidly and starts them laying weeks ahead of the time when they would do it without lights. However, these pullets, if too rapidly forced, will not stand up under lights as well as those that have reached normal growth and maturity. Lights appear to affect the egg organs more than they affect growth, and their use on immature birds, especially in connection with a ration high in animal protein, is quite apt to bring them into laying before the birds are fully grown. As Professor James E. Rice has said: "You want to turn the lights on when the birds are ripe." It is true that a moderate use of lights in connection with nonstimulating rations will hasten the "ripening" of pullets, but they must be used with care or the birds, after laying for a short time, will break down.

What has just been said in regard to the importance of having the pullets thoroughly ripened applies equally to hens. In other words, it is not wise to force these for eggs until they are entirely through the molt, have grown a new coat of feathers, and are in good flesh.

Morning and Evening Lighting

There is no reason to believe that there is any actual difference in the value of lights used at any particular time in the day. Choice in this respect is determined chiefly by the convenience of the caretaker. One advantage of turning the lights on in the morning is that the birds will come down at once and may readily be kept down thereafter until the usual time for going on the perches in the evening. However, if there is no feed for them when they come down and if they must go without water for two or three hours, or must drink ice water left over from the day before, there may be little gain from the use of lights.

The chief reason why lights in the evening do not always give good results is that the poultry keeper fails

POULTRY FARM WHERE ARTIFICIAL ILLUMINATION IS IN REGULAR USE
See night scene at foot of page. Photo from Storrs (Conn.) Agricultural College.

to adapt his feeding methods to conditions. It is quite common on going into lighted poultry houses at dusk, to find half of the birds on the perches and the remainder standing around more or less idle. It is useless to expect much benefit from the use of lights under such conditions. The caretaker must have a feeding schedule that will offer some inducement to the birds to keep reasonably active during the lighted hours. With proper attention to this, lights in the evening will be just as satisfactory as in the morning.

When to Turn Lights Off

Many poultry keepers have found that while they have had good results with their lighted flocks during the winter and early spring, the egg yield quickly falls off when lights are discontinued. This often is accompanied by a molt. Apparently the explanation lies in the sudden change in conditions, something that always unfavorably affects fowls even though it may seem to be of slight importance. No matter when the lights are to be turned off, do not discontinue them all at once, but reduce the lighted period gradually, 10 minutes or so per day, thus introducing the change by degrees.

Many poultry keepers report that they get few more eggs in a year's time from lighted pens than from pens without lights, the advantage in their use, in this case, being that the eggs are transferred from the low-price spring months to the high-price period of early winter. Some however, contend that this is simply evidence of poor management. They maintain that the extra eggs laid in the winter under lights are not necessarily deducted from spring production. On the other hand, they believe that since lights rightly used improve the hen's physical condition, the extra number of eggs secured during winter months should be a definite addition to the number she would produce in the year without lights.

NIGHT SCENE ON POULTRY FARM SHOWN AT TOP OF PAGE, WITH LIGHTS ON
Photo from Storrs (Conn.) Agricultural College.

The reader will find complete information in regard to this subject in "Use of Artificial Light to Increase Winter Egg Production," published by Reliable Poultry Journal Publishing Company.

CHAPTER VI

High Egg Production by Individuals and Pens

Fowls Differ From Each Other in Marked Degree, and to Insure Maximum Production Should Have Close Individual Attention—If Kept in Pens Should Be Carefully "Matched" as to Feeding Habits, Etc.—Special Methods of Handling Fowls for Pen Records Briefly Outlined

EVERYONE who has had occasion carefully to observe fowls, even of the same breeding and age, has noted that, far from being "like peas in a pod," each one has a distinct individuality and that they differ from each other, often to a marked degree. Just how far it will pay to go in recognizing these differences and in catering to them in everyday work is not clear. In some careful studies of individual hens at Pennsylvania State Experiment Station important differences were found to exist in feeding habits and preferences. In these experiments various feeds were placed before the birds, in separate hoppers, and they were allowed to help themselves at will. It was quickly observed that each hen had a feeding formula of her own. In many cases this was so clearly defined that the attendant was able at any time to identify a given hen by the proportions in which the different feeds were consumed. Some ate practically no grain but corn; others ignored corn and demanded wheat; some ate liberally of dry mash while others practically rejected it. The difference extended even to green feed, some hens quickly picking their small grass runs bare, while the amount consumed by others was scarcely noticeable. In a number of cases it was found that some of the birds would not, under the conditions of the experiment, make up for themselves a ration sufficiently well balanced to enable them to be productive or even to keep in good health for more than a short period.

It is hardly to be expected that maximum records will be secured where fowls with marked differences are bunched together and all handled alike. Under such conditions the strongest and most aggressive always have the best chance, and many layers capable of making extremely good records are too high strung to reach the best production of which they are capable, when subjected to the jostling, crowding and possibly the abuse of other members of the flock. Without doubt those who aim to secure highest possible egg yields must adopt individual penning or its practical equivalent. This plan makes it possible to vary rations and methods exactly to meet individual requirements, and it enables the operator to observe closely the conduct and condition of the birds from day to day—details of the greatest importance where production is being pushed up to the limit of the birds' physical endurance. To date, few if any efforts have been made to carry methods of high egg production to this extreme, but it is reasonable to expect that when this is done marked increases over the best present records will be secured.

Reference has already been made to the fact that birds are confined in individual pens in several laying contests in Australia, but there is nothing about their management to indicate that any special effort is made to take advantage of the opportunity thus afforded for applying highly intensive methods. As a matter of fact, some of the American contests appear to go a good deal farther in efforts to get high records in pens than is done in the case of any of the Australian single tests. D. F. Laurie, government poultry expert in South Australia, appears to have been the first to adopt the use of single-hen pens, doing so chiefly, as it would seem, to secure individual records without the use of trap nests. He provided each bird with a house pen 3 feet square and a yard 3 by 20 feet.

The ration recommended by Mr. Laurie is as follows—the names of the feedstuffs being changed to the American equivalents for the convenience of the reader: The mash, fed early in the morning, consists of 1 part of bran and 2 parts of middlings, exact proportions depending upon the amount of flour left in the middlings. To this is added one-third by bulk of short-cut, steamed alfalfa or clover hay. Meat is supplied in the form of a soup made of either fresh meat or meat meal, which is used to moisten the mash. In addition to this, three or four times a week a little meat scrap is given when it seems to be required, taking particular pains, however, to avoid feeding it to excess. Mr. Laurie is of the opinion that to feed meat in the proportions recommended in the average American laying ration would speedily prove disastrous. At midday a little succulent green feed is given, and an hour before dark a handful of grain, usually wheat, is scattered in the straw litter with which the 3 by 20-foot yard is covered to a depth of several inches.

Mr. Laurie says: "On no account should any attempt be made to force egg production. You wish to ascertain what the hen will do under suitable conditions and on normal feeding and you also hope to have a sound, healthy hen at

VINELAND CONTEST HOUSE AND PEN OF R. I. REDS
Note open front with hood affording protection from rain and sun without obstructing ventilation. A muslin shutter is provided for closing front in extreme weather. Courtesy of N. J. Experiment Station.

the termination of your test. Give what variety of food you can afford or obtain, but remember egg production depends on the use of food having the necessary constituents."

At the Bendigo Contest in Victoria, a semiofficial contest, where some of the highest Australian records have been secured, including the 339-egg record referred to in Chapter I, the ration is as follows: The wet mash, which is fed every day of the year, consists of 3 parts of middlings, 1 to 1½ parts of bran, ⅛ part of oatmeal. This is mixed with 50 per cent of chopped green feed, usually alfalfa or clover. The mash is mixed with a hot liver soup, secured by boiling liver with which also are cooked various vegetables. Four times a week liver is minced and added to the mash at the rate of 1½ ounce per bird. The mash is mixed in a crumbly condition and fed warm early in the morning. In addition to the moist mash a dry mash is kept before the birds, this consisting of 2 parts of middlings, 1 part of bran, ¼ part of alfalfa chaff, ⅛ part of oatmeal, 1/29 part of peameal and 1 per cent of brown sugar. At noon chopped clover or other succulent green stuff is given with minced liver. The evening meal, which is thrown into the scratching litter, consists of 4 parts of wheat, 1 part of oats, 2 parts of corn.

The International Laying Contest at Victoria, B. C., Canada, represents the only step so far taken in American contests in the direction of concentrating attention on individual birds. Beginning in the fall of 1920 the plan was adopted of placing only two birds in a pen, one belonging to a breed laying white eggs and the other laying brown eggs, thus making it possible to keep accurate laying records without the use of trap nests. Prior to 1920 the pens at this contest consisted of six birds. J. R. Terry, poultry expert in charge of contest, supplies the following data showing production of the 2-hen pens, from November 1 to April 1, as compared with production in 6-hen pens for the same period the preceding year:

Comparison Between Six-Hen and Two-Hen Pens

Duration of comparison, winter months.................5
Trap-nested birds (six birds) average eggs per bird......38.6
Two-bird pens, average eggs per bird (both classes)....65.0
Trap-nested birds, lightweight, average number of
   eggs laid per bird.................................39.3
Two-bird pens, lightweight, average number of eggs
   laid per bird.....................................67.
Trap-nested birds, lightweight, average number of
   eggs laid per bird.................................37.8
Trap-nested birds, heavyweight, average number of
   eggs laid per bird.................................63.

The marked difference in favor of the hens in the present contest cannot be attributed entirely to the difference in treatment, as the winter of 1920-21 was much more favorable to production than the previous one. However, it seems entirely probable that individual treatment has had a highly favorable influence.

By way of showing how moderate are the efforts made to secure high egg yields, even in laying contests, it is interesting to note that the highest individual record secured in a laying contest in this country, also the highest pen average (with one exception) and the highest average for all fowls entered in contest, have been secured at the West. Washington Contest where the Leghorns entered are housed in flocks of 54 and the American-class birds in flocks of 36 (see Chapter I, Part III, for details).

As fowls approach their natural limit in production their records become more and more a test of the caretaker's skill, and there probably is no fixed point beyond which not another egg can be secured. Whatever a hen's performance may be under given conditions she will cer-

SECOND PEN AT SECOND ALL-NORTHWEST CONTEST, WASHINGTON
Average production 251.6 eggs per bird. Photo from Professor R. V. Mitchell.

tainly respond with increasing production to suitable improvement in rations or methods. Thus a hen that lays 275 eggs in 12 months can lay 300 in the same period if her caretaker is able to feed her to a little better advantage—and beyond the latter figure production can be further advanced, up to the limit of human knowledge and skill.

"Forcing" Egg Production

No arbitrary limit can be established for "natural" production beyond which it can accurately be said that the fowls are being forced. That can be determined only by the physical condition of the birds. Perhaps as good a definition of "forcing" as can be given is to describe it as the point beyond which production can only be secured at the expense of the fowl's general health or vigor. That, however, is extremely vague, because a rate of production that would be forcing under some conditions—that would cause the hen to fall off in weight or would promptly result in various disorders of the egg organs—would be sustained indefinitely on a better ration or with better methods of care.

Correctly speaking, no hen can be "forced" to lay. If she is in the right physical condition and other things generally are favorable she will lay without forcing; and where her requirements are not met there will be few eggs no matter what measures may be adopted. It is possible, it is true, to use rations that will temporarily speed up the egg organs beyond what is natural, but in most instances this will be followed sooner or later by a slump that may leave the average for the year at a lower point than would have been attained without interference.

Securing Maximum Production

Maximum production of individuals is secured under the same general conditions that any market producer seeks to provide, but carried to extremes that would not be possible nor practical in commercial production. For example, the egg farmer feeds meat, and so does the test feeder. But the former fixes upon a proportion that is safe and that can be fed week in and week out without special attention; the latter aims to keep the proportion at the limit of what the birds can assimilate as established by daily observation and adjustment. Because of its great palatability (if for no other reason) the test feeder will use fresh meat freely, even though it may cost decidedly more. Milk will be supplied almost regardless of expense or trouble, because within suitable limits there is no better egg-producing feed known.

Practically every experimental test that has been made indicates that a regular daily feed of moist mash will increase production over that secured with dry mash. And incorporated in this mash will be a reasonable proportion of cooked feed. Just how much of this to supply and in what form are questions that have never been thoroughly investigated but, without doubt, easily or partially digested feed will prove a highly important factor in feeding for maximum production.

WATER VESSEL, MASH, GRIT AND SHELL FEEDERS USED AT VINELAND CONTEST
Located on a roomy feeding platform and in a good light, which encourages frequent visits. Courtesy of N. J. Experiment Station.

The appetite of fowls is subjected to extreme fluctuations, and while these no doubt can be greatly modified by careful feeding there will always be occasions when the question of an egg or two more in a given period will be determined by whether or not the hen's appetite is tempted by something extra palatable, this "hand feeding" taking whatever form and being carried to whatever length may seem necessary.

There is good reason for believing that maximum production demands moderate warmth and protection from extreme cold snaps in winter. This is suggested, among other things, by the fact that, almost without exception, winter yields are highest at the laying contests located in sections where winters are not severe.

The use of artificial light in winter is, of course, imperative, and one who is in charge of only a few fowls can go to extremes in this respect that the commercial producer would be unwise to attempt, because he can check up closely on results and make prompt modifications where required.

Many more details in methods could be added to the foregoing, each with its definite reaction upon production. How far it may be practical to go in introducing these is for each poultry keeper to determine for himself. But the advertising value of an exceptionally high record or one exceeding all others is so great that the breeder, at least, can well afford to incur considerable trouble and expense in giving his birds an opportunity to demonstrate what they can do when they have the best possible chance.

High Production in Pens

Outside of experimental work and laying contests where economy in cost is not directly an object, single penning is not practical, but the poultry keeper who is earnestly striving for high averages will find that he can go a long way toward meeting individual requirements by carefully grading his birds and keeping in small or medium-sized pens those that most closely resemble each other in their needs or preferences.

While the term "pens" is used to designate a small flock, it has no arbitrary limit. Speaking generally, it may be said that a "pen" becomes a "flock" when the fowls are too many in number for even matching, or for close individual observation. However, the term is most commonly applied to groups of twenty-five or less and so is used in that sense in this book. Inside of that limit there should be no difficulty in providing reasonably favorable conditions for individuals or in giving to each a measure of personal attention; beyond it "wholesale" methods usually are applied to varying extent.

TYPE OF NESTS USED AT VINELAND CONTEST
A very satisfactory trap nest for general use. Courtesy of N. J. Experiment Station.

Within moderate limits it is doubtful whether size of pen is directly a limiting factor in production; whether, in other words, a group of ten hens will necessarily lay fewer eggs per bird than one of five, all other conditions being equal, or more than one of twenty. On the contrary, it seems probable that careful matching of the birds will prove much more important than the actual number kept together—that twenty or even forty layers evenly matched in their physical characters and feeding habits may make better average records than five among which marked differences in individuality exist.

The chief recent demonstrations in what can be accomplished in pen production have been staged at various laying contests, the best known of which are the "International" at Storrs, Conn. and the "Vineland" in New Jersey. Here ten birds constitute a pen and while extreme methods have not been adopted in either case, in fact have been carefully avoided, the production averages secured have set the pace for poultry keepers the country over. At Storrs the highest pen average secured was in 1917-18 where the leading pen produced 2,352 eggs, while the highest average for all pens in the contest (1,634 eggs) was in 1916-17. At the Vineland Contest, 1918-19, the best pen record was 2,431 eggs, or 243 per bird—and the average per pen for the entire number entered in the contest was 1,790. These averages are much above those secured by the great majority of poultry keepers and perhaps there is no better way of showing how such records are reached than by stating briefly the rations and methods adopted at these two institutions.

Storrs Contest Method

In the Storrs Contest the birds are fed dry mash to which they have access at all times, and a grain mixture the major part of which is fed in the afternoon from three to five o'clock, depending somewhat upon weather and season. More grain is fed at this time than the birds will consume so that there may always be a little left in the litter on which the fowls can go to work early the next morning. On cold days and in damp weather, when the birds are disposed to be less active, grain is fed in small amounts two or three times during the forenoon just to keep them busy. About four or five inches of litter is provided and changed as frequently as necessary to keep it dry and reasonably clean. The houses have dirt floors and are 12 feet square, being divided into two pens of equal size, thus providing 7.2 sq. ft. of floor space per bird.

The grain ration used is composed of equal parts by weight of cracked corn and wheat, while the mash consists of equal parts by weight of wheat bran, corn meal, ground oats, flour middlings and meat, the latter being a mixture of equal parts of meat and fish scraps. The average consumption of feed per pen of 10 birds, for a period covering five years, was 544 lbs. of mash and 406 lbs. of grain. Mangel beets supply succulence during the winter, and about the first of May the birds are allowed to run out in the yards in which rye, clover and grass are growing. The yards supply sufficient green feed as a rule, until about midsummer, after which rape or Swiss chard is fed.

Vineland Contest Method

The houses in the Vineland Contest are 8 by 10 feet and the pens consist of 10 hens, thus giving each bird 8 sq. ft. of floor space. (In the 1920-21 contest the number was increased to 20, reducing the floor space to 4 sq. ft. per bird.) Each house has a yard 36 by 80 feet, to which the birds are given access from about the first of May. The remainder of the time, or from November 1 to May 1, they are confined constantly to their houses which are well lighted and ventilated, even when the muslin curtains must be closed on account of cold or severe storms. A further provision for the comfort of the fowls in houses where Leghorns are kept is a curtain of muslin or burlap that can be let down in front of the perches in extremely cold weather. The hood also helps to keep out rain and shades the interior in summer, without obstructing ventilation. Both the Storrs and Vineland houses provide the fowls with as comfortable quarters in winter and in summer as can be supplied at a moderate expense.

The grain ration fed consists of 100 pounds each of cracked corn, wheat and clipped oats. In the wintertime an extra 100 pounds of cracked corn is added. The dry mash consists of equal parts by weight of wheat bran, white or flour middlings, ground oats, corn meal and meat scrap. This is fed in open, circular earthen hoppers with a loose-fitting grid resting on the mash to prevent its being wasted. The birds are given access to the

THE BACK-YARD POULTRY KEEPER HAS AN EXCELLENT CHANCE TO GET GOOD PEN RECORDS
Practically all the birds in this village pen of Barred Rocks (Parks' strain) laid 200 eggs or over in pullet year. Photo from Professor E. F. Grundhoeffer, State College, Pa.

dry mash at all times and the grain is fed in limited amounts, morning, noon and night, straw litter 3 to 6 inches deep being provided and frequently renewed.

At this contest the birds are encouraged to eat a larger percentage of dry mash than at Storrs, the proportion in the Vineland Contest being in the ratio of 1.4 of mash to 1 of grain. Based on the results secured at this contest the New Jersey Station has worked out the following table showing the amount of grain to feed to layers each month in the year, the expectation being that they will consume enough dry mash in addition to give them the correct ration for their probable production in the different months.

Amount of Grain to Feed Layers Each Month in the Year

Months	Pounds per Day per 100 Birds	Pounds for Each Feeding A. M.	P. M.
November	12	4 and	8
December	12	4 and	8
January	12	4 and	8
February	12	4 and	8
March	12	4 and	8
April	10	4 and	6
May	10	4 and	6
June	8	3 and	5
July	8	3 and	5
August	6	2 and	4
September	5	2 and	3
October	5	2 and	3

The feeding and general care are comparatively simple at both contests. There is nothing in either method that differs materially from the practice of commercial poultry keepers generally. Without doubt, a more elaborate feeding schedule, the use of artificial lights, etc., would have had a marked influence on the records secured, and those who are working primarily for high production will find many modifications or additions that can be advantageously adopted.

CHAPTER VII

High Egg Production by Flocks

Flocks May Vary in Size According to Convenience of Owners, but High Records Can Be Secured Even in Flocks of a Thousand or More—The Commercial Egg Farmer in Particular, Must Keep Production Well Above the Minimum Required to Meet All Costs—Special Suggestions Are Here Given on How to Do This

BY "flock" is meant a comparatively large group of fowls, as "pen" means a small group. For the purpose of this book, twenty-five is accepted as the dividing point. From this minimum flocks range in size up to several hundred. While pens are generally accepted as offering better opportunities for securing maximum production per hen, flocks of good size are held to afford economies in labor and equipment costs that more than offset whatever loss in egg yield may result. The size of the flock is determined partly by breed and partly by the extent of the individual poultryman's willingness to sacrifice production averages to reduced labor costs. Apparently, no one has ever taken the trouble to determine definitely by careful experiments the influence of numbers upon production, so as to be able to say with certainty where, in practical egg production, the line of improving egg yields crosses the line of decreasing labor cost, and beyond which point better averages can only be secured at disproportionate expense. As a result, there is no agreement among large-scale producers with respect to this detail, flocks of 50, 100, 250, or 500 or more being maintained, according to individual preference. In general, 100 appears to be recognized as the practical limit in the case of fowls of the larger breeds while some operators regard 1,000 as practical in the case of Leghorns, though most prefer not to exceed 200 or 250.

THREE-YEAR LAYING RECORD OF HEN NO. 88-8, AT WEST. WASHINGTON EXPERIMENT STATION

This recopied record of Hen No. 88-8 is here reproduced to show the daily performance of the long-cycle layers being developed at this institution. This hen began laying on July 26, went through a partial neck molt in January that checked production for a little over a month and then laid continuously until November, a record of 370 eggs in 442 days. After about two months' rest she began laying again in January, 1920, but did not settle down to steady production until in March. This year she reached a total of 189. She molted earlier in the fall of 1920 and began laying earlier (December 15), reaching a total of 137 by June 21 of her third year—the date on which this record was made out—and still laying.

Good Records Possible in Extra-Large Flocks

An especially interesting illustration of the fact that high egg yields can be secured in extra-large flocks is afforded by the success of Mr. and Mrs. Geo. R. Shoup at the West. Wash. Experiment Station, in producing an exceptionally large number of 200-egg records and better, with the layers kept in flocks of 1,000. The management of the birds while making their pullet-year record under such conditions is thus described in substance by them.

"The caretakers are on the job at 7 a. m. when all frozen feed or icy water is removed. Enough fresh kale is supplied to last until noon but no longer. If kale is not available sprouted oats are fed instead. At eight o'clock water is supplied; at 8:30, clabbered milk or buttermilk in earthenware vessels. At 3:30 mixed grain is fed and in the winter artificial light is turned on about this time. Fresh water is again supplied at five o'clock, and at 5:30 a moist mash is fed. At seven o'clock kale is supplied. Dry mash is kept before the birds in self-feeding hoppers all the time.

"This institution now owns whole flocks of breeders, up to 400 or more, that are all of 200-egg grade or better. Our birds spend their pullet year in flocks of 1,000 and only by their trap-nest records are they identified as extra-high producers. At the end of the first laying period all birds that are physically fit and have records of 200 or more eggs are placed in the breeding pens."

Mr. and Mrs. Shoup have made a specialty of extra-early hatching and have been successful in producing a number of January and February-hatched pullets that have remained productive through the fall and winter without going through a "false molt," the result being that the first laying season of these pullets totals 13 to 16 months instead of being less than 12 months as is commonly the case with birds that are hatched at the usual time—March to May.

The table presented herewith shows the records of a number of these long-cycle layers and is arranged to show their records for the first 12 months and for the entire pullet year; also their subsequent performance. We re-

they run, with reference to the false molt and its effect on production, Mr. and Mrs. Shoup have furnished complete monthly records for September, October and November for the first 99 pullets to begin laying in the summer of 1918. This may be summarized as follows: No. 1 began laying on June 10 and No. 99 on July 26. Individual production prior to September 1 ranged from one up to 66 eggs, on which date 88 of the 99 birds were laying. Total number of eggs for September was 1,758 but toward the last of the month production fell off sharply. On October 1 57 were laying, though several stopped on that day or the following one and the number of layers was down to 31 on October 20, the total pro-

Production Records of Long-Cycle Hens Bred and Owned by the West. Wash. Experiment Station

Band number	224-5	118-6	74-7	4-7	94-7	70-7	71-7	88-8	350-8	199-9	306-9
First egg	7/31/'15	10/29/'16	8/6/'17	6/10/'17	8/13/'17	8/4/'17	8/4/'17	7/26/'18	10/29/'18	7/27/'19	8/16/'19
Last egg of pullet yr.	10/21/'16	11/1/'17	9/14/'18	10/15/'18	9/25/'18	9/29/'18	9/18/'18	11/10/'19	11/5/'19	11/28/'20	11/27/'20
12-Month record	231	295	289	240	254	232	272	287	315	290	292
Pullet-year record	306	298	319	333	286	275	307	370	322	365	363
Second year	197	169	216	190	216	208	194	189	231		
Third year	167	162	151	158	188	189	169				
Fourth year	200										
Fifth year	173										
Total production	1044	785	686	691	690	672	670	559	553	365	363

Twelve-month records up to 315 eggs and pullet-year (13 to 16 months) records up to 365 eggs are presented in above table. The dates when first and last eggs were laid show the exact length of each hen's pullet year. It will be noted that these winter-hatched pullets began laying—each in the summer—and continued to lay until the fall of the following year. Thus, the pullet with the highest record began laying July 26, 1918, and stopped November 10, 1919—producing 370 eggs in 642 days.

gard these records as truly remarkable and as pointing the way to a practical method of increasing first-year production and profits that no commercial poultry keeper can afford to ignore. This effort to develop a strain of nonmolting, early-hatched pullets will be watched by practical poultry keepers with great interest. These investigators are emphatic in their belief that the long-distance layers should not be used for breeders, "as the fertility of the eggs is very poor, the chicks are quite weak and the pullets are far inferior to their parents. Breeders should have at least sixty days' rest and an opportunity to store up a new supply of yellow pigment and physical vigor generally, which nature demands for the production of normal offspring."

The reader should note that this heavy and long-continued laying does not detract from the ability of the birds to produce heavily in their second year, as is indicated by second-year records running as high as 231, while even in the third year some closely approximate 200 eggs, and one hen reached 200 in her fourth year.

The hens represented in the foregoing table are selected birds, of course. In response to a request for information as to what early-hatched pullets do, just as

duction for October being 900. Fifty-three were laying the last of the month, however, and by the 10th of November only 11 were nonproductive. The total production for November was 1,898. All of this indicates that the average rest period did not exceed 3 or 4 weeks.

In the September record only one molter is mentioned and one broody; in October two partial molts were recorded and none at all in November. With reference to this it is stated that "when they rest three weeks or less they do not show any molt, but when they rest longer there is either a neck or a fluff molt, or both. No attention is paid to those birds that are resting. They stay in the flock, receive the same feed, light and care, and return to production according to their capacity. However, if the birds take frequent rests, molt their flight or tail feathers, go broody and lay one day and rest two as a habit, they are marketed as soon as the defect is noticed—the sooner the better."

Cost of Production

While the feed bill is the largest single item of expense on the poultry farm, there are many other costs that must be reckoned with, though they often are over-

SOME HIGH-RECORD, LONG-CYCLE BIRDS AT THE WESTERN WASHINGTON EXPERIMENT STATION

Reading from left to right the first hen is No. 88-8, which has the remarkable record of 315 eggs in her first 12 months, and 370 in 442 days of continuous laying. The second bird is No 71-7 and the third, No. 350-8. Production records of all three birds will be found in table herewith. No. 88-8's daily record for three years is reproduced on opposite page.

looked or underestimated. Egg production, to justify itself as a truly profitable industry, must show a profit over all items of expense including not only feed, but also labor, interest on investment, depreciation, etc.

To indicate the expenses regularly associated with egg production as a business enterprise, the accompanying table has been made up from data given in New Jersey Bulletin No. 329, the items listed representing averages on 100 poultry farms stocked chiefly with Leghorns. If fowls

END VIEW OF LAYING HOUSE, WEST. WASH. EXPERIMENT STATION

This illustration shows type of house in which were produced the records presented in table on page 55. Front is equipped with roll curtain (not shown), and moist mash is fed in trough running entire length of front of building.

of the larger breeds are kept, most of these items will have to be increased somewhat, though there does not appear to be any definite basis for figuring the increase except in the case of feed, where about 10 per cent should be added for fowls of such breeds as Plymouth Rocks, R. I. Reds, etc.

It should be understood, in a study of this table, that a poultry plant fully meeting this schedule of costs is self-sustaining—that is, it not only meets feed and all other costs but it pays its operator laborer's wages for all work done and returns interest on investment, the income being sufficient, with the help of a good fruit and vegetable garden, to afford a modest living. On such a plant (again using data from Bulletin No. 329) approximately two-thirds of the total income is from market eggs sold and one-third from stock, eggs for hatching and miscellaneous sources. Depending upon market eggs then for two-thirds of the income it is clear that to meet the total annual expense of $3,953, 1,000 hens must average to lay 78 eggs each at an average price of 40 cents per dozen, or 104 at 30 cents per dozen. Everything above these averages of production or price —provided that costs are not further increased in securing the advance— goes toward the "comfortable" living that is the ideal of everyone who engages in the work.

In the light of the foregoing it would seem that the importance of securing a comparatively high average of production would be self-evident, but that it is imperfectly understood is evidenced by the general low average secured the country over, and even on the plants of specialists. For example, the

Estimated Cost of Egg Production on a Poultry Farm of 1,000 Layers

Items of Cost	Amount
Grain	$1,726.
Succulence (beets, cabbage, etc.)	17.
Human labor	620.
Hired labor	126.
Horse labor	32.
Litter and use of range at 6 per cent	101.
Seed for range	9.
Use of building—depreciation at 10 per cent	155.
Interest on buildings at 6 per cent	90.
Use of equipment, at 12.95 per cent	89.
Interest on equipment, at 6 per cent	18.
Interest per hen, at 6 per cent	81.
Use of cockerels	97.
Hatching eggs used	128.
Hatching eggs bought	12.80
Day-old chicks bought	17.
Pullets bought	13.
Yearlings bought	12.
Crates, cost	24.
Spray materials	8.
Deaths	99.
Fuel and oil	30.
Depreciation	497.

In this table, cost of feed is estimated on the 1915-16 basis, the average price per hundred pounds being $2.07. This probably is a fair average under normal conditions, though much below the average for exceptional years like 1918, '19 and '20.

bulletin just referred to, shows that in New Jersey, where commercial egg farming is generally regarded as quite highly developed, the average per hen on 100 egg farms was but 109 eggs. A similar survey of 42 poultry farms in Connecticut revealed an average of only 97.

The causes for these low averages will be found in readily corrected conditions, in almost every instance. There is not a low-producing flock, anywhere, that cannot be put on a paying basis simply by comfortable housing, good feeding and culling. The latter may result in reducing the size of the flock to a level much below what is desired, but if the alternative is to continue producing eggs at a loss, why not face the fact and take steps to add to the income in some practical way instead of continuing straight toward bankruptcy? That there is nothing impossible or even difficult about getting good, even high, averages in flocks is proved by the numerous

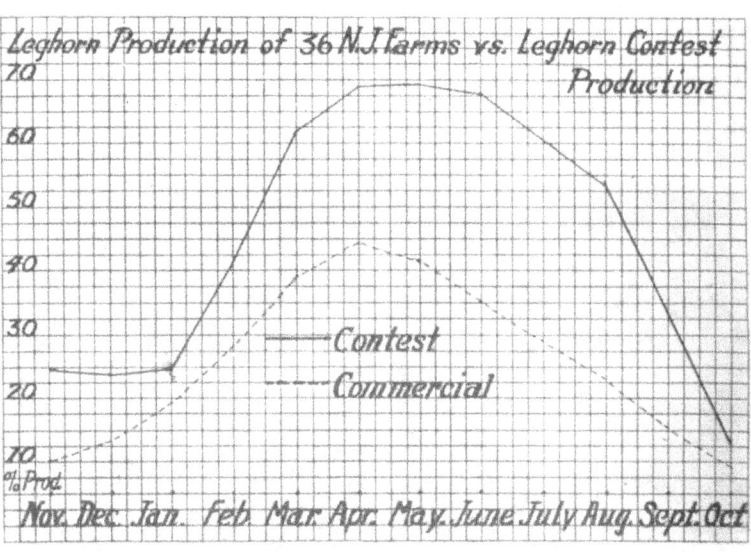

HIGH AND LOW-LEVEL EGG PRODUCTION CONTRASTED

Average production per bird in New Jersey is about 10 eggs higher than in the country at large, but the dotted line representing the monthly percentage of production of 36 Leghorn egg farms in that state, when contrasted with production secured with Leghorns at the Vineland Contest, shows that even in this prosperous poultry center averages are far below what may readily be secured with better stock, feeding and management. Courtesy of N. J. Experiment Station.

illustrations given in more or less detail throughout this book.

Increasing Flock Production

Each Connecticut farmer with his average of 97 eggs per hen, each New Jersey egg farmer with his average of 109, and all other commercial egg producers, no matter in what part of the country located, whose production is around these low levels, will find that vastly better results than they are now realizing are readily within their reach. To secure them it is only necessary to meet the reasonable conditions outlined in previous chapters, adapting details to the practical requirements of commercial production where, of course, they must be more or less general in application. Some methods of doing this, also some of the commercial egg farmer's special problems, may be here mentioned, though necessarily with great brevity.

It goes without saying that the results secured by Wyckoff, Atkinson, etc. (see Chapters I and IV, Part II), can only be secured by having first-class, bred-to-lay stock. Comparatively few of those who are engaged in commercial egg production care to undertake systematic breeding for high egg production, but those who do not can and should regularly patronize those who do, to the extent, at least, of purchasing from them all males used in the breeding pens from which their annual supply of pullets is to come. This step alone will, in a year or two, make all the difference between a flock that barely pays its way and one that provides a comfortable net income to its owner.

The feeding of the commercial flock must be on a wholesale basis as compared with the elaborate attention often possible in the case of small pens, but it should be carefully adapted to the requirements of the birds, nevertheless. Whether the commercial flock is to be fed on dry mash or to have a moist mash each day will depend on a number of conditions. The additional labor cost of feeding moist mash is slight, however, and will be covered by a comparatively small increase in production—an increase that will be realized in a majority of cases if the moist mash is properly fed. The precise formula of the mash, whether fed moist or dry, is not so important so far as results are concerned, as making sure it is palatable, so that the birds will eat the proper amount, though it should, of course, be well balanced. A num-

A FLOCK OF LEGHORNS THAT HAVE AVERAGED TO LAY OVER 200 EGGS EACH IN THEIR PULLET YEAR
Photo from C T. Patterson, Springfield, Mo.

ber of good rations are given at various places in this book and with any of them satisfactory results become simply a matter of management.

The commercial poultryman, particularly, must give close attention to costs. The use of a grain or particular mill feed the price of which is out of proportion to other feeds that could be substituted for it, makes a noticeable difference in the expense of feeding the birds and in the net results secured. It is important to bear in mind however, that while economy is highly important, cost should be measured not by the cost per pound, but by the nutritive value of the feed. The poultry keeper who selects a grain or a mixture chiefly on the basis of price almost invariably supplies his birds with a low-value feed that, in the long run, costs him a good deal more in waste resulting from the excess fiber or in reduced production due to lack of palatability, than the added expense involved in using a much better ration.

The large-scale poultry keeper is especially apt to err in the direction of treating his flock as a unit and giv-

VIEW ON QUEENSBURY FARM, TOMS RIVER, N. J., SHOWING TYPICAL LEGHORN POULTRY PLANT IN THIS SECTION

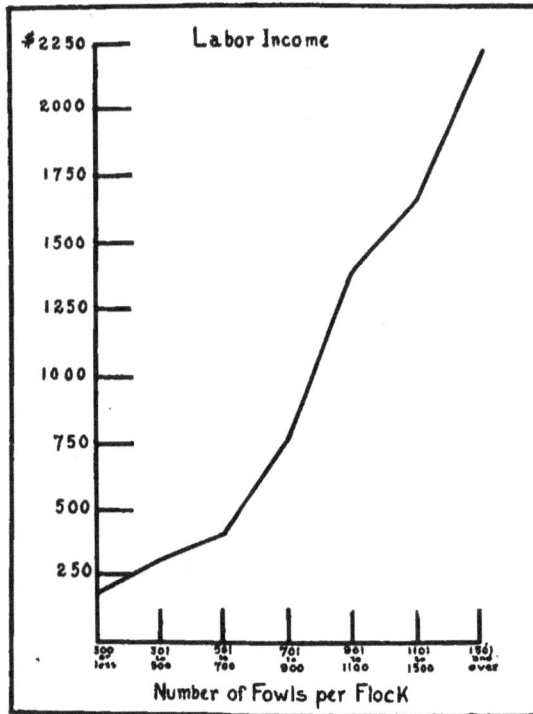

GRAPH SHOWING RELATION OF NUMBER OF FOWLS TO LABOR INCOME

This graph shows that on the exclusive poultry farms of New Jersey 900 hens or over are required to provide a fair income. It is true that these figures are based on a low average of production (109 eggs per bird) but, whether the average is low or high, true efficiency on the poultry farm requires that relatively large numbers of fowls be kept. Reproduced from New Jersey Bulletin No. 329.

ing scant consideration to individuals, even though he may know that many birds thus are placed at a disadvantage. It is not necessary to spend a great deal of time in watching the fowls, but a sharp eye for those that are not in good health, that are falling off in condition, or that are being mistreated by other members of the flock, and their prompt removal to more suitable quarters, will be well repaid in the better results secured. This also applies to the prompt weeding out of birds that are evidently inferior producers. The commercial poultry keeper, especially, should make it a practice to do more or less continuous culling. The presence of ten per cent of unprofitable birds in a large flock means serious and entirely unnecessary financial loss. The methods of culling described in Chapter V are sufficiently definite and easily mastered so that the wide-awake egg farmer will have little difficulty in weeding out any birds that become nonproductive.

The house in which the laying flock is kept may be made an important aid to securing increased production, though there are few things besides sufficient room, good ventilation and reasonable protection that are regarded as absolutely essential. The chief gain resulting from the careful planning and construction of laying houses will be found in reduced labor cost, enabling the attendant to take care of a large number of birds in a given time. In this chapter are illustrated several types of houses that are in use on commercial plants, and though they differ to a marked degree in their architectural features they all meet the foregoing essentials. It is impossible, within the limits of this book, to go into details of poultry-house construction, but those who are interested in this subject are referred to a new book on "Poultry Houses and Fixtures," published by Reliable Poultry Journal Publishing Company, in which complete plans are given, also much helpful information in regard to location, purchase of materials, construction, etc.

It is doubtful if any commercial poultry keeper can afford to ignore the practical benefits which grow out of the use of artificial light in short winter days (see Chapter V). In the last few years this method has come into great popularity. Thousands of persons in all parts of the country have adopted it as a permanent aid in winter production, and in no instance has it failed to give good results where properly used.

INTENSIVE POULTRY KEEPING AND WHERE TO PRACTICE IT

Unquestionable Evidence That Intensive Poultry Keeping Is Profitable—Both the Advantages and the Disadvantages of the Method Frankly Presented—Intensive Poultry Keepers Are Mostly Town Raised, Which Explains Why They Rarely Attempt Other Lines of Production

BY intensive poultry keeping is meant the method of keeping fowls in comparatively close confinement, as distinguished from EXTENSIVE poultry keeping, where an abundance of room is available so that the birds can have free range, or its practical equivalent. The relative merits of the two methods have always been more or less in dispute. The former has a number of theoretical disadvantages, but it has features that make it highly attractive to many persons. The latter method has important theoretical advantages, though they do not always develop in actual test. Moreover, it carries with it conditions that most of those who want to keep fowls on a large scale simply refuse to meet. Aside from farmers who specialize in poultry, it is safe to say that nine-tenths of those who are considering going into this line of work will do it on the intensive plan or not at all. It is important therefore, to consider carefully just what is

TYPE OF LAYING HOUSE LARGELY USED IN EAST

This house is well adapted to the requirements of all poultry keepers who have good-sized laying flocks. Hood shades the front in hot weather and affords some protection from storms. Photo from N. J. Experiment Station.

practical in this method and the conditions under which it is advisable to undertake it.

It is difficult, if not impossible, to draw a clear-cut comparison between the requirements of intensive and extensive methods because, in the last analysis, local conditions and individual preferences are the real determining factors. An effort to compare them, based on surveys of single plants or averages for whole communities, will always be more or less indecisive because it will never be quite clear how much of the success achieved in either case is to be attributed to the method itself, how much to the skill of the poultry keeper, and how much to conditions that have no direct bearing whatever on the subject.

Advantages and Disadvantages of Method

The more important theoretical advantages and disadvantages of intensive poultry keeping may be listed as follows:

Advantages
- Small amount of capital required.
- High-priced land available.
- Ability to concentrate on single line of production.
- Limited exposure in work due to having fowls in close quarters.
- Community life.

Disadvantages
- Increased cost of feed.
- Increased labor cost.
- Risk associated with a single source of income.
- Danger of loss from disease.
- Monotony of work.

INTERIOR OF LARGE-SIZE, CLOTH-FRONT COMMERCIAL LAYING HOUSE

This scene shows a single flock of 1,500 pullets. Muslin shutters are hooked up on sunny days and closed for warmth at night or when the temperature drops extremely low. This is a very popular type of laying house on eastern commercial egg farms though, as a rule, it is not considered desirable to have so many birds in a single flock.

These advantages and disadvantages alike are subject to decided modification in practice. The advantages can readily (though quite unnecessarily) be nullified by bad management, while the disadvantages can to a great extent be offset by proper planning. The fact of the matter is that a reasonably competent and industrious worker can succeed in intensive poultry keeping almost anywhere, meaning by this that he can pay expenses and "make a living." But if he wants a "good living"—the kind of a living that is demanded by most city men who move to the country—it should be understood that it is only realized under reasonably favorable conditions.

Taking up this list of advantages and disadvantages in some detail, no doubt one of the most important of the former is the small amount of land required, which enables the poultryman either to reduce the amount of capital invested, or to utilize relatively high-priced land. Bearing in mind what has already been said, that the average poultry keeper has had no experience in other lines of agricultural production and wants none, it is clear that a method requiring the use of a large acreage must be undesirable, as the land, if unused or cultivated in an indifferent manner, will be a source of expense rather than of income. Moreover, the city-bred poultryman does not take at all kindly to the isolation of country life, but wants to be in a comparatively well-settled community and preferably not over 30 minutes from Broadway or its local equivalent; to meet which condition comparatively high-priced land must be used.

A good market is vital to the success of the intensive poultry plant. Other things being equal, the man who is closest to his trade will get the highest price and realize the most profit. However, closeness to market is quite largely a matter of transportation. With a large volume of production within a limited radius, good organization among growers, direct railroad connections, refrigerator car service, etc., miles from market are comparatively unimportant, as is proved by the fact that eggs are regularly shipped from the Pacific Coast to New York City, and grade as extras, at some seasons bringing the same prices as Vineland, N. J., eggs. The marketing advantages just mentioned, however, are available only where centers of production have been highly developed. Outside of these it usually is unwise to locate an intensive poultry plant where the product cannot be sold to a fancy retail trade or readily shipped direct to a high-priced city market.

Convenience in caring for fowls where they are all housed within a limited radius is a practical advantage, though its importance is not great except in sections where the winters are quite severe. Generally speaking, and with the foregoing exception in mind, where extensive plants compare unfavorably with intensive ones in point of convenience and labor cost, it is because the former have not been wisely organized.

A much more important advantage to most persons is that the intensive method permits of concentrating one's efforts entirely upon a single line of production. The average man who is interested in poultry keeping

OPEN-FRONT LAYING HOUSE WITH COMBINATION ROOF

For houses exceeding 16 feet in length a combination roof, such as here illustrated, is practical. This house is 20 by 240 feet, the front wall is 6½ feet high, the rear wall, 5 feet, the ridge, 9 feet from the floor and distance from ridge to front, 7 feet, giving good head room with a minimum of enclosed air space and cost for material.

on a large scale does not care to divide his attention, but wishes to specialize exclusively on fowls. And as a matter of both convenience and economy he aims to make each acre carry the maximum fowl population. As to the wisdom of applying oneself to poultry keeping to the exclusion of everything else there is room for argument, but final decision on this point will rest largely on individual preference, so the subject need not be considered here.

One indirect advantage of intensive poultry keeping that is more important than appears on the surface is the opportunity it affords for community life. Small farms mean close neighbors, and if these are mostly engaged in the same line of work a number of practical benefits accrue. These are so important that the average beginner should think long and seriously before deciding on an isolated location instead of settling where others are engaged in the same line of work. In the latter case he has the advantage of coming in daily contact with persons who have had more experience, of watching the results of others' experiments, of sharing the better facilities and markets that always develop in important producing centers or may be developed with proper cooperatives.

The poultry keepers of the Vineland district, for example, market their eggs through an association. Through the same medium they do more or less wholesale feed buying, contract with farmers for the growing of special crops, such as mangels, and in various ways realize advantages that the isolated operator does not have. Moreover, since the intensive poultry keeper almost invariably is city bred, this community life supplies some of the social advantages to which he has been accustomed, and the absence of which in an isolated or less congenial environment often makes irksome what would otherwise be a pleasant occupation.

The Disadvantages of the Method

While, as has already been said, the theoretical disadvantages can be offset to a considerable extent by proper management, higher feed cost is a factor that the intensive poultry keeper must always negotiate. He raises no grain, often not even the winter's supply of green feed and, as a rule, intensive producing centers are not located in grain-growing sections, so that all feed must be shipped in. Something can be done to reduce feed cost by wholesale buying, but the poultry specialist's profits depend more on his getting top prices for his product than on buying economies.

Theoretically, keeping fowls in confinement calls for more care and attention and demands more labor per fowl than smaller, well-managed flocks on the broad acres of a farm. But if the methods adopted by the latter are not down to date it may develop that the intensive poultry keeper has a lower labor cost and higher labor income than the farmer keeping an equal number of fowls, simply because the latter has failed to realize to the full the advantages that his opportunities afford.

Danger of losses from disease is one handicap that the intensive poultry keeper must constantly guard against. It may not be necessary for him to suffer heavier losses than those who have more room, but he is extremely liable to do so.

Because sandy soil gives good drainage and does not retain impurities to as great an extent as heavy soil, it is looked upon as practically essential in locating for intensive poultry keeping. A great deal of New Jersey's popularity as a poultry-producing center is due to its light, sandy soil. With or without a clear understanding of the reason for doing so, it will be found that those who keep fowls on heavy clay soil rarely persist long in efforts to follow highly intensive methods.

In the Petaluma district, where distinctly different classes of soil lie side by side, it is said that the poultry plants stop short where sandy soil gives way to that of a distinctly heavier nature. Sandy soil does not give immunity from disease, as a great many California and New Jersey poultry keepers have discovered to their sorrow, but it unquestionably is an important advantage.

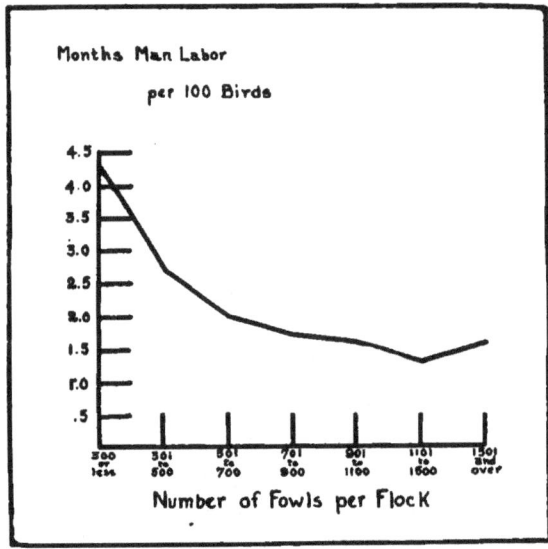

RELATION OF SIZE OF FLOCK TO LABOR COST

Where fowls are kept in comparatively small numbers, the labor cost is apt to be disproportionately high. This graph shows that on 150 poultry farms in New Jersey the greatest labor efficiency was secured in flocks of from 900 to 1,500. The months of labor per 100 birds where 300 or less were kept, amounted to double the time required for the same number where 500 to 700 are kept, and three times as much as in flocks of 1,000 to 1,500. Reproduced from N. J. Bulletin No. 829.

More of the failures that occur in poultry keeping are due to the operator's inability to stand the strain of long hours of monotonous employment than to any other one cause. Few persons accustomed to living in town or city and working at some occupation which calls for comparatively short and regular hours have any appreciation of the strain involved in being "on the job" every waking hour, week in and week out. When to this is added a more or less mechanical round of duties day after day, it is easy to see why only the elect survive. If only as "relief work" the average poultry keeper should develop one or more interesting side lines—fruit, flowers, garden crops or whatever most interests him. If he will do this he will find that he will enjoy the work more and in the long run will be more permanently successful.

FARM FLOCKS AS A SOURCE OF INCOME

The Farm Flock Has Always Been Profitable, No Matter What Its Size, But There Are Sound, Practical Reasons Why Enough Fowls Should Be Kept to Make the Poultry Income Worth While

PROBABLY no one will seriously challenge the statement that, labor and investment considered, no branch of general farming pays better than the poultry flock, and this statement is true not as applied to present conditions, but it has been so for an indefinite number of years past. There are economic reasons which we cannot here discuss, to explain why, in spite of the foregoing statement, the average flock remains comparatively insignificant as a source of income. However, the careful student of farm poultry production cannot escape the conviction that an important change in this branch of the industry not only is due but probably is already well underway.

Conditions affecting commercial poultry and egg production have been so changed in recent years through increased average productiveness, greater labor efficiency and higher prices obtainable for the product, that poultry keeping as a specialty is now on as sound a financial basis and the probabilities of success in well-considered undertakings as great as in any other branch of agriculture. The better understanding of poultry problems and the improvement in methods that have made specialized commercial poultry keeping so profitable clearly warrant increased attention to poultry keeping on farms. In practice, however, there appears to be a distinct tendency for farmers to divide into two classes: those who follow the line of least resistance and keep only the number of fowls that can practically maintain themselves (less than the present general low average rather than more) and those who definitely specialize in this branch of farming, keeping large flocks so that they can produce eggs efficiently as regards labor, and in sufficient numbers so that they can be marketed to best advantage.

Specializing in Farm Poultry

It is to be expected, of course, that where poultry keeping is considered as a source of income it will have to stand or fall in the estimate of the individual farmer, depending on how it ranks in comparison with other lines of farming open to him. Doubtless there are many who, because of local conditions or personal preferences, will do much better in other lines. But doubtless, also, there are others who would realize larger profits from a good-sized flock of hens than they are ever apt to secure in any other way.

The farmer who proposes to specialize in any line of farming, regardless of what it may be, should carefully study his conditions and be sure that he is working with the forces about him, not against them. Where climate or general environment is unfavorable, where local markets are poor and shipping to a distant market involves heavy expense for transportation or deterioration in quality, where there is no one to take a personal interest in the work—under these and other similar conditions, it clearly would be unwise to specialize in poultry. So, in recommending poultry as an important source of income, in stating that thousands of farmers are overlooking the most promising branch of agriculture open to them, in pointing out the practical advantage that specialized farm poultry keeping offers, it is always with the understanding that reasonably favorable conditions exist.

Assuming that conditions are favorable, it only remains to satisfy the individual that this branch of farming is practical under his conditions, and that it will be

A FARM POULTRY PLANT IN EAST TENNESSEE
It is not necessary, nor is it economy, to use a medley of nondescript houses and sheds for the farm flock. The plant here illustrated shows all the building equipment needed to accommodate a flock of 500-600 hens, and brooder capacity to renew half the flock with early-hatched, winter-laying pullets each fall. Building in foreground is brooder house equipped with two coal-burning colony hovers—capacity 600 to 800 chicks. Feed room is in center of laying house, with incubator cellar underneath.

profitable in direct proportion to the good judgment and sound business methods applied to the work. In doing this there are three general difficulties that usually have to be met. One of these is the idea already referred to—that the poultry flock is "too small business" to claim the attention of a real man; another is the common belief that the farm poultry flock is only profitable when it is more or less self-supporting; and another is lack of help.

Possibilities for Profit in Farm Flock

With reference to the first objection it may be said that poultry keeping is "small business" only when it is made so. The statement that the farm flock pays as well as, or better than, any other branch of farming will stand a lot of repeating, and it applies to well-managed flocks of any size. More than that, up to a reasonable limit the larger the flock the better it pays as regards labor income. For this there is ample proof in all sections of the country. As a single illustration, selected from numerous ones available, two Nebraska farms, reported by the College Extension Department of that state, are especially interesting: Farm No. 1, with 72 hens, gave a total return of $365 over feed cost, in 1919-1920, the hens averaging to lay 126 eggs each. But Farm

No. 2, with 209 hens and only a little more labor, returned $779 over feed, the hens averaging to lay 152 eggs per hen.

An illustration, on a larger scale, of what may be accomplished with good-sized farm flocks is given in the annual report of the Missouri Farm Flock Laying Contest for the year ending November 1, 1920. The different flocks entered were classified in the report under breeds, and the following condensed table shows that a labor income of $2.44 to $4.07 per hen was readily secured.

TWO HIGH-RECORD BUFF LEGHORNS OF EXHIBITION BREEDING
Bird on left laid 215 eggs in 11 months at National (Mo.) Contest; one on the right laid 204 during the same period. Bred by Dr. L. E. Heasley, Holland, Mich.

Summary of Reports from Demonstration Farms in Missouri for the Year Ending November 1, 1920

Breed	W. Wyan.	W. Leg.	Br. Leg.	R. I. Red	B. Rock
Number of flocks	6	22	10	13	26
Number of hens per flock	101	186	197	139	109
Number of eggs per hen	132.2	127.8	113.8	109.3	103.1
Total income per farm	$655.07	$1089.51	$978.46	$777.75	$600.35
Labor income per hen	$3.76	$3.16	$2.74	$3.48	$2.44

This contest demonstrates beyond question the practicability of handling fowls in good-sized flocks on the farm, and of realizing a profit which, for a side line, certainly is attractive.

How Fowls Rank with Other Classes of Farm Animals

The idea that to make the farm flock profitable it must be maintained on a scavenger basis is widely accepted and often hard to overcome though, as a matter of fact, the farm flock is exceptional in its ability to turn low-priced feed into high-priced finished products. So far as known nobody has ever disputed the statement made many years ago by Dr. Jordan, of the New York Experiment Station, that the hen is the "most efficient transformer of raw material into the finished product that there is on the farm." The table on next page will illustrate this:

In this table the relation of feed consumption to egg production is based on the low average of 100 eggs per hen, per year, though no one who is in earnest need fail to secure a decidedly higher average than this, in which case—also where a better average price is secured—the comparison will still further favor the hen. If the object in feeding grain is to make money, why not feed it to that class of animals from which the largest cash returns are to be expected?

How the Labor Problem Affects the Farm Flock

A common objection encountered just now in regard to increasing the farm flock is that help is difficult to obtain and poultry keeping for that reason is impractical. Under some conditions this undoubtedly is true. Where the farmer has a large acreage to handle, where all the working force available is utilized and perhaps overtaxed in order to handle the land efficiently, the practical thing to do may be to cut down the farm flock rather than to increase it. But, on the other hand, there are large numbers of farmers who probably would find it to their advantage to reduce other lines and increase their farm flocks, instead of cutting down this, the most profitable branch of their work, investment and labor considered.

For example, Professor Elford, Dominion Poultry Husbandman of Canada, writes: "We find that many farmers are becoming more interested in their poultry flocks and are quite satisfied that, if properly handled, poultry is a paying proposition—so much so that in some cases where labor is scarce, farmers are increasing their poultry, though they may have to cut down other work on the farm." This is not a matter of sentiment nor need it be one of guess work. The different lines of activity have been so generally analyzed and it is possible for the farmer so readily to determine in just what branches of the work he can realize the largest labor income that decision becomes a matter of simple comparison.

MISSOURI FARM POULTRY HOUSE
This well-lighted, gable-roof house is most practical and convenient. Affords excellent control of ventilation, and is provided with straw loft which makes the house warm in winter and cool in summer. Photo from University of Missouri.

What Size Should the Farm Flock Be

The exact size of the income-producing farm flock will be determined by a number of factors and it has no arbitrary limits. It must be of fairly good size, however, for the simple reason that it is not practical to provide the equipment and learn the methods that make commercial poultry keeping efficient and then apply them to a

Relative Feeding Efficiency of Farm Animals
100 pounds of digestible matter produces—
  139 lbs. milk, worth @ 3 cents per pound $4.17
  13.5 lbs. steer (live weight) worth @ 9c per pound 1.215
  13.9 lbs. sheep, worth @ 7c per pound .... .973
  30.4 lbs. swine, worth @ 9½ cents per pound 2.888
  19.6 lbs. fowl, worth @ 20 cents per pound 3.92
  22 lbs. eggs, worth @ 27 cents per lb. (40c a doz.) 5.94

flock so small that the income from it cannot amount to a substantial sum no matter how skillfully it is tended. It is possible to get excellent returns per hen this way, but the labor income (meaning the net income after allowing for feed and all other cash expenditures) will be small— and in commercial poultry keeping that is what counts.

Making due allowance for the numerous exceptions that will always exist it is doubtful if any farmer who aims at making money from fowls can afford to stop short of around 200 layers. This is because the equipment and methods that should be adopted for a flock of 75 or 100 will provide almost as readily for a flock of 200. This latter number of good, well-bred fowls makes direct shipment of eggs in case lots practical, thus freeing the poultry keeper from the limitations of his home market and it affords a labor income that almost anyone would regard as worth while.

Just how far one can profitably exceed 200 fowls is largely an individual matter. However, it is a fact that the additional equipment required to raise the 200-hen flock to 400 is much smaller than is needed to increase the average flock of 46 to 200, and the extra labor involved is also proportionately less. With good organization and a 10-hour day, it is reasonable to estimate that a 200-hen flock on the farm should not occupy an able-bodied person more than three hours daily, the year around, as a general average, while 4½ hours daily should meet all the requirements of a 400-hen flock. On a labor income of $2 per hen that would mean an annual income of $400 for a 3-hour day, or $800 for a 4½-hour day. The figures on labor income per hen, as secured in the Missouri Farm Laying Contest already referred to, prove that this is a conservative estimate.

MODERN HOUSE FOR COMMERCIAL FLOCK OF 800 TO 1,000 LAYERS
This illustration shows house with windows raised for ventilation in extremely warm weather. Is built 48 feet deep by 60 feet in length and has a good-sized gable room through the center for feed and general storage purposes. One-third of the floor on either side of the center section is covered with poultry netting or light boards, and this is covered with several inches of straw to afford ventilation and to insulate the lower room against heat in summer. Photo from American Poultry School, Kansas City, Mo.

CHAPTER VIII

The Production Possibilities of Different Breeds

Ability to Produce Large Numbers of Eggs Is not Peculiar to Certain Breeds but May Be Developed in Any—Leghorns Have Had Special Attention in This Respect But They Have no Monopoly on High Production and Excellent Records Have Been Made in Numerous Other Breeds

HIGH egg production is in no sense a breed character but one which is already well developed in a large number of breeds and which no doubt is capable of being well developed in other breeds that up to this time have not been conspicuous producers. In dealing with some five-year averages secured at the Storrs Laying Contest, Storrs Bulletin No. 100 states: "When one considers five years' contest records as a whole, there appears to be nearly an even chance for any breed to win. The 1913-14 contest was won by a pen of White Leghorns that belonged to Francis F. Lincoln, Mt. Carmel, Conn. Their record for the year was 2,088 eggs. The next, or fourth, contest was won by a pen of White Wyandottes entered by Tom Barron, Catforth, England. These 10 hens laid 2,072 eggs. In 1915-16 Obed G. Knight of Bridgeton, R. I., furnished a pen of Wyandottes that were winners with a fine score of 2,265 eggs. In the sixth contest a pen of Barred Rocks owned by Fairfield Poultry Farms, Short Falls, N. H., outlaid all their competitors, finishing the year with 2,119 eggs. In the seventh competition, which closed October 30, 1918, a pen of Oregons won first place with a total of 2,352 eggs. Moreover, the Oregons established a new high record for a pen of ten birds in any American contest.

"If one considers individual records, the best bird in the 1913-14 competition was a White Wyandotte that belonged to Merrythought Farm, Columbia, Conn. This hen laid 265 eggs during the year. In 1914-15 a Rhode Island Red owned by Hillview Poultry Farm, St. Albans, Vt., made the best individual record with a yield of 257 eggs. In 1915-16 individual honors went to a White Leghorn owned by A. P. Robinson of Calverton, N. Y., with the excellent record of 286 eggs. In the sixth competition a Barred Plymouth Rock pullet entered by Merrit M. Clark of Brookfield Center, Conn., won first place among individuals with a yield of 277 eggs. In the contest now being reported, a White Wyandotte entered by Obed G. Knight of Bridgeton, R. I., made the very exceptional record of 308 eggs."

Since annual laying contests were inaugurated in 1901, fowls of the following breeds and varieties have made official or semiofficial records of 200 eggs or over:

Barred, Columbian and White Plymouth Rock
Buff, Columbian, Silver and White Wyandotte
S. C. Black, R. C. and S. C. Brown, S. C. Buff and S. C. White Leghorn
Rhode Island Red
Rhode Island White
Black, Buff and White Orpington
Ancona
Black Minorca
Oregon
Black Langshan
Red Sussex

In addition excellent records are reported for Light Brahmas, Campines, etc.; also some crossbred birds.

It must be remembered that there are two types of fowls being bred generally, to which the names of "utility" and "fancy" are popularly applied. These distinctions are purely artificial, being based simply upon the primary breeding objective, whether that is greater productiveness, superior table quality or some particular and often exaggerated character relating to size, shape, plumage, etc., and it so happens that in some cases the characters that are most highly valued in certain breeds are such as are distinctly unfavorable to production, or their development has been carried to such an extreme that to maintain them it is necessary to concentrate attention chiefly upon them. Doubtless any of these breeds, if bred primarily for production and modified in type where necessary to favor abdominal capacity, would promptly respond with marked improvement in production.

Breeding for production need not in anyway interfere with standard quality—at least within the limits of what are known as the popular breeds. It is probable that the breeder who concentrates his attention solely upon production will make more rapid progress for a time, but he will always have an inferior flock and one that will be extremely difficult to bring up to any breed standard.

Repeated charges have been made that standard fowls have been bred so intensively for exhibition that productiveness had been sacrificed, but egg-laying contests in which standard-bred, crossbred and mixed or mongrel fowls have been entered have almost invariably demonstrated the superiority of standard birds, even though comparatively little attention may have been given to breeding for production. Moreover, their use as breeders on low-producing stock regularly results in increasing egg yields, other conditions being favorable.

Use of Standard-Bred Males Increased Production

A striking illustration of what can be accomplished

TWO HIGH-PRODUCING OREGON HENS
This new breed developed by Professor Dryden has made an excellent showing as to productiveness. Hen on left has a record of 303 eggs; the one on right, 257. They are somewhat larger than average Leghorns but are white and have single combs.

in increased production by introducing blood of standard-bred stock into a mongrel flock is afforded by an experiment at the Manitoba Experiment Station. Reporting this experiment in "The Nor'-West Farmer," Professor Herner, who conducted the experiments, writes:

"The value of the male bird in the flock should not be underestimated. On him depends the revenue just

PORTION OF MONGREL FLOCK USED IN GRADING EXPERIMENT
These were farm-raised birds and fairly representative of the general appearance and quality of farm flocks in Manitoba. Average production in pullet year, 76 eggs. Photo from Professor M. C. Herner.

about as much as on the females. Our own experience in grading up a farm flock of mongrels has afforded striking object lessons. We started out with a flock of 100 mongrel pullets three years ago. These were mated with six purebred Barred Rock males of such type and quality as any farmer could buy at that time for about $3 each. These pullets were certainly a nondescript flock. These birds laid an average of 76 eggs each for the year. The chickens hatched from the eggs from these mongrels mated with the above males were used the second year. We killed off all the original mongrels. In the second year of the experiment the grade pullets averaged 90 eggs each for the year, or an increase the second year, due to the use of the purebred males, of an average of 14 eggs per bird. The progeny from these birds were in turn mated with Barred Rock males, with the result that they gave us pullets that averaged 111 eggs for the year, or an increase of 15 eggs over the previous year. This gives us an average of 76, 90 and 111 eggs respectively for mongrels the first year, and grades the next two years."

While increased egg yields may thus sometimes be secured by using birds of standard quality even when not known to have been bred for production, marked improvement demands not only standard-bred birds but such as are from lines specifically bred for production. This is well illustrated by an elaborate experiment in the improvement of mongrel flocks undertaken in 1912 at the Kansas Experiment Station. In this test four pens of birds were used, these being from a local poultry packer just as they were received from the farmers and without any culling or selection, the object being simply to secure a representative mongrel flock. Forty birds were purchased and divided into four pens of 10 each. One pen was mated with a White Orpington male from a high-record hen, another with a Barred Plymouth Rock and another with a White Leghorn, while a fourth was mated with a mongrel male.

The graph on next page shows in striking manner the results of this method of mating continued for three years. Steady and marked increase in production was realized in the Leghorn and Rock grades but, presumably through an unfortunate selection of males, there was a marked falling off in the Orpington grade pens after the first year. There is no apparent reason why the Orpington mating should not have shown as good progress as in the case of the other breeds if good birds had been secured, but it is quite plain that, at least in mongrel matings, the breeder is at the mercy of the males used which, if not strongly prepotent for the character of heavy production, may completely fail in raising the flock to a high level in this respect.

Crossbreeding Not Generally Desirable in Practical Poultry Keeping

These results of experiments in crossbreeding are presented to show the benefit to be derived from the introduction of the blood of standard fowls in any low-producing flock, but should not be accepted as indicating the general desirability of crossbreeding. While this practice clearly may result in marked improvement in production, the poultry keeper will always have crossbred fowls. These are better than mongrels, it is true, but they fall far short of standard-bred stock in money-making possibilities. As compared with grading up a flock, the additional cost of starting a small breeding pen of standard fowls of good quality and gradually increasing the number of these, at the same time reducing the graded flock so that at the end of a few years the entire flock is of standard breeding, is slight, little more effort or expense is involved, and the final outcome will be vastly more satisfactory from every point of view.

Comparison of Breeds

While high egg production can be developed in any breed, generally speaking, to date greater progress has been achieved in some than in others, and the poultry keeper whose primary object is market egg production and who wants immediate results, will be wise in selecting a breed, and a strain in that breed, that has been systematically bred with this object in view, thus to enable him not only to start at a higher level but more quickly to advance his average by careful breeding within his line.

Where market egg production alone is the object White Leghorns will be selected under most conditions. It is not by accident but because of extremely good practical reasons that the White Leghorn has become the almost universal egg-farm breed. The merits of Leghorns

THIRD-YEAR GRADE IN MANITOBA BREEDING EXPERIMENT
After using standard-bred Plymouth Rock males for three years this flock of pullets was producing an average of 111 eggs. Photo from Professor M. C. Herner.

are briefly set forth in the accompanying extract from Bulletin No. 338 of the New Jersey Experiment Station:

"It is interesting to note the efficiency of the Leghorn breed. The material increase in the profit realized from this breed is due to three direct causes: first, a greater egg production per bird; second, a greater selling price per dozen and third, a marked decrease in the amount and cost of feed consumed.

"From a study of the data we find that because of heavy production, low feed cost and increased selling price for white eggs, the Leghorns far outstripped the American breeds. We find also that the Plymouth Rocks, Wyandottes and Rhode Island Reds are in a class by themselves in respect to this profit factor. The first year the Leghorns paid a net return above feed of $4.34, while the American breeds paid a net return of $2.90. The second year the Leghorns paid a net return above feed of $3.41 while the American breeds averaged a little less than $1.50. In computing these results no consideration is given to the greater weight and value of the carcass from the heavy breeds, which is about offset by the increased cost of producing the pullet."

Financial Returns by Breeds—Second or Yearling Year, 1917-1918

	Plymouth Rocks	Wyandottes	Rhode Island Reds	Leghorns
Amount of mash consumed per bird (lbs.)	61.10	55.70	59.00	52.20
Amount of grain consumed per bird (lbs)	27.50	24.70	27.50	27.70
Cost of mash per bird	$2.22	$1.94	$2.03	$1.73
Cost of grain per bird	$0.97	$0.99	$1.01	$1.14
Total cost per bird	$3.19	$2.93	$3.04	$2.87
Number of eggs per bird	119.10	115.40	117.20	137.60
Average per cent production	32.60	31.60	32.10	37.70
Average price per dozen	$0.46	$0.46	$0.46	$0.522
Value of eggs per bird	$4.58	$4.44	$4.54	$5.98
Returns per bird over cost of feed	$1.39	$1.51	$1.50	$3.11

It will be noted here that the argument in the foregoing quotation is based chiefly upon greater egg production per bird and a greater selling price per dozen. The application, however, is local rather than general, as these two advantages are by no means peculiar to this particular breed. This is proved by the fact that in other contests, comparisons between breeds have not been so favorable to Leghorns. At the Storrs Contest, for example, there is practically no difference between Wyandottes and Leghorns with respect to net returns realized, value of eggs alone being considered.

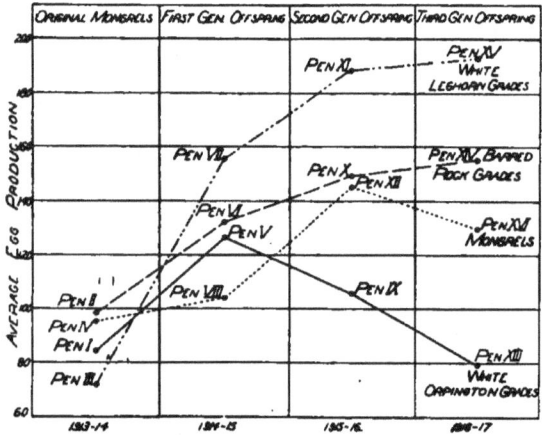

GRAPH SHOWING RESULTS OF EXPERIMENTS IN CROSS BREEDING AT KANSAS EXPERIMENT STATION

A flock of 40 mongrel hens was divided into lots of 10 each and mated with Barred Plymouth Rock, White Leghorn and White Orpington males of high-producing ancestry. One pen was mated with a mongrel male. Curve shows marked improvement in Plymouth Rock and Leghorn crosses. Through unfortunate choice of males the Orpington cross fell short even of the production of the mongrel pen. Reproduced from Kansas Bulletin No. 223.

Under Connecticut conditions Leghorns have proved to be much more seasonal in their production than Wy-

A HEAVY-LAYING CAMPINE

Laid 216 eggs (averaging over 26 ounces to the dozen) in one year at the National Egg-Laying Contest. Owned by M. R. Jacobus, Ridgefield, N. J.

andottes. In other words, fowls of the latter breed were found to be more uniform in egg yield the year through and they laid more eggs at the season when prices were high. Bulletin No. 100 states in regard to this:

"A few outstanding points about these curves should perhaps be mentioned (this relates to illustration shown on page 67). The first point to be observed is the decided drop in egg production which the Leghorns have suffered from the sixth to the tenth week after the beginning of the contest. As has been pointed out in previous reports, this drop is very largely due to temperature. Combined with the effect of low temperature is also the fact that Leghorns which reach the competition in laying condition are practically certain to go through a partial molt and thus suffer a setback in egg yield. Considering the breed as a whole, they make up for the eggs lost at this time by their higher production during the summer months. At the end of the laying year, when all birds are dropping off in egg yield, the Leghorns seem to have an ambition to be first in the matter of reaching minimum production. Almost in direct contrast to the seasonal performance of Leghorns is that of Wyandottes. When the Leghorns are suffering their first drop in egg yield, the Wyandottes are outlaying all other breeds; similarly at the end of the year when the Leghorns are dropping rapidly, Wyandottes are at the head of the list."

As contrasted with Leghorns and other small breeds, fowls of the larger breeds are at a disadvantage in that they require on the average about 10 per cent more feed to produce a given number of eggs in a year's time. As usually bred they are from one to two months longer in maturing, which not only increases the cost of raising the pullets but often gives the Leghorns a practical advantage in the start in the fall. The other breeds also undoubtedly lose out somewhat on account of broodiness. Some of these disadvantages are offset by greater market value of eggs, greater profit in sale of surplus cockerels, capons, etc., and better winter production.

Losses due to broodiness can easily be overestimated, hence comparisons between the larger breeds and Leghorns cannot always be accepted at their face value. For example, the Leghorn that stops laying for a two-week rest without going broody loses just as much time as a Plymouth Rock or a Wyandotte that goes broody but

THE PRODUCTION POSSIBILITIES OF DIFFERENT BREEDS

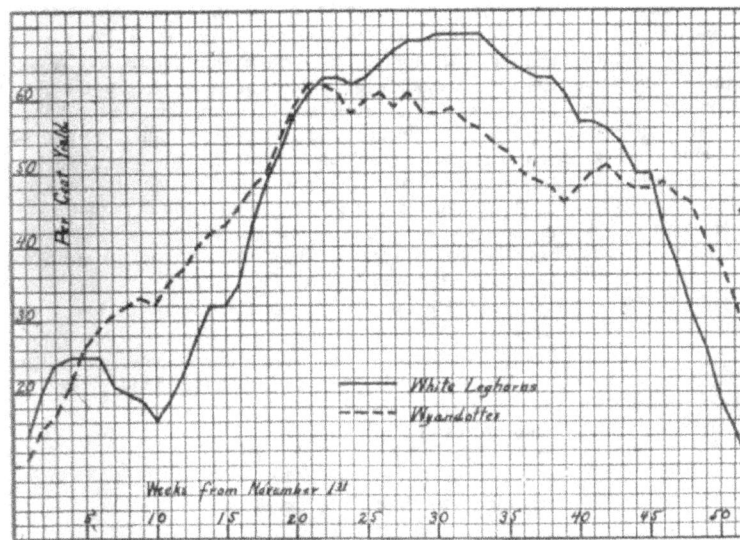

GRAPH SHOWING PERCENTAGE OF PRODUCTION OF WYANDOTTES AND LEGHORNS AT STORRS LAYING CONTEST

These curves indicate production percentages. The weekly records of all Wyandottes and White Leghorns in the contest for a period of 5 years were used in preparing this graph. Average number of eggs produced for the year was almost the same for each breed, but Wyandottes led in production during the high-price months.

is promptly broken up and brought back into laying within the same period. It is also possible that the broody period bears some relation to the bird's ability to lay later in the fall. This is indicated by the graph reproduced herewith, in which the production line of the Wyandottes and White Leghorns at the Storrs Contest is indicated. It will be seen here that while the Wyandottes fell off to a marked extent during the warm months of the year, toward the latter end of the year (September and October) their production was quite a little better than that of the Leghorns. That the loss through broodiness is more nominal than real also is suggested by Storrs' Bulletin No. 100, where it is shown that the average production of Wyandottes for a five-year period was 162.6 per hen, while the average for White Leghorns was 162.4.

HIGH EGG YIELD OF MEDIUM AND HEAVY-WEIGHT VARIETIES

Leghorns, Anconas, Etc., Have Distinct Advantages as Egg Producers for Table Use, but the Heavier-Weight, General or Dual-Purpose Breeds, Such as the Plymouth Rocks, Wyandottes, Rhode Island Reds, Sussex, Orpingtons, Etc., Also Have Points of Superior Value

By Grant M. Curtis, Editor of Reliable Poultry Journal

AMONG the outstanding advantages possessed by the lightweight breeds, such as the Leghorns and Anconas, are these, in brief: they come into egg yield one to two months earlier on the average than the heavier-weight breeds and are brought to this point at considerably less cost; they lay as many and as large (as heavy) eggs during their production period and eat less food in doing it; they can be kept to advantage in larger flocks and require less housing and yardroom per fowl; up to twelve weeks old the surplus cockerels that usually are sold as broilers, grow more rapidly than do the cockerels of the heavier-weight breeds.

Reprinted from Reliable Poultry Journal, issue of April, 1920.

On the other hand, the medium and heavyweight breeds (most varieties of these breeds) enjoy the following advantages, comparatively speaking: they are better for table use, especially as roasting chickens, etc.; are heavier when sold as hens for use as human food; they excel as capons; they stand cold weather better in northern latitudes and average to lay more eggs during severe winter months; they are more easily confined, are excellent mothers and have better grain and flavor of flesh than the smaller, more active breeds and varieties.

For good enough reasons far more attention to date in the history of poultry culture, especially in the United States, has been given to the development of the Leghorns along high-egg-production lines, notably the White variety. For these same good reasons they now "have the call" in the great egg-farming districts, such as Vineland, N. J., and Petaluma, Calif. Also at a number of agricultural colleges considerably more attention to date has been given to the Leghorns, with a view to developing their ability as prolific layers, than to the heavier-weight breeds. Nevertheless, as one journeys northward on a tour of investigation, it will be found that the medium and heavier-weight breeds soon "come into their own" in the colder districts, doing so in common practice largely on account of a reversal of natural reasons that

WHITE ORPINGTON WITH OFFICIAL RECORD OF 301 EGGS

This exceptional hen, owned by Wilburtha Poultry Farms, Trenton Junction, N. J., made a record of 301 eggs in her pullet year (1916-17), at Vineland Laying Contest, and 177 in second year. Is still in Wilburtha breeding pens and the manager, M. L. Chapman, writes that her daughters lay around 200 eggs on the average and splendid results are being secured from her granddaughters.

AUSTRALIAN BLACK ORPINGTON WITH RECORD OF 335 EGGS
This is the highest official record produced to date.

have given the lighter-weight breeds the lead in southern latitudes. It appears, therefore, as a general proposition, that the lighter-weight breeds fare best under warm skies, taking the full calendar year into account, whereas in much colder sections the larger-bodied, better-fleshed Plymouth Rocks, Wyandottes, R. I. Reds, etc., forge to the front. The Leghorns deservedly have a good deal said in their behalf, both in the poultry press and elsewhere, as prolific layers, but we have felt for a long time that far more should be known about the high egg producing power or ability of the medium and heavyweight breeds. The one right or best way for this information to be given to the interested public is for poultrymen to CREATE THE FACTS. By this we mean actual production—the trap-nest performances of these heavier-weight, general-purpose or dual-purpose breeds and varieties. Of course mere guesses are not sufficient. The trap nest is the proper "recorder" and the three hundred and sixty-fifth day may be regarded as the individual hen's day of judgment! These trap-nest records, furthermore, should not be made solely at agricultural colleges or on state experiment stations, but ought to be made generally, as common practice, by student-readers who have the good judgment to breed "on scientific lines," which means simply in an intelligent, systematic way for definite results, based on accurate records of performance.

That 365-Egg Hen or Duck

Writer of this article is on record as holding the belief that before long individual domestic fowl, both chickens and ducks, will be laying to exceed 365 eggs in 365 consecutive days. This does not mean necessarily that some particular hen (chicken or duck) will lay an egg a day for 365 days at a stretch. So far as our claim is concerned—our belief—Miss Biddy or Mrs. Biddy (or her duckship) is privileged to lay two eggs a day for as many days of the three hundred and sixty-five as she prefers and we shall not object, because our requirement is that she is to lay 365 eggs within one year. And some alert, up-to-date hen or duck will be doing this almost before we know it! No doubt of this exists in our mind and we confidently await well-authenticated reports of such performances. Years ago eight Indian Runner Ducks averaged 320 eggs each AS A FLOCK in 365 days and an individual duck of this breed at Malvern, Pa., laid 358 eggs in 365 consecutive days. All varieties of chickens, therefore, may well look to their laurels.

On this question of laying more than one egg a day we refer the interested student to Chapter I for numerous illustrations of what has already been done along this line, these being well authenticated in every case.

However, the main object of this article—necessarily brief—is to direct attention to the surprising ability of the medium-weight and heavier-weight breeds to lay eggs, as has been proved by recent records. Only a few years ago there was much said in the poultry press, at poultry shows, etc., about the "200-egg hen," but she is now a back number—decidedly so. Many people disputed fifteen to twenty years ago, the arrival of the 200-egger, and a good many still claim that a 200-egg strain is a misnomer—a misrepresentation. It may be that they are right, as regards a 200-egg "strain," but it is certain, as we believe, that the 200-egg FLOCK is here—that numerous flocks of this kind have existed in the last five years and that quite a few now exist; how many of them we do not know, nor does anyone else, so wide is the field, domestic and foreign, and so limited are the records.

A 261-EGG R. I. RED HEN
Photo from N. J. Experiment Station.

Two-Hundred-Egg Pens and Flocks

It is an established fact, nevertheless, that flocks ranging up to 200 birds and more (Leghorns) have averaged to lay over 200 eggs per bird in one year. Many small pens, ranging from five to ten birds each, have far exceeded an average of 200 eggs per bird—have reached 235 eggs per bird and upwards. This is true, not alone of the Leghorns, Anconas, etc., but includes also the Plymouth Rocks, Wyandottes, Rhode Island Reds, Rhode Island Whites, Sussex and Orpingtons. These records, many of them, have been made officially at egg-laying contests conducted at agricultural colleges or under their supervision as at Storrs, Conn., at Vineland, N. J., at Mountain Grove, Mo., etc.

And now unquestionably we have the 300-egger with us! Better still,

BLACK LANGSHAN, RECORD 221 EGGS   WHITE WYANDOTTE, RECORD 226 EGGS
Photo from National (Mo.) Egg-Laying   Photo from Prof. J. Holmes Martin, of
Contest at Mountain Grove.   the University of Kentucky.

THE PRODUCTION POSSIBILITIES OF DIFFERENT BREEDS

S. L. WYANDOTTE, RECORD 227 EGGS
Bred and owned by Ira C. Keller, Prospect, Ohio. Weight 7½ lbs. Winner of 1st at Marion and 2nd at Columbus.

BROWN LEGHORN, RECORD 215 EGGS
Record made at National (Mo.) Egg-Laying Contest. There are many heavy-laying commercial flocks of this breed.

she is with us quite numerously. Fortunately also, she is not limited to the Leghorns, Anconas, etc. On the contrary, she exists in the form of Plymouth Rocks, Wyandottes, Rhode Island Reds, White Orpingtons, Black Orpingtons, etc. It is truly wonderful what a little three-and-one-half or four-pound Leghorn or Ancona can do as an "egg machine," and the wonder need not grow less because this "machine" weighs two or three pounds more and wears a different coat of feathers.

Frankly, the writer of this article, as a long-time advocate of high egg production, on the basis of utility and beauty combined, was surprised—very agreeably so—by the remarkable performance of five Morris-strain White Orpingtons (blue ribbon winners) that were entered at the American Egg-Laying Competition, Leavenworth, Kansas, November 1, 1918, three of which, during the ensuing 365 consecutive days, laid 241 eggs, 245 eggs and 303 eggs. In far-off Australia and New Zealand the Orpingtons, notably the Black variety, have been doing splendidly in prolific egg yield, one Black Orpington reaching the high mark of 339 eggs, but it is conceded that these birds were not of exhibition quality; were rather inferior specimens, except for their ability to "shell out the eggs." But the Morris Poultry Farm White Orpingtons were not of that kind. Each of these three birds, including the one that had laid 303 eggs in trap nest, was up to standard weight and had won highest honors in strong competition. Repeatedly in the last five to ten years we have received letters reporting egg production on the part of Buff Orpingtons that ranged from 200 to 260 eggs by individuals in 365 consecutive days. So much for the Orpingtons, within the limits of this article.

Plymouth Rocks as Layers

It has long been known that Plymouth Rocks occupy the front ranks as layers. Eight or ten years ago W. R. Graham, Professor of Poultry Husbandry at the Ontario Agricultural College, Guelph, Ont., reported the performance of a Barred Rock hen that laid 282 eggs in 365 consecutive days. A year or two later T. E. Quisenberry, then in charge of the National Egg-Laying Contest, Mountain Grove, Mo., conducted under the auspices of the Missouri State Poultry Board, told the public about "Lady Show You," a White Plymouth Rock that laid 281 eggs in one year. Still later at the International Laying and Breeding Contest, Vineland, New Jersey, a White Rock hen laid 301 eggs within a year—a hen now owned, along with her relatives, by Wilburtha Poultry Farms, Trenton Junction, N. J., where this line is being carefully bred. We shall not attempt in this article to list all the "300-eggers," even of the medium and heavyweight breeds, but are aiming to report only enough of these cases to show students of the subject what has been done, as a fair index of what CAN BE DONE. Referring again to the Barred variety of the Plymouth Rock breed, we present a photographic reproduction of a "bred-to-lay" specimen that was bought as a baby chick, at a nominal price, from J. W. Parks, Altoona, Pa., by E. F. Grundhoeffer of the Engineering Department, Pennsylvania State College, State College, Pa. In

PEN OF BROWN LEGHORNS WITH AVERAGE RECORD OF 126 EGGS EACH IN 150 DAYS AT MISSOURI LAYING CONTEST
Photo from H. V. Tormohlen, Portland, Indiana.

RHODE ISLAND WHITE WITH RECORD OF 291 EGGS
Was an alternate bird in All-Northwest Laying Contest. Photo from Professor R. V. Mitchell.

the possession of Mr. Grundhoeffer this bird laid 313 eggs within one year, producing 203 eggs in 210 consecutive days, an exceptional record to date for any breed or variety. (See "Miss Graduate," on page 96, Chapter III, Part II. On the same page will be found a photographic reproduction of another J. W. Parks hen, "Miss Smarty," that appears to have established a new world record for this variety by laying 325 eggs in twelve consecutive months.)

The Wyandottes have been good layers ever since the breed was originated in the form of Silvers, thirty to forty years ago. The Silver variety has produced numerous layers that have passed the 200 mark and the Whites have done still better—have passed the 300-egg mark in a number of instances, notably in the case of the Storrs Contest 309-egg hen, illustrated herewith. Many hundreds of White Wyandottes in recent years have passed the 200-egg mark, while scores of them have reached and passed the 250-egg mark. Eight to ten years ago the Cyphers Incubator Company on its poultry farm at Elma, N. Y., had numerous Wyandottes that laid upwards of 220 eggs in 365 days, one bird of this strain reaching the high point of 288 eggs within one year.

Said to Have the "Egg Type"

The Rhode Island Reds always have been good layers. They got their start that way! Moreover, the "lay" has not been bred out of them.

In the case of Rhode Island Reds, H. W. Sanborn, at that time proprietor of the West Mansfield Poultry Farm, Attleboro, Mass., built up a strain that produced a world-record layer for this breed with a yield of 309 eggs in 365 consecutive days (see Chapter V, Part II). How many other Rhode Island Reds have reached the 300-egg mark we do not know, but at the Seventieth Anniversary Boston Poultry Show, December 29, 1919-January 3, 1920, we saw on display a Single Comb Rhode Island Red, owned by Lester Tompkins, Concord, Mass., that holds the world record, so far as our knowledge goes, for egg production within a period of three years—and each of the three years was a good record indeed, taken by itself. A small-sized circular was handed out at the Boston Anniversary Show by Tompkins, which contained the following facts regarding the standard or exhibition qualities of this bird, also her three-year egg yield:

"This hen won as pullet at the Flower City Show at Rochester, N. Y., January, 1917, first prize, shape and color specials and champion female of the show. She laid in 1917, 298 eggs. She laid in 1918, 311 eggs. She laid in 1919, 289 eggs; for the three years, 898 eggs. She was sired by the first pen cockerel at Grand Central Palace (New York City), 1915, said by our best judges to be the best Rhode Island Red male ever exhibited. Her dam was one of the third pen females at New York, 1914."

Sussex Also Are Good Layers

In this article we can refer only briefly to the Sussex breed (made up of three varieties, Speckled, Red and Light), today the leading market fowl of England, but comparatively newcomers in this country. However, they certainly are entitled to favorable mention in this connection; therefore we present herewith a report on the performance of three Speckled Sussex hens formerly the property of Jas. A. Lawrey, Carson City, Iowa, but purchased later by the well-known firm of A. & E. Tarbox, Yorkville, Ill., foremost breeders for twenty years of Silver Wyandottes and who took up the breeding of Speckled and Red Sussex ten or twelve years ago. During the time this egg record was made by the three Speckled Sussex hens, their owner wrote us several times. Following is his last letter, giving their yearly production:

"Carson, Ia., Oct. 4, 1919.

"Editor, R. P. J.:

"Following up our recent correspondence, I am giving in this letter the information you asked for about my three Speckled Sussex hens. They have now completed the year and during the 365 days they laid 795 eggs, or an average of 265 eggs each, which I think is very good.

"Yours truly,
"JAS. A. LAWREY."

208 EGGS FROM SEPTEMBER TO JUNE 1
Lady Curdmore, with above record, is of exhibition breeding. Sired by 1st cock, Illinois State Show, 1920; dam, 1st pullet, Chicago Coliseum, 1918. Photo from Mrs. W. G. Curd, Saverton, Mo.

The remarkable performance of the S. C. Rhode Island Red owned by Lester Tompkins, as here reported, brings up another highly interesting question in this connection, to wit: how many eggs can a well-bred or bred-to-lay domestic hen or duck lay in her natural lifetime? Repeatedly they have laid 1,000 eggs and upward, doing this in six, seven or eight years. The first case of this kind ever recorded in R. P. J. was that of a

309-EGG WHITE WYANDOTTE
This neatly built, standard-bred White Wyandotte made the best record so far secured at the Storrs Laying Contest.

standard-bred Buff Wyandotte owned by Dr. N. W. Sanborn, then living at Hudson, Mass., now in the employ of the U. S. Government as a field extension poultry instructor and demonstrator. Numerous other bred-to-lay standard fowl have passed the thousand-egg mark, notably among the White Leghorns. Birds of this variety laid as many as 1,200 eggs in a lifetime—one or two at Cornell University and an equal or greater number in the hands of Jas. Dryden, Corvallis, Oregon, Professor of Poultry Husbandry at the Oregon State Agricultural College. Many experimenters, comparatively speaking, especially at our agricultural colleges, are now studying egg production for two-year periods, three-year periods and longer—a most promising field for careful research work.

A FINE RECORD MADE BY CANADIAN PEN OF BARRED ROCKS
Pen illustrated above is owned by Passmore's Ranch, B. C., and made the following record in 11 months: 253, 256, 264 and 267—a total of 1,049 eggs. At the end of 12 months three had increased their records to 284, 286 and 291, respectively (final record for fourth bird not given). These are "Ringlet" Rocks, and were bred by E. B. Thompson, Amenia, N. Y.

Brown and Buff Leghorns

Brown Leghorns, while distinguished for their beauty and for that reason highly popular with fanciers, have never been greatly in demand among commercial poultry keepers. In the aggregate, however, quite a large number of egg farms are stocked with Browns. Where they have been bred for increased production they have promptly responded and many first-class records are held by representatives of this variety. H. V. Tormohlen, Portland, Indiana, the well-known Brown Leghorn breeder, in supplying the photos from which the large illustration on page 69 was reproduced, gives the following information, in substance, regarding them:

"These five birds were entered in the American Laying Contest and while too immature at the start to have much of a chance to make an exceptional year's record, they showed marked high productive capacity, two pullets laying 29 eggs each in 31 days; two, 30 eggs each in 31 days and 1, 31 eggs in 31 days. These five pullets averaged 126 eggs each in 150 days. One of these pullets laid 137 eggs in these 150 days. Each of the birds was sired by a Madison Square Garden winner and they thus combine laying and exhibition qualities to a marked degree."

Buff Leghorns, while less generally bred than the Whites, have carried out the traditions of the Leghorn family by running up excellent records at laying contests when given the opportunity. At more than one contest the Buffs have made better winter laying records than the Whites, while in 1914, at the Missouri Contest, a Buff owned by Dr. L. E. Heasley of Holland, Michigan, made a record of 95 eggs in 4 winter months, being only 4 eggs behind the highest hen among all the breeds represented. Other contest records made by Dr. Heasley's birds are 230 in 11½ months, at the Storrs Contest, 215 and 204 in 11 months and 211 in 9½ months, at the Missouri Contest. See half-tone photo-engraving of two of these birds on page 62.

Anconas and Minorcas

Anconas have proved to be exceptionally good layers and there are many good-sized commercial flocks of this breed that are giving the best of satisfaction. Owing to the comparatively small number entered in laying contests not many official records are available, though as far back as 1912 fowls of this breed were conspicuous for their productiveness at the National (Mo.) Contest.

A Sheppard-strain Ancona was recently reported to have made the exceptional record of 331 eggs in her first year, 266 in her second and 206 in her third—a total of 802 in three years—an extraordinary instance of sustained high production. Writing under date of April 12, 1921, H. B. Weaver of Morrisonville, N. Y., states that eight pullets raised from chicks hatched from eggs purchased of H. C. Sheppard, the well-known Ancona breeder, "laid an average of over 200 for the first year. One made a splendid trap-nest record of 285. I think they are wonderful layers."

Black Minorcas have also distinguished them-

HEAVY-LAYING EXHIBITION-QUALITY S. C. RHODE ISLAND RED
Has a record of 298 eggs in first year, 311 in the second and 289 in the third—a total of 898 in three years. Won first prize, shape and color specials, and was champion female at Rochester Show in 1917. Owned by Lester Tompkins, Connecticut.

BARRED ROCK PULLET LAYS 166 EGGS IN 182 DAYS
This bird, the property of the Poultry Department at Purdue University, Indiana, was sired by a Thompson male from a 200-egg Thompson hen. She made the above record in 6 months—December to May inclusive—and not under forcing conditions.

A HEAVY-LAYING PEN OF WHITE PLYMOUTH ROCKS
Some exceptionally good records have been made by fowls of this breed. Pen here illustrated averaged to lay 188.8 each, at the National (Mo.) Egg-Laying Contest.

selves by some excellent records and are especially popular among producers catering to extra-fancy trade, as Minorca eggs are exceptionally large in size, as well as chalk-white in color. C. G. Pape, the well-known Minorca breeder of Fort Wayne, Ind., states that eggs laid by his fowls average to weigh 32 ounces to the dozen and his flock averages, when entire-year records were kept, ranged from 165 to 175 per hen. He reports his highest trap-nest record as 262 eggs, with several up close to 250.

A Remarkable Light Brahma Record

The Asiatics are not generally credited with capacity for high egg production, but it is quite probable that with the same attention that other more popular breeds have received an equally good showing would be made. Attention is called in Chapter I to the fact that one of the first 300-eggers on record was a Light Brahma and the following report from I. K. Felch, the veteran poultryman and breeder, presents a recent record that has been exceeded but a few times by hens of any breed. Following is Mr. Felch's first report:

"From 1876 to 1916, Rebecca, a Light Brahma hen with a record of 313 eggs in 333 days, was never beaten till now. Mollie Wellington, with her 325 eggs in twelve months, has stepped to the front. Six days in the year she has laid two eggs each day. Both these specimens have been of the Felch-Chamberlain strain. Mollie's eggs weigh thirty-two ounces to the thirteen eggs. She herself in her best laying condition weighs eight pounds, but so long as she keeps on laying it will be hard to put her in exhibition weight.

"This remarkable hen was the daughter of a pair of $10 specimens sold by me from a pen of 8 pullets that laid 88½ average for four months and 200 eggs in the month of May."

Later Mr. Felch supplied the following additional information:

A RECORD-BREAKING LIGHT BRAHMA
Owner of this Felch-strain Light Brahma has made affidavit to the fact that she laid 325 eggs in 365 consecutive days. She continued to lay after the year was up, reaching a total of 405 eggs in 15 months and 6 days.

"Since writing you I have learned more about Mollie Wellington and am enclosing the facts with her picture. Her owner has furnished an affidavit that she laid 325 eggs in 365 consecutive days. This is what he says:

" 'Mollie was hatched from the first sitting of eggs from the birds I bought of you, the rats getting all but her. Later I had eight more hatch and the nine birds were kept together in a yard. Mollie was selected before she began to lay to breed back to her father, because she was the exact type of a hen that I owned that was said to have come from stock from your yards. This hen laid 37 eggs in 37 days during the molt. She laid 200 eggs in a year, weighed 13¾ pounds and was accidently killed.

" 'Mollie laid two or three eggs while she was in the yard with her sisters, but I did not credit them to her because I was not positive. I then made a coop 2½ feet square, with a yard 4½ by 2½, the tops and sides being covered with wire netting, so that it was impossible for another hen to get in, and there Mollie lived. She had the same care and feed that the rest of the flock had. I think her sisters will all go 200 eggs or better, so she is not to be considered a freak or an exception. She has shown no desire to sit and has laid 80 eggs since finishing her year, or 405 eggs in 15 months and 6 days.' "

ANCONA WITH RECORD OF 285 EGGS
Bird shown above is one of eight pullets raised from eggs purchased from H. C. Sheppard, the well-known Ancona breeder. This individual made a trap-nest record of 285 eggs in her pullet year, while the average for the entire pen of eight was over 200.

CHAPTER IX

Diseases of Egg Organs: Their Cause, Prevention and Treatment

Lack of Exercise, Improper Feeding, Forcing Methods and Accidents Are Most Common Causes of Derangements of Egg Organs—A Large Percentage of Losses from These Causes Could Readily Be Avoided by Better Methods of Management—Laying of Abnormal Eggs Can Largely Be Prevented in the Same Way

IT is not at all strange that these delicate organs (ovary and oviduct), functioning as they do at so rapid a rate, should be subject to numerous disorders—derangements rather than diseases for the most part. Bulletin No. 338 of the New Jersey Experiment Station, summarizing the causes of mortality in the First Vineland Laying Contest, covering the first and second laying years of the birds, reports that in a total of 101 deaths among the 1,000 entrants, 19 were directly due to disorders of the ovary or oviduct in the first year, and 16 in the second. In addition to these, however, there were 24 deaths due to abdominal dropsy, hemorrhage, peritonitis and tumors in the pullet year, and 30 from the same causes in the second year, many of which no doubt were directly caused by derangements of the egg organs.

It will be seen from this report that there is here an important source of loss, and one which in less competently managed flocks may grow to serious proportions. Many of these diseases may develop directly as a result of constitutional weakness, infection or accident, but it is no exaggeration to say that at least three-fourths of the losses due to such diseases could have been prevented by better management. In feeding for maximum production, as in laying contests, it is to be expected that the pace will prove too swift for the weaker birds, but in the average commercial flock the occurrence of a high percentage of such losses is unnecessary and should be accepted as evidence of incompetence or at least of extreme methods injudiciously applied. Among the more common causes of egg-organ diseases are the following:

Insufficient Exercise. This not only results in the birds becoming overfat on the highly concentrated rations which usually are fed to layers, but makes the muscles of the abdomen weak and flabby. The proper functioning of the digestive organs as well as the egg organs is conditioned quite largely upon the activity of the abdominal muscles. With fowls on open range or fed in deep litter so that these muscles, through constant use, are kept vigorous and firm, there will be comparatively few abdominal diseases of any sort. It is possible to overdo the matter of exercise and thus reduce egg production, but the danger of underexercise is vastly greater. The poultry keeper who finds that the percentage of diseases of egg organs is increasing should know that his most effective and promptest means of reducing losses is to provide more exercise.

Improper Feeding. By this is meant not so much wrong methods of feeding, which are largely corrected by making proper provision for exercise, as poorly balanced rations which do not adequately nourish the birds, result in an injurious accumulation of fat or stimulate the egg organs to a degree of activity that the supply of nutrients is not adequate to meet or in excess of the fowl's physical ability to endure. Particularly where maximum production is sought the caretaker should keep in close touch with his fowls, noting whether they are gaining or losing in weight, whether their appetites are capricious, and whether they are properly digesting their rations. It is reasonably certain that when high-producing birds depart far from what is normal with respect to any of these particulars the ration needs revision. Failure to do this is to invite a prompt increase in the percentage of ovarian and intestinal disorders, few of which can be successfully treated in advance stages.

Forcing, or overstimulating of the egg organs, can be induced in various ways, such as feeding rations carrying an excessive proportion of meat, the use of condiments and stimulating drugs, the excessive use of lights, etc. It is a comparatively simple matter, by any of the foregoing means, to bring about a prompt increase in the number of eggs laid, but unless the caretaker is experienced and skillful he is apt to find that the nice physical balance that must be observed is upset, with a rapid increase in ovarian diseases as a direct result.

Accidents, fright and rough handling are responsible for many losses. Allowing fowls to fly down from high

A HOLTERMAN BARRED ROCK WITH RECORD OF 262 EGGS IN 11 MONTHS
Holterman-strain Rocks are reaching high levels in production. Above record was made at American Egg-Laying Contest. The bird is above standard weight and an excellent breeder.

perches, particularly when they must alight on hard floors, frightening them so that they fly wildly about the house or yards, catching them and holding them head downward, are all fruitful sources of trouble.

Constipation. Fowls on highly concentrated rations are subject to constipation, which may in turn cause inflammation of the oviduct. The straining to evacuate the cloaca is a common cause of prolapse.

BUFF LEGHORN WITH RECORD OF 231 EGGS IN 8 MONTHS AND 14 DAYS
Is granddaughter of 283-egg hen. Photo from R. E. Sims, Little Rock, Ark.

General Treatment

The poultry keeper should aim constantly at prevention rather than cure, as the presence of diseased conditions is rarely discovered until too late for effective treatment. Many of these disorders can only be identified by post-mortem examination. Avoidance of the aforementioned causes of disease and prompt removal of any birds that seem to be in any way affected, so that they will not be annoyed by other members of the flock, a dose of Epsom salts, followed with light feeding to check production, and an abundance of succulent green feed to keep the blood cool and the bowels active, are steps that should be taken in every case. Usually there is little to be gained by medical treatment, though some special measures, as herein suggested, may prove helpful in the case of extra-valuable birds or in first stages of the disorders, where prompt attention may prevent many losses.

Notes on the More Common Diseases

Internal Laying. One of the most common ovarian troubles is the miscarriage of yolks or partially formed eggs, resulting in their escape into the abdominal cavity instead of being passed on through the oviduct and regularly excluded. In many instances internal laying proceeds indefinitely with nothing to indicate to the casual observer that anything is wrong, except the hen's unproductive visits to the nest. The escape of yolks into the abdomen instead of into the oviduct may result from the failure of the funnel of the oviduct to function properly at the time the yolk sac is ruptured. The normal peristaltic action of the oviduct may also be reversed after the egg is partially formed, forcing it back and out at the upper end. This sometimes happens even after the shell has been formed. Internal laying also occurs where the eggs are diverted into a sac which develops at some point along the oviduct, or the eggs may escape into the abdomen through the rupturing of the wall of the oviduct.

A BUFF PLYMOUTH ROCK WITH A GOOD RECORD
This Buff Rock owned and bred by C. R. Baker, Abilene, Kansas, made a record of 229 eggs at the National Egg-Laying Contest.

It is possible for internal laying to go on for a long time without any noticeable effect on the health of the bird and in some instances accumulations of egg material weighing over one pound are found. Dr. Thompson of New Jersey Experiment Station has reported to the author finding such a mass weighing three pounds, in a hen submitted for post-mortem examination. Under some conditions the presence of this foreign material in the abdomen may set up an inflammation resulting in peritonitis. In other cases the bird dies without any apparent unfavorable symptoms aside from the mere presence of the foreign matter. A case of internal laying is illustrated on next page. Here the yolks are encysted or enclosed in a membranous sac formed on the sidewall of the oviduct.

Rupture of the oviduct may result from any cause which interferes with the passage of the egg. An oversized egg, an inflammatory condition of some section of the oviduct as a result of which it is not normally lubricated, stricture of the walls of the oviduct (also prob-

A HIGH-PRODUCING PEN OF R. C. RHODE ISLAND REDS
These birds laid 1,042 eggs—an average of 208.4 each, at the National (Mo.) Egg-Laying Contest.

ably due to inflammation), are common causes. If production ceases and the wound has a chance to heal the bird may not be permanently injured.

Prolapse, or eversion of the oviduct, may be caused by constipation or by the efforts of the bird to lay an extra-large egg or one that for some reason does not readily pass through the oviduct. Frequently all that is necessary is to push the protruding parts back into place, put the bird in quiet, comfortable quarters away from all other fowls, and feed light. As constipation is often associated with the trouble, the bird should have a dose of Epsom salts at once, and a ration containing a liberal allowance of succulent green stuff should be supplied. If the trouble is not noticed until the tissues have become badly inflamed or other fowls have begun to pick at them, treatment is a waste of time. A small percentage of cases is to be expected in any heavy-laying flock but if numerous there is something decidedly wrong with the ration or method of feeding.

Peritonitis, or inflammation of the peritoneum, the membrane lining the abdominal cavity and covering the intestines, is caused by the presence of foreign matter, hence frequently develops in connection with internal laying. It may also result from the extension of inflammation of the oviduct. There are few external symptoms by which this disease can be identified, though the presence of egg material or an excess of fluid usually can be detected by handling the bird. When opened, the abdominal cavity will usually be found to contain liquid matter, free or in a sac resembling a water blister, such as the one illustrated on this page. Small masses of cheesy material also will be found adhering to the intestines. There is no practical remedy for peritonitis and the only preventive measures are such as are designed to keep the bird in good health generally.

Egg Bound. Fowls suffering from this trouble have an egg in the oviduct which, for some reason, they are unable to lay. It may be an extra-large one or, through an inflammatory condition of the oviduct, the natural lubricant may be insufficient to enable even a normal-sized egg to pass readily. The first step should be to put the bird in a quiet place away from the rest of the flock and see whether she will not presently succeed in dropping the egg without help. If it appears necessary to give assistance, try injecting a small quantity of olive oil. By coating a finger with vaseline and inserting in the vent the egg can sometimes be worked loose. The danger of breaking the egg in doing this, however, is so great that it is doubtful whether any measures except injecting a little olive oil are practical. Holding the abdomen of the bird over a steaming kettle for half an hour or so as a means of relaxing muscles is recommended by some.

Bagging Down. This condition is common in fowls of the larger breeds, and particularly so among birds in their second year or older. It results directly from the accumulation of excess fat, or from the weakening of the abdominal muscles, due to heavy feeding and underexercise. Fowls having short keel bones are especially apt to bag down, for which reason this tendency should be avoided in selection of breeding stock. Professor Graham of Ontario Agricultural College, in discussing this subject with the writer, stated that he had regularly bred his birds for long keels and, as a result, bagging down rarely ever occurs in his flock. There is no remedy for this trouble when it develops and no practical form of treatment. By way of prevention, select breeders with good long keels, make the hens scratch for the grain part of their ration unless they are on open range, and feed so as to avoid the accumulation of excess fat.

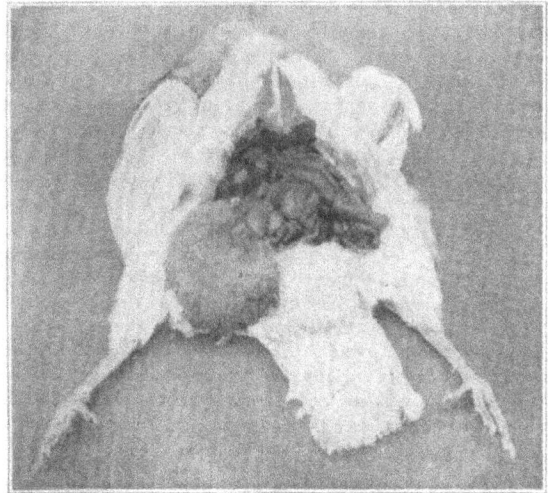

A CASE OF INTERNAL LAYING
Post-mortem photo shows large sac formed on wall of oviduct, about halfway down, and containing 16 fully formed yolks and albumen. Ovary was in condition of hen in full lay. All other conditions were normal and healthy. Photo from N. J. Experiment Station

Vent Gleet. Vent gleet is a contagious disease, indicated by the inflammation of the skin around and below the vent, the frequent voiding of excrement and by a discharge, watery at first and later white and foul smelling. Birds so affected should be removed from the flock, as the disease is highly contagious. It is comparatively infrequent, however, and its cause is not clearly understood.

Uncommon Diseases. There is quite an array of diseases of the egg organs that must be mentioned where an attempt is made to cover the subject with textbook thoroughness, but they hardly need be discussed in a brief chapter designed to include only the more common disorders. The percentage of birds affected by such diseases as atrophy of the ovary, ovarian tumors, gangrene, etc., is comparatively slight and specific preventive measures and treatment of affected individuals are alike impractical. The poultry keeper must aim at keeping his birds in good

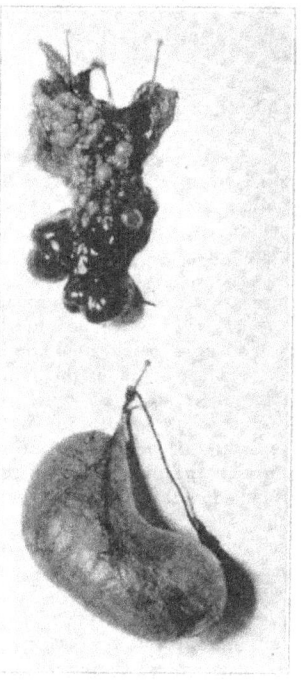

A COMMON SYMPTOM IN PERITONITIS
The disease is caused by entrance of foreign matter into abdominal cavity and often accompanies internal laying. Illustration shows ovary and large water sac found formed between the intestines as a result of inflammation. Photo from N. J. Experiment Station.

general health, avoid measures that are known to have an injurious effect upon the egg organs, and when isolated cases develop, include them in the moderate percentage of loss that is to be expected in any flock. Neither need he become unduly exercised over the danger of infection. With the exception of bacillary infection, meaning by this the particular type of ovarian infection which later causes bacillary white diarrhea in chicks, and vent gleet, probably none of the diseases of the egg organs is infections. Sick birds should be isolated however, for comfort.

Abnormal Eggs

A considerable percentage of abnormal eggs will be produced in almost any flock, these being especially noticeable in the season of heaviest production, and particularly so where the birds are not properly fed. While abnormal eggs sometimes result from organic defects, usually they should be taken as evidence that there is something radically wrong in the ration or in the treatment of the birds.

Blood Clots. Blood clots in eggs are caused by an inflammatory condition of the ovary or the oviduct, resulting in the escape of small quantities of blood from the bursting of minute blood vessels. This may occur in the ovary when the follicle ruptures to allow the egg to escape into the oviduct, or at any point along the oviduct before the soft shell is formed. Blood clots are most apt to be found when the birds are laying heavily and, when numerous, the practical thing is to put the flock on a lighter ration, thus to reduce production slightly and allow the inflamed organs to recover their natural tone. Frequently the inflammation which causes blood clots is brought on by constipation, the remedy in this case being to give a dose of Epsom salts and to see that the ration contains a liberal percentage of succulent green feed. There is nothing unwholesome about eggs containing blood clots, but customers naturally object to them and for that reason preventive measures should promptly be taken when they appear.

Bloody Eggs. Eggs with more or less blood on the shell are often found. They are most apt to occur when pullets are just coming into laying. Special treatment seldom is needed as the slight tearing of the tissues about the vent will cease after the bird lays a few eggs.

Extra-Large Eggs. These are produced, as a rule, by birds that are overfat, or whose organs, for some other special reason, are not functioning normally. Double-yolked eggs result from the entrance of two yolks into the oviduct at about the same time. Occasionally, eggs after receiving their coating of shell material, are returned to the upper end of the oviduct, through some reversal of peristaltic action, and again pass down through the oviduct, in which case the eggs will be surrounded with additional layers of albumen and another shell, thus resulting in abnormally large eggs which sometimes cause a great deal of trouble in exclusion. An egg so formed is shown in the illustration on this page.

Soft-Shelled Eggs. Soft-shelled eggs that are not caused by a lack of shell material usually are laid by hens that are overfat or that are laying so rapidly that the eggs pass through the oviduct without having sufficient time in which to acquire their normal shell covering. If a number of such eggs are produced in flocks having ac-

DEVELOPMENT OF SECONDARY SEXUAL CHARACTERS
Ovaries of hens sometimes become atrophied for one reason or another, after which the birds often develop combs, hackle and tail feathers, like those of the male. Above bird was photographed in the yard of a back-lot poultry keeper in Quincy, Ill., and shows remarkable development of secondary sexual characters.

cess to oyster shell they should receive a less-fattening ration. As some fowls do not appear voluntarily to eat oyster shell enough to meet their requirements, it is a good plan to add one or two per cent of bone meal to

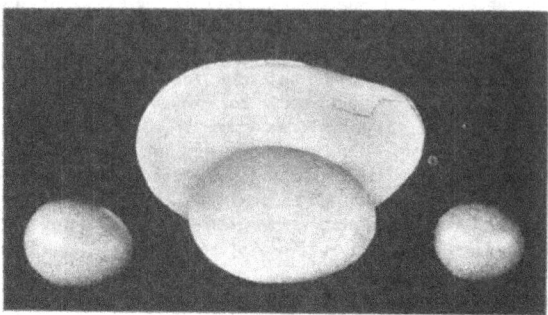

ABNORMAL EGGS
After the average-sized egg in center had received its hard shell it was returned to the upper end of the oviduct and then again passed down, receiving another layer of albumen and a second shell as shown. The extremely small eggs shown on the sides were laid by the same hen.

the mash mixture when the shells are not heavy enough. Sometimes underconsumption of oyster shell is due to the fact that the supply is stale or dirty. A lack of shell material is not always indicated by soft-shelled eggs; in some instances this results, instead, in a decrease in production or in small-sized eggs. There is no economy in depriving the birds of this important part of the ration, and a fresh, clean supply should always be available.

Rough Shells and Irregular Shape. These are most often caused by an inflamed condition of the oviduct, though occasionally there are birds who habitually lay misshapen eggs, due to some organic defect. Misshapen eggs are most common at the height of the laying period just as are soft-shelled and double-yolked eggs, and they probably result from about the same general causes. A less-forcing ration and a dose of Epsom salts will usually correct the tendency, except in the case of a few chronic offenders, which, if they can be detected, should be removed from the flock.

Discolored Eggs. Discoloration of yolks and albumen usually is traceable to the ration, though in rare cases it may be due to bacterial infection. It is stated in "Diseases of Domesticated Birds" that the use of cottonseed meal is particularly apt to result in discolored yolks. It may also be caused by turnip leaves, frosted rape, etc. Discolored albumen usually results from an excess of green feed or from supplying the wrong kind, which fact suggests the remedy.

Poor Flavor. Fresh eggs of distinctly inferior or unpleasant flavor usually result from feeding something of an unwholesome nature or with a strong odor, such as onions, turnips, etc., or any kind of decaying vegetation. There will be practically no danger of this so long as such material is avoided or fed only in a limited quantity. Very rarely strictly fresh eggs show noticeable deterioration in quality due to their being retained in the oviduct for hours after they are partially or wholly formed.

PART II

CHAPTER I

A Visit to Wyckoff's Grandview Poultry Farm

The Principles and Methods that Make This Farm One of the Simplest Yet Most Successful Poultry Establishments in Existence—Wyckoff's Leghorns Averaged to Lay Around 200 Eggs Each Over a Quarter of a Century Ago—Practical Methods and Substantial Results Are Life Work of Men of Energy and Brains

By GRANT M. CURTIS, Editor of Reliable Poultry Journal

MOST of this chapter consists of an interview the writer had in October of 1920 with C. H. Wyckoff, senior member of the firm of C. H. Wyckoff and Son, proprietors of Grandview Poultry Farm, located on the east shore of Cayuga Lake, less than a mile from the village of Aurora and twenty-seven miles north of Ithaca, on the Lehigh Valley Railroad. At that time Mr. Wyckoff senior was in his seventieth year and quite probably was the only man living who had been engaged for forty years in breeding White Leghorns—and almost continuously, the only break being a period of about a year in 1899-1900, just after he sold his original plant at Groton. Grandview Poultry Farm was established in 1900-1901 and has been in continuous successful operation since then, under the personal management of C. H. Wyckoff and his son, E. L., who at this time (1920) was forty-five years old.

Writer had the good fortune to visit Mr. Wyckoff senior at the Groton plant, summer of 1896, a report of which visit was published in the September, 1896, issue of the Reliable Poultry Journal and attracted wide attention, because at that time this undoubtedly was the most successful breeding and egg-production poultry establishment in America, devoted to the S. C. White Leghorns and conducted on practical, money-making lines. In this chapter will be found a number of references to that visit of a quarter of a century ago, also a photographic view of part of the old plant. Back there, especially during the early career of the Groton plant, Mr. Wyckoff sold eggs largely to the daily market for table use and therefore was deeply interested in prolific yield—in a high flock average, as every additional egg meant increased cash income; but as the years went by, the name and fame of the "Wyckoff White Leghorns" traveled far, with the result that before long he was favored each year with a demand for breeding stock and layers, also for hatching eggs in season, that took all his surplus, that broke into his flock almost continuously and that called for more eggs at hatching time than the fowls could produce, even when held back from full average production earlier in the season.

Practical common sense from the first guided Mr. Wyckoff in all he did with his fowls. A mechanic or machinist by trade (repairing and rebuilding railway locomotives), he was methodical, systematic, thorough. He had a trained mind and made use of it, spurred on by necessity, because when he went on the sixty-acre, run-down general farm near the village of Groton, Tompkins County, N. Y., he had no capital. To enable him to secure possession of the place, his father had to sign the papers and advance some of the cash for a few garden and farming tools. The first year or two he found that his few fowls were the best-paying "crop" he had, and for that reason he gave them more and more attention, soon replacing the mixed flock with purebreds—with Standard S. C. White Leghorns "of that day and generation," now practically forty years ago.

It is believed that Mr. Wyckoff senior was the first man to give S. C. White Leghorns the now popular title, "Business Hen of America." To bestow on him further well-deserved credit, we quote here a tribute paid him lately by James E. Rice, head of the poultry department of the New York State College of Agriculture (Cornell University), Ithaca, and also publish "An Appreciation" by John H. Robinson, for years the editor of "Farm Poultry," Boston, Mass., now an associate editor of the Reliable Poultry Journal. Said Professor Rice to us, October, 1920:

"Mr. C. H. Wyckoff was a generation ahead of his fellow poultrymen on lines of actual achievement in high-egg production. Yes, I have visited his poultry plant almost every year for the last thirty years or more and know about his splendid results."

Following is the statement of Mr. Robinson:

"For more than a score of years my personal opinion has been that the greatest single positive factor in giving the S. C. White Leghorns their great popularity was the influence and wide distribution of the Wyckoff strain.

"Since 1897 I have been in a position to learn and have generally made it a point to learn something of the origin of the large stocks of White Leghorns on commercial farms and also of the origin of small flocks making remarkable egg records.

C. H. WYCKOFF, AURORA, N. Y.
Originator of the Wyckoff strain of S. C. White Leghorns and the Wyckoff system of flock breeding in limited space for prolific egg yield.

"In the first ten or twelve years of this period I do not think a single such case came to my knowledge when the stock was not either wholly or principally from Wyckoff foundation stock. In all the stocks that I saw then the Wyckoff type was conspicuous.

"Later there was more evidence of the influence of the popular exhibition strains, and a tendency toward smaller birds. The fact that the Wyckoff Farm is still one of the striking successes, with the same type of Leghorn that made its reputation, speaks volumes for the value of the type."

Minimum Cost—Maximum Returns

The Wyckoffs never have used trap nests nor given special consideration to the egg yield of individual birds. Their dominant idea has been—above and beyond good size and great vigor in the stock—maximum yield and cash returns from moderate-sized flocks, with the minimum of labor and expense. That plan, for them, has been "the middle of the road," although as the success of their methods and the money-earning value of their stock became widely known, the demand for their products shifted largely from the daily market to that from their fellow poultrymen, as above briefly set forth.

Of recent years, therefore, the Wyckoffs have not found it practical nor to their special interest, as they believe, to keep a yearly egg record on the basis of the number of hens on hand because, in filling orders and making shipments to all parts of the civilized world, they constantly are disturbing the flocks, first in one house, then in another, with the result that during probably no two weeks of the year do they have the same number of layers on the farm. They do keep a book record of the number of eggs gathered each day, also of shipments of eggs, both for hatching purposes and to the daily market, but at Grandview Farm they have not tried to keep accurate and continuous daily records,

Reliable Poultry Journal on the flock production of the above-mentioned 600 S. C. White Leghorns, we said:

"We should say it was!" referring thus to our remark of twenty-five years ago that an average of 194 eggs per bird in one year was "a very good record for so large a number of hens." Continuing in the January, 1921, issue, we said:

"At that time editor of R. P. J. did not at all comprehend how good a record that was, nor did anyone else, so far as we know—and we make this statement from the vantage-point of the present, looking backward over the last quarter of a century. To have secured such an average that long ago was truly remarkable, no doubt about it, and the pity is that poultrymen in general of that day did not realize what it meant—did not study more carefully Mr. Wyckoff's methods and benefit to a far greater extent by his extraordinary achievement.

"And still another remarkable fact is that Mr. Wyckoff kept those 600 hens—had them housed and yarded on a little over one acre of ground, and he is doing the same thing today so far as his laying houses are concerned, including the runways connected therewith. At Groton the runways or parks were 33 by 108 feet in size, or 3,564 square feet to each park. In each half of the 12 by 40-foot house he kept 50 to 70 hens or pullets—70

PART VIEW OF ORIGINAL C. H. WYCKOFF PLANT NEAR GROTON, TOMPKINS COUNTY, N. Y.
Photograph kindly furnished by James E. Rice, head of the poultry department, New York State College of Agriculture (Cornell University), Ithaca, N. Y. Was taken by Professor Rice some thirty years ago, "about 1890." Shows early low-cost type of 12 by 40-foot Wyckoff laying and breeding houses and style of yards, 33 by 108-feet, with wooden picket fences; also cornfield range in background, and new plum orchard in ploughed field at right in foreground. A cornfield range for young stock has been popular with Mr. Wyckoff many years.

based on any given number of hens and pullets, either for certain flocks or in an attempt to cover all birds on the plant.

At this point let us quote from the printed report of that visit to the original Wyckoff Poultry Plant, near Groton, N. Y., as same was published in the Reliable Poultry, September, 1896:

"ONE YEAR HIS 600 HENS AVERAGED 194 EGGS PER HEN, A VERY GOOD RECORD FOR SO LARGE A NUMBER OF LAYERS."

The 600 birds above referred to were kept in comparatively small yards, in flocks of about 60 to each flock. Herewith is shown a photographic view of the Groton plant, with its 12 by 40-foot laying and breeding houses, each having a partition across the middle and each half section (quarters for 50 to 70 birds) opening into a yard or park, 33 by 108 feet in size, enclosed with a low-cost picket fence. Commenting in the January, 1921, issue of

early in the season and about 50 later on after he had sold off what he regarded as surplus, disposing of them as layers or breeders. An acre consists of 43,560 square feet, therefore it is easy for the reader to arrive at a near enough knowledge of the number of layers Mr. Wyckoff was able to keep on one acre of ground at Groton years ago, and now keeps at Grandview Farm by the same system.

"But it should be well understood that while Mr. Wyckoff, in the early days near Groton and now at Grandview Farm, could keep and does keep as many as 600 layers on what is equal to about an acre of ground, all his laying and breeding stock is raised on range, from chickhood to maturity. It is entirely practical, in other words, to keep well-bred, range-raised, vigorous fowls in comparatively limited quarters for egg production after they have matured but, as a rule, it is both difficult and expensive to raise chicks in too limited quarters where they do not get sufficient exercise and the ground is sure to become contaminated unless great care is taken to prevent this fatal condition. Now, as in the old days, Mr. Wyckoff is a firm believer in having some form of helpful vegetation or at least untainted ground under the feet of his fowls, also in a liberal daily use of green food. Said he, in our interview with him twenty-four years ago:

"'It seems to me that I would almost rather stop feeding grain than green food. That is of course an extreme statement, as green food is mainly an appetizer and bowel corrective, but I could not do business without a daily ration the year round of green food."

Location and Size of Grandview Farm

Associated with Mr. Wyckoff as a partner is his son, E. L. Wyckoff, and both families live at Grandview Farm in a fine colonial-type house with broad veranda looking out on spacious, well-kept lawns, embellished with shrubbery and fine old trees. Thought and care were used by the Wyckoffs in selecting this location. About five hundred feet away and perhaps seventy-five to one hundred feet below the average level of the farm, is the widest part of Cayuga Lake—four and one-half miles wide and over six hundred feet deep somewhere near the middle. Cayuga Lake is so wide and deep opposite Grandview Farm that in the twenty years the Wyckoffs have lived there the lake at this point has frozen entirely over twice only. Six degrees below zero is the coldest it has been at Grandview Farm in the history of this plant. The large body of water tempers the atmosphere both winter and summer and has made the east shore in this section one of the best fruit-growing localities in central New York—in our whole country. Moreover, twelve to fifteeen miles farther west, beyond a single range of hills, is Seneca Lake. Both these lakes are practically forty miles long, the two largest among those that form what are known as the Finger Lakes, five in number.

At Groton Mr. Wyckoff used plum trees extensively in the poultry yards and during the earlier years at Grandview he had peach trees in the poultry yards and on the ranges, from which he gathered numerous large crops, but when these trees became old and less profitable, he replaced them with English walnut trees,

E. L. WYCKOFF, AURORA, N. Y.
Son of C. H. Wyckoff and partner with his father in conducting the business of Grandview Poultry Farm. Has been "at it" twenty years and is making the poultry business a life work.

doing this on account of the somewhat moderated climate and because there were several trees in the neighborhood, old in years and large in size, that have borne good crops of walnuts time and again since Mr. Wyckoff senior selected this locality for his present poultry plant. All told, Grandview Farm consists of but nine acres—not a very large area for carrying 1,400 layers and breeders, also for raising 2,500 to 3,000 birds each season for replacement purposes and to sell, yet sufficient after one has learned how to do it, it, as the success of the Messrs. Wyckoff amply proves.

Personal Interview, October, 1920

Next, for the information and benefit of the readers of this book on "High Egg Production," we are pleased to quote at length from an interview we had with C. H. Wyckoff in October, 1920, at Grandview Poultry Farm, telling of the origin of the Wyckoff strain of S. C. White Leghorns, his theories and practices in developing this strain, what his views are about standard weights, correct type, selection, inbreeding, culling, season of molting, etc., and also giving the Wyckoff method of feeding the layers and breeders, together with the profitable results they have secured at Grandview Farm by the use of artificial lighting to increase egg production during the short-day period of the fall and winter months.

Origin of the Wyckoff Strain

Asked about the origin of what later became known as the Wyckoff strain of S. C. White Leghorns, Mr. Wyckoff senior, gave us these facts:

"I got my stock originally at the New York State Fair, from a man named George Weed, who then lived at or near Poughkeepsie, N. Y. This was along about 1880 or 1881. Weed had quite an exhibit at the Elmira, N. Y., fall fair, and I liked the looks of them. On second thought I am not sure this was a state fair, but I know it was held in Elmira. Have since lost all trace of Mr. Weed. Never heard of his exhibiting after that. I re-

PARTIAL VIEW OF BUILDINGS ON GRANDVIEW POULTRY FARM, AURORA, N. Y.
Showing at left, stable and storehouse for garden and farming tools; right of center (foreground), work shop, feed and root cellar, shipping rooms and storage loft; two rooms of the 16 by 40-foot breeding and laying houses; a peach orchard range in background, and the character of well-kept lawns and driveways. Might well be called "The Home of Spick-and-Span," everything is so orderly and neat.

call that his birds were larger, sturdier and appeared to be more active and vigorous than any other Leghorns I had seen. I bought ten or twelve—do not now recall the exact number—out of his exhibit at the fair and from the first they proved to be unusual layers. After that it was a matter of study and selection—a study of common-sense methods and of selection according to egg yield—what is now called 'performance.'

"Before the time you visited me at Groton in 1896 we had developed our methods to a high degree, so far as heavy egg production and stamina were concerned, and had reached almost the 200-egg per hen mark for a flock average covering 600 birds. These days we do not work that hard for big flock averages. We let the eggs come right along and breed for maximum vigor, but it pays us better to cater to the eggs-for-hatching trade, this and the production during March and April each year of the 2500 or more head of young stock we need for replacement and to sell."

Weights and Type of Wyckoff Leghorns

In this connection, let us report on the size or weights of the Wyckoff strain of S. C. White Leghorns, which represent what the Messrs. Wyckoff prefer, after their long years of experience. The pullets, as they approach maturity, "weigh up close to 4 lbs.," as Mr. Wyckoff senior expressed it, "and the hens 4½ to 5 lbs." Said he: "Birds of this kind will stand up better to heavy egg production." In males, the cockerels weigh 5 to 5½ lbs. and the cock birds "6 lbs., or a little better." Asked to describe the type, size, etc., of birds they prefer Mr. Wyckoff said:

"In males we want them of good sound color as to plumage, with bright yellow, sturdy legs set well apart; want them long in body and with good upright combs, red eyes and well-developed lobes and wattles, also with tails carried fairly well down—not upright nor squirrel. I favor the present standard type with full weights and maximum vigor. We never use undersized, ill-developed birds or birds with thin, long heads and smallish or undersized combs.

"In females we want good length and depth of body—every breeding bird to be full breasted, with long back, tail carried at moderate angle, fully developed head points and expressive face, showing signs of a gentle, intelligent disposition—not wild nor scarey. They must have good leg bones, not thin and spindly. Combs of males should be medium large—not too large, but set well on heads and carried upright. They also should have intelligent faces, of which the eye is the best sign. We prefer what might be called medium-length legs in both males and females, rather than thin or spindly shanks and legs.

"No, in making our selections, we do not go by finger measurements. We invariably select birds that have good depth of body, also good length and strong, well-furnished heads. A poor layer will show it in her head quicker than any other place. Always avoid small, weak-looking heads, also weak-appearing beaks and you will find that the combs on such birds usually lack color. Avoid also small, undeveloped or shrunken combs. Select birds with broad skulls, full face and eyes and stout, strong beaks. Good breadth birds behind will have good finger capacity, which is what we all want, I guess. But the difference of a finger or two isn't very definite, as I see it. It will be found that pigmentation supports the above description of the poor producer or nonlayer and therefore is a help in picking out the drones at a time when they should be laying."

Outside the eggs-for-hatching season, Grandview Farm sells new-laid eggs at premium prices to a hotel in Auburn, N. Y., and to a commission merchant in New York City, for which they obtain "top prices." The past season, on October 29 and for some time previous to that date, they were getting $1.10 per dozen in case lots. Their eggs range in size or weight from twenty-five to thirty ounces to the dozen. Said Mr. Wyckoff senior:

"They are almost too large for the daily market, but the hatching-egg trade demands good size—in fact, large-sized eggs. Shipping them to market in the ordinary egg case we often have trouble to get them into the egg fillers, without danger of undue breakage."

Not Worried About Inbreeding

"No, we do not use trap nests, nor do we worry about inbreeding. When the birds of a flock have proper vigor and the chicks are well reared, I believe that much of the talk about the dangers of inbreeding is mere theory. In our case we have introduced no outside blood since twelve years ago, at which time the experiment was not a success. The eggs from those crossbred birds did not hatch as well, the chicks did not grow and develop as well and the hens and pullets from that mating were not as good layers. We made careful comparisons with our own stock, and with those results. We select closely season after season, keeping near to our choice of type—good length and depth of body, full-breasted, long backed, stout-legged, moderately low-tailed specimens of

NEW-STYLE POULTRY HOUSES AT GRANDVIEW FARM

Are of the semimonitor type, each 16 by 24 feet in size. Capacity, 120 hens or pullets with five or more male birds during hatching season. Also shows uniform standard type of world-renowned Wyckoff strain of S. C. White Leghorns.

THE WYCKOFF TYPE OF COLONY HOUSES FOR PULLETS

View of long row of the 4 by 8-foot chick-growing coops, twenty-four in number, used for young pullets at Grandview Poultry Farm (after sexes are separated), showing how fronts are constructed, the raised floor, etc., with two "sun shelters" between each two houses, where feed troughs and water fountains are located for use of the pullets. Houses are 4 feet high at rear and 6 feet high in front. For detailed plans and specifications suitable for building this style of colony house, see R. P. J. book, "Forty Years with the White Leghorns."

both sexes, then we mate them as we like, regardless of family ties or relationship.

"Frankly, we pay no attention to whether our mated birds, when they finally get into the laying houses after we have selected or culled them carefully, are or are not brothers and sisters. I have studied this matter a good many years and cannot see that it makes any difference. We select very carefully according to our experience and best judgment for size and vigor, then we aim to house and feed properly—do this and you will never know any difference, is my belief. All these years it has been to my interest, also to that of our thousands of customers, to learn about this matter and I know that today our birds are better than ever as to size, vigor, health and egg production, and I claim this is the answer and proof, in combination with the fact that they never have had disease and now lay better than at any time in the past, according to our requirements in yearly egg production.

Birds Speak for Themselves

"Just go out and look our birds over for yourself. You certainly should know by this time when you see healthy, vigorous birds—and they are out there to speak for themselves. However, at one time we reached a state of mind—about twelve years ago—when we thought we should have some new blood. But every time we tried it we got a set-back from these new matings that took several years to overcome. Meanwhile, those of our flock or strain into which we did not introduce new blood, but depended solely on selection for improvement —selection based on vigor of the individual specimens— kept on doing better and laying better than anything we could build up alongside of them. I know that this practice of indiscriminate mating, as to sires and dams and brothers and sisters, is contrary to the theory or teaching of numerous poultry experts, but it is the truth about my long experience, and I understand it is facts that you want me to give you.

"As regards trap nesting, we do not get that close to individual production. On the other hand, we work for pen or flock averages, on the percentage basis. As an example, a year ago last spring, from 1,100 birds our daily egg collection went up to 830 per day and better for quite a period in the flush season. This was without the help of artificial lights. Year after year in the flush season, production with us runs from sixty-five to eighty per cent and a little better, at the time we want our largest yield for the hatching-egg season.

"Back at Groton, at one time, I catered to the market egg trade and worked for maximum annual production, with special efforts for fall and winter yield to catch the high prices, but for years now we have been primarily a breeding plant, our main object being to produce strong, vigorous breeding stock and eggs to be used for hatching purposes, both to be ready for our customers when they can receive and use them to best advantage. On this account we are not today specially interested in high egg records, by individual birds, but work for standard type, proper size of frame, full weights, sound plumage color, outstanding vigor and good pen or flock averages in all-the-year-'round egg production.

Molting of Hens and Pullets

"You ask about molting. We do not want to see our hens molt before late August and during September and October. However, we do want a full new coat of feathers by the time real cold weather arrives. Birds of our strain that are of the same age, partly as the result of long years of selection and breeding, now molt uniformly —at about the same time and they require about the same length of time to complete their new coat of feathers. The poultryman who observes well will notice that any poor-type, weakly, undersized specimen, or any bird of poor physical condition will molt earliest. Year after year we have taken pains to get rid of such birds, keeping at it right along, so as not to be penalized by these loafers or slack producers, let alone breeding from them.

"We like to have our pullets go through a light molt the latter part of September and during October, so that by the last of January and in early February they will get into full lay of hatchable eggs after their midwinter rest. These pullets, as a rule, are hatched the latter part of March and third week of April—which is our preference for this latitude and climate. Early fertility from these well-matured pullets usually runs better than from hens, and eggs for hatching from such pullets give very good satisfaction.

"In each half of our large breeding houses we place 70 to 75 hens or pullets and use four to five males in each pen. We start usually with five and wind up with four, as a rule. Four or five males will get along better than two. We find that it is not wise to make changes in males during the season. We put them in the pens early in the winter and after three or four days they will not bother one another. They may or may not be brothers. This makes no difference, if they have been raised on range together and are acquainted."

Asked if he understood that he was first to call S. C. White Leghorns "The Business Hen of America," Mr. Wyckoff replied:

"Yes, I am responsible for that. And it seems a long time ago! Am now in my seventieth year and have been breeding and studying the White Leghorns for practically forty years. At the Groton farm 700 to 800 breeder-layers were our limit, but here we are carrying 1,200 to 1,400 birds, and with proportionately greater profits and success.

Hens' Eggs and Cockerels

"We prefer to raise our breeders from hen eggs, but we pay no attention whatever to relationship of males

POUND-AND-A-QUARTER, SIX-WEEK OLD COCKEREL
Sample of six-week-old, Wyckoff-strain S. C. White Leghorn, weight, 1¼ pounds. Note vigorous-looking head, substantial body and sturdy shanks.

and females. In the past we have retained our best breeders only two years, but in future, on account of the demonstrated benefits of artificial lighting, we are going to hold them over the third year, and I believe we can do it to advantage.

"Yes, we like to use cockerels, as a rule, for breeding, selecting early-hatched, fully-matured birds and thus avoiding the expense and inconvenience of carrying older male birds over till they are two and three years old or older. We invariably find a good demand for the cock birds and the supply of course is limited.

Feeding the Breeding Stock

"How do we feed the layers or breeders? There is nothing complex about it. We give them a light scratch feed of whole oats for exercising purposes, doing this as soon as they can see to eat each morning. It doesn't pay to fill them up and make them logy. Dry mash is kept before them all the time, in a simple form of non-waste feed hoppers. For years we have used high-grade commercial laying mash, with good satisfaction. This season we are using the Ful-O-Pep brand, and with fine results—never better.

"Then there is the night feed of scratch grains, consisting of cracked corn, wheat, barley and buckwheat in this proportion: corn 1,400 lbs., with 1,000 lbs. each of wheat and barley and 600 lbs. of buckwheat, thoroughly mixed. We give this feed early enough before dark so they can pretty well fill their crops. All scratch feed is fed in the deep litter, to make them work for every particle they get. And they work—believe me! All these years it has been a constant delight to me to see those sturdy, vigorous, good-sized birds kick the straw about to find their daily ration. If I were to see a hen or male bird 'loaf on the job' at this time, out she or he would go EVERY TIME. Such a bird is either sick or of no real value.

"Our green food for the breeder-layers consists in the early winter of cabbage; later on of mangel-wurzels. In the summer we use lawn clippings (you can see we have lots of them) and Swiss chard. Cabbage and wurzels are fed in suspended net bags. Clippings also are fed in bags. Chard is thrown in on the litter.

"Yes, you can overdo the green food proposition. Sometimes the birds will overeat and 'leave off' on grain food and dry mash, then they soon drop off in laying. We found this out several times by experience. Green food should be used to help balance the ration, not as the whole ration, nor even a main part of it."

Tested Lantern Lights in Stable Basement

The installation recently of electric lights, notably in the breeding and laying houses at Grandview Farm, is partly the result of what the Messrs. Wyckoff had read on the subject, but probably was mainly due to an experiment they made last winter. Note herewith on page 79, the well-built stable at left of picture. This building has a large basement with stone walls and has been used for years as overflow quarters for two or three hundred hens, to carry them partly through the winter. Size of this basement is 28 by 40 feet, with an entrance and three or four fair-sized windows on the north side, to the left in picture.

Last fall and winter (1919-1920) the Wyckoffs had about 300 hens in this basement, most of them three years old—a few two years old. Prior to December 1 the weather was cold and they "got no eggs at all." December 1 Mr. Wyckoff junior hung up three gasoline lanterns, suspending them from the joists seven or eight feet above

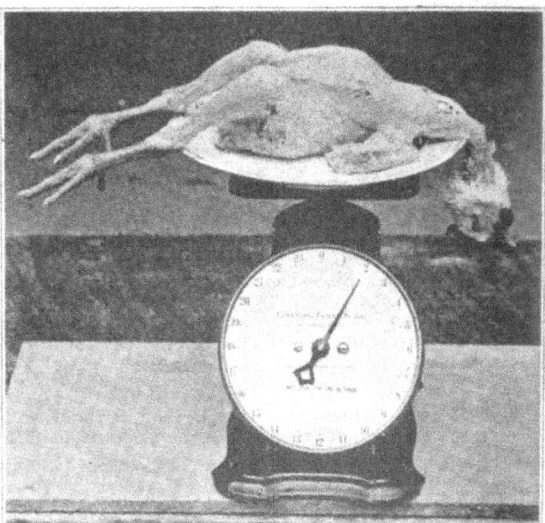

SAME COCKEREL, WHEN DRESSED
Bird was bled and dry picked; then weighed a trifle more than one pound, exclusive of plate. Shows rather unusual growth, as one pound at six weeks old is truly good weight. This toothsome-looking squab broiler was not special fed in any way—in fact, was one of a brood of 400 brooded under a coal-burning colony hover.

the fowls. These were lighted about 6:00 a. m. and kept going till daylight, then were again lighted at 3:30 to 4:00 p. m., and kept burning until 7:00 p. m. This was on December 1, 1919, at which time they were getting only two to three eggs per day from this flock of hens.

"By the eighteenth of December, or the eighteenth day under lights," said Mr. Wyckoff junior, "the production was up to 33 1/3 per cent. This certainly was a revelation to us! And it was wonderful how those hens would at once brighten up as soon as the light was turned on in this stable basement. They would scratch, eat, sing and lay and we soon were convinced of the remarkable benefits. Eggs from these hens we sold to the hotel in Auburn or shipped to our commission merchant in New York and the high prices helped a lot in paying for feed. In late December, however, we had to quit, because the lamps 'acted up' and we were afraid of fire—thought we might lose the stable and the hens, too. But this experience—this demonstration convinced us that we should equip our entire plant with electricity, which we are now doing."

The forepart of February, 1921, writer again visited Grandview Poultry Farm, in quest of further information and found the Messrs. Wyckoff truly enthusiastic about the installation and use of electric lights to increase egg production during the short-day period of the year. Briefly stated the lights were turned on two days before Christmas—December 23, 1920, at which time about 1,500 hens and pullets (hens practically all yearlings and early-hatched pullets—hatched the last week in March and third week in April, 1920) were in the eight laying houses. On that date, December 23, these birds produced 174 eggs. By the first week of January there had been a notable increase in egg yield and on January 23, exactly one month after the lights were turned on, the Messrs. Wyckoff gathered 926 eggs from these hens and pullets. Said Mr. Wyckoff senior to writer, at the time of this visit, the forepart of February, 1921:

"The 'lights' in our case certainly have done the work claimed for them, and I am speaking on the basis of comparison. To me the results have been little less than astonishing. Believe you will agree with me. Of course market eggs have sold at unusually high prices this season—somewhat higher even than last winter, but the fact is, as shown by our records, that it cost us a little over one thousand dollars to install the electric lights on the entire plant, including our residence, the work shop, etc., and during last month alone the increased cash receipts from eggs sold by us for table use—to Auburn and New York City—have paid for the entire installation, with a few dollars to spare. If anyone had told me such a thing a few years ago, it would have been hard for me to believe it."

Book, "Forty Years with the White Leghorns"

The Reliable Poultry Journal Publishing Company, in April, 1921, obtained the consent of C. H. and E. L. Wyckoff to furnish complete, detailed information about their experiences and success with Single Comb White Leghorns during the long period of years they have handled this variety of standard-bred fowl, which information is to be compiled by writer of this chapter into a book to bear title, "Forty Years with the White Leghorns," contents of which are to go back to the early start made by C. H. Wyckoff at the Groton plant forty years ago

SHOWS SIZE, UNIFORM SHAPE AND AVERAGE WEIGHT OF WYCKOFF-STRAIN WHITE LEGHORN EGGS

The dozen eggs at left were laid by yearling hens, week of April 25, 1921, and eggs at right were laid on same date by pullets that were one year old that month. Middle picture shows that shallow pan weighed 6 ounces; hence these eggs from hens nearly two years old weighed 28 ounces to the dozen and the pullet eggs, 26 ounces to the dozen. Average-sized eggs were selected in this case by writer, not extra-large ones. These weights were approved by the Messrs. Wyckoff as meeting their requirements for the Wyckoff strain.

and will aim to embody all facts of importance, including illustrations of breeding and laying houses, brooding houses, colony coops, feed shelters, etc.; methods of cropping and cultivation of the soil to keep it free from contamination; the home growing and practical use of green foods of different kinds; plans and draughtsman specifications of the present buildings, colony coops, chick shelters, labor-saving fixtures, etc., at Grandview Poultry Farm; also a detailed account of the efficient and economical methods now employed by the Messrs. Wyckoff in hatching with incubators, brooding the chicks by artificial means, self-taught roosting, single and double yarding, separation of sexes, etc.; also the building of shipping coops, their methods of packing eggs to ship to market and for shipment by parcel post as hatching eggs; also the use of shade trees for poultry, including plum, peach, English walnut, etc. Book will contain upwards of 150 illustrations made from photographs taken at Grandview Poultry Farm, 1920-1921, expressly for this purpose. For further description of book, "Forty Years with the White Leghorns," see annual book catalogue of Reliable Poultry Journal Publishing Company, which will be mailed free to any address on request.

CHAPTER II

Where Exhibition Quality and High Egg Production Are Combined

Origin of Morris-Farm Strain of White Orpingtons and Methods Used with Noteworthy Success in Its Development
—Line Breeding Practiced and Careful Records Kept so that Pedigrees Can Be Supplied with Birds of
Exceptional Merit—Have Lengthened Backs and Bred Away from Excess Fluff

By GRANT M. CURTIS, Editor of Reliable Poultry Journal

IT is quite generally understood—naturally enough so—that the smaller breeds of domestic fowl, such as the Leghorns, Anconas, etc., are our best layers, as to the number of eggs individual hens can be brought to lay in a given length of time, also as to the number of eggs they can be induced to lay as an average in pens and flocks. Fact is, however, that this claim has not yet been established, at least not to the satisfaction of all concerned. Of late years it has become widely known that the mediumweight breeds, embracing the Plymouth Rocks, Wyandottes, Rhode Island Reds, Orpingtons, etc., also are capable of remarkable egg production, especially so in sections where cold winters are the rule, and it is stoutly claimed by devotees of the heavyweight Light Brahmas that they, likewise, must be taken into account in determining and crediting the wonderful power of these desirable, easily managed, convenient-sized creatures to produce unexcelled human food in truly liberal fashion.

An Australian Black Orpington now holds the world's record for annual egg yield by an individual hen—having laid 339 eggs in 365 consecutive days. In this country the honor of being the highest producer is held by a 336-egg S. C. White Leghorn, bred and owned by Hollywood Farm, M. E. Atkinson, part owner and manager, Hollywood, Washington. Readers of this book have further been informed that in a steadily growing number of cases, other mediumweight birds have passed the 300-egg mark in 365 days, both officially at egg-laying contests and in trustworthy private hands, including Barred and White Plymouth Rocks, White Wyandottes, Rhode Island Reds and White Orpingtons. It would seem, therefore, that T. E Quisenberry is right—that egg production "has to be bred into them, then fed out of them." Furthermore, no less an authority than John H. Robinson stoutly claims that prolific egg yield on the part of domestic fowl is not a question of standard size or weights, within practical limits, but is a question of breeding, housing, feeding, right care, etc.

Another interesting and vitally important question that here presents itself is: whether or not standard fowl, bred to an advance degree of excellence for exhibition purpose under the requirements of the "American Standard of Perfection," can also be developed into prolific layers at one and the same time—that is, is this possible or practical and if so, what is the method or system of breeding to achieve such results in combination? It is fortunate that the answer to this inquiry exists in tangible form—in actual results achieved and publicly recorded. This chapter sets forth an outstanding example. Of late, readers of the poultry press of the United States and Canada have repeatedly been told about the remarkable egg performance of the prize-winning strain of S. C. White Orpingtons bred by Morris Poultry Farm, Lebanon, Ohio, at the American Egg-Laying Contest, Leavenworth, Kansas, in the contests of 1917-1918, 1918-1919, 1919-1920. During that period Morris Farm entered a total of fifteen birds, everyone a first-prize winner at some leading poultry show and also a high-scoring specimen, five of which birds in the hands of strangers several hundred miles from home made records as follows, in 365 consecutive days: "Princess Pat," 303 eggs; "Pauline," 245; "Peggy," 241; "Polly," 226 and "Pansy," 216. These birds, after they arrived at Leavenworth and before they were placed in the contest pens, were scored by E. C. Branch, licensed A. P. A. judge, and were given scores of 94½ to 96 points. For example, the five birds entered as far back as October, 1917, scored as follows: 95½, 95¼, 94½, 94½, 94¼. "Pauline," with an egg record of 245 eggs in 365 consecutive days, scored 96 points, fall of 1919, and as late as January, 1921, this same fine specimen was in the first prize old pen at Madison Square Garden—a combination record in high egg production and exhibition quality probably never before equaled in the history of poultry culture.

Obtained Facts First Hand, for This Book

For some time the author of this chapter had wanted to obtain first-hand, reliable information about these Morris-strain White Orpingtons—to learn their origin, how they were bred to this high state of egg production and the breeding methods employed to combine in this manner and to such extent the exhibition qualities of standard fowl, along with prolific egg yield; also we wanted the information for this book, "High Egg Production by Individual Hens, Pens and Flocks." Frankly, writer is among those who believe that standard requirements, as they apply to most of our popular breeds and varieties, are identical, generally speaking, with vigor, stamina

MOARI QUEEN
Owned by Morris Poultry Farm, Lebanon, Ohio. English born and New Zealand bred. Was brought to the United States seven years ago. Nine years old this spring. Was in good health February 18, 1921, on which date this photograph was taken. Long life is an indication of unusual vitality.

and prolific egg yield, provided the birds are bred that way and properly cared for, but such a belief should be supported by definite facts, leading up to positive results, as disclosed and proved in matters of this kind by the trap nest; therefore, in February, 1921, we spent the necessary time at Morris Poultry Farm and obtained from J. S. Morris, proprietor, and Harold Rawnsley, manager, the information desired. Said information is presented herewith, mainly in the actual words of Messrs. Morris and Rawnsley.

First, here are the offical records of the five blue-ribbon prize winners, in the form of Morris Farm S. C. White Orpingtons, above mentioned:

Princess Pat—First prize pullet, Ohio State Fair, 1918; laid 303 eggs from November 1, 1918, to October 1, 1919, in the American Egg-Laying Contest, Leavenworth, Kansas.

Pauline—First hen in single class at Cincinnati, December, 1919, and was in first prize old pen at Madison Square Garden, January, 1921; laid 245 eggs in American Contest, November, 1917-October, 1918. Hasn't yet "gone to pieces," as is proved by her triumph at Madison Square Garden this year—January, 1921.

Peggy—First pullet, Ohio State Fair, 1918; laid 241 eggs in American Egg-Laying Contest, November, 1918-October, 1919.

Polly—In first young pen, Ohio State Fair, 1918; laid 226 eggs, November, 1918-October, 1919, American Egg-Laying Contest at Leavenworth.

Pansy—First prize pullet, Michigan State Fair, Detroit; laid 216 eggs, November, 1918-October, 1919, in American Egg-Laying Contest, Leavenworth.

(Note:—Herewith is reproduced by photographic process—much reduced in size—the reading matter of a placard displayed by Morris Poultry Farm in connection with their exhibits of White Orpingtons at Madison Square Garden Poultry Show, New York City, January 18-28, inclusive, 1921, which is supposed to represent a "little speech" made by Pauline, one of Morris Farms' justly famous White Orpington hens—a bird that for the year ending November 30, 1918, laid 245 eggs in trap nests and that, when more than four years old, was a proud member of the Morris Farm exhibition pen that won first prize as old fowls at Madison Square Garden, January, 1921—truly a remarkable double achievment by a wonderful specimen of standard-bred fowl.)

Started with White Orpingtons in 1910

Writer spent most of three days at Morris Poultry Farm, Lebanon, Ohio, in early February, 1921, on which occasion Messrs. Morris and Rawnsley answered all questions promptly and freely. Asked when he started with standard-bred fowl, how he happened to decide on the White Orpingtons as his choice, etc., Mr. Morris said:

"The Reliable Poultry Journal really got me into the poultry business. For many years here in Lebanon I conducted a book

WHITE ORPINGTON WITH UNEQUALED COMBINATION RECORD

Was first hen in single class, Cincinnati, December, 1919, and in first-prize old pen at Madison Square Garden, January, 1921; laid 245 eggs in American Egg-Laying Contest, November, 1917-October, 1918. Received a score of 96 points from F. O. Branch, licensed A. P. A. judge.

store, selling magazines, newspapers, etc.; sold the 'Reliable' and became a reader of it. Saw several pictures of White Orpingtons, was impressed by their size and color and noted that the Orpingtons were rated as 'good winter layers,' particularly in northern latitudes where cold winters occur. In 1910, while living in town where I had a moderate-sized city lot, I got together four high-class White Orpington females, three of them from E. B. Rogers of Cincinnati, and an Owen Farms' pullet out of 100 eggs bought by a friend. A little later I bought two pullets from the Aldrich Poultry Farm at Columbus, this state, and next bought a cockerel from Mr. Rogers, which he had imported from Murray Linder of England. This mating produced, right from the first, a wonderful line of females. My Linder cockerel won second at the Chicago Coliseum that year. This can truly be said to be the original foundation of our present line of females, though we now have several families, as I'll explain later.

"That second year I had the good fortune to get acquainted with Len Rawnsley, brother of Harold Rawnsley, who recently had moved to Canada from New Zealand. Learned at our first meeting that he had been breeding White Orpingtons in England and had developed there a strain of successful exhibition birds from a line of high-record layers. This interested me very much. Still later, he had moved to the United States. As a result of this acquaintance we sent to New Zealand for the best birds he had left there, eight in number, two males and six females, which arrived in due course—good-sized, healthy, vigorous birds and unusually white in plumage.

"After the arrival of these additional birds, I bred along that way, using my Rogers-Owen Farm-Aldrich line of females and the New Zealand birds three years, during which period Len Rawnsley helped with mating the birds, doing this until his brother Harold came to me as manager five years ago, at which time he took charge of the mating. In 1918, as you will recall, we bought out the Aldrich Poultry Farm, obtaining their prize-winning stock which also had been bred several years for prolific egg yield, and as soon as Harold Rawnsley came on the job he got right after the egg yield, I am glad to state. Let me add further that these New Zealand birds, from the first year, showed themselves to be exceptional layers. As you know, down there in New Zealand and Australia they go in strong for egg production—also in England, the native land of the Orpington breed, which was originally intended as winter layers—as a practical combination of good table qualities with plenty of eggs.

Decided to Work for the Combination

"When Harold came with me, I said to him: 'Those New Zealand birds certainly

---

I AM PAULINE

THE

$1000 White Orpington Hen

Bred and Raised at The MORRIS POULTRY FARM, Lebanon, Ohio

I WAS ENTERED IN AN OFFICIAL EGG-LAYING CONTEST AT THE AMERICAN SCHOOL of Poultry Husbandry, Leavenworth, Kansas, conducted by the noted Prof. Quisenberry, November 1, 1917, to November, 1918, and during one of the severest winters on record I defeated 551 Pullets for 6 months, all varieties being represented, laying 141 eggs, and finishing the year with 245 eggs to my credit. Four of my sisters averaged 110 eggs each for the 6 months in the same contest.

But my younger sister, PRINCESS PAT, this year established a WORLD'S RECORD. In the Leavenworth Contest, November 1, 1918, to November 1, 1919, she laid 303 eggs. I feel a little bit peeved because she has beaten my record, but then you know blood is thicker than water, and I am really proud of Pat because she has made a World's Record, and all of us Orpingtons are really proud of it.

I am also an exhibition Orpington. Judge Branch scored me while I was in the contest, recording me at 96 out of a possible 100. I was exhibited at the Cincinnati Poultry Show last September, winning 1st Hen. You know our family is really the aristocratic family of the White Orpington World. Besides making famous Egg-Laying Records, our family has won more prizes at Chicago Coliseum and Madison Square Garden, New York, than all our competitors combined. So you see we are a real classy dual purpose fowl.

My owners have never regretted that they did not sell me for $1000. Sister Princess Pat, no doubt, is worth much more than $1000, for she laid 303 eggs in the contest this year. Peggy, who also has an official record of 241 eggs and was 1st Pullet at Ohio State Fair, is worth as much as I am.

Secure some of our family; they will deliver the goods in meat and eggs and will win for you in the Show Room. Really, you can not go wrong with the Morris White Orpingtons. We are bred to lay, win and pay. Very Truly Yours,

PAULINE.

Reproduction of 12 by 18-inch placard displayed by Morris Poultry Farm at the Chicago Coliseum Poultry Show, December, 1920, and at the Madison Square Garden Poultry Show, January, 1921, in connection with their prize-winning exhibits of "Proven Leader" strain, Single Comb White Orpingtons.

are remarkable layers,' and right there we talked it over and decided that we could have and ought to work for or breed for a combination of exhibition qualities with heavy egg yield. Harold was of the same opinion and at once made a thorough study of the Hogan system. The next mating season we began faithfully to Hoganize our breeders and now the poultry world knows the results. Our last and perhaps most gratifying triumph and demonstration is found in the fact that Pauline, our 245-egg hen by official record, was in the first-prize pen at the world's greatest annual poultry show, Madison Square Garden, New York City, in January of 1921, or 'right down-to-date.' Also the two male lines that originated with those eight New Zealand birds have won highest honors year after year at our two greatest annual winter exhibitions, Madison Square Garden and the Chicago Coliseum,—and to win at these two shows is proof positive that our birds will win first honors anywhere, under the 'American Standard of Perfection,' as it now describes and illustrates the modern S. C. White Orpington, which I claim to be our foremost table-meat and winter-egg breed, par excellence.''

In this connection Harold Rawnsley—who was present at this interview with Mr. Morris—said:

"My personal preference? Yes, indeed, I prefer the present-day American type of White Orpington over the former shorter-bodied, rounder-bodied, heavier-fluffed, creepy type that was copied in the earlier days from the Black Orpingtons. This new or American standard type with its increased length of body is a bird of greater vitality and activity. The males especially 'have more daylight under them,' as the saying is, possess greater vigor and fertilize the eggs better. Also the females lay better—no doubt of it. The increased length of body gives them all the room needed for large, healthy egg organs and ample food capacity. As they are now bred, their backs in the rear are well covered with thick mats of protective feathers that keep warm the vital organs."

At this point we quote as follows from the 1921 catalogue and mating list of the Morris Poultry Farm:

Combination Exhibition, Egg-Laying Strain

"The Morris record layers and first prize showroom winners are line bred from the same stock. We have no utility egg-laying line and a separate exhibition showroom line of White Orpingtons. They are one and the same line, a combination exhibition, egg-laying strain. We have scattered to the winds the old theory that show birds are simply beautiful, exhibition samples without utility qualifications. Those who have attended any or all of the poultry shows, Madison Square Garden, Chicago Coliseum, Kansas City, Boston, Cincinnati—where our record hens, Pauline, 245 eggs per year; Princess Pat, 303 eggs; Peggy, 241 eggs; Pansy, 226 eggs; Polly, 216, have been exhibited—will attest to the fact that they are all real show birds, capable of being shown in single class or pens wherever exhibited and taking a good position among the winners. Every one is a real exhibition hen."

"At the Madison Square Show, January, 1921, our first old pen was a center of admiration for those who admire beauty, coupled with utility. The male that headed this attractive pen was admired for his noble style, broad back, elegant tail, wide, rounded front and beautiful snow-white plumage. The four females were large hens with not a break in their broad backs anywhere. Their heads and tails were alike, in fact it was a well-matched pen. Every individual in this mating was a beauty. This Garden pen was the same pen that made up our Chicago Coliseum first pen, with one exception. Pauline, our $1,000 hen, displaced the poorest one of the Coliseum hens and went to the Garden in the pen. Pauline was the queen of the pen, for besides ranking as the best of them as an exhibition individual, she had her pullet official egg record of 245 eggs at Leavenworth. Another of these hens, leg band No. 61, unofficially, for us, laid 215 eggs in her pullet year. Pauline won first hen in the single class at Cincinnati, December, 1919.''

Mr. Rawnsley Gives Credit to Walter Hogan, Deceased

Harold Rawnsley, at the time of our interview with him and Mr. Morris, February, 1921, said:

"I am strong for giving due credit to Mr. Hogan. His work for greater egg production is at the bottom of the whole thing. He started this work before the colleges took hold and had it well developed in practical, successful use before other students of the problem woke up to the possibilities.

"On arriving here in the fall of 1915, I started in at once to Hoganize the birds and found good material to work with. Among other points, we decided to lengthen the backs, doing this without losing width. We did it by selection of the breeders, going through the entire flock. Here at Morris Farm we do not have what some call utility stock, and never have had. In other words, to get the results desired, we worked on our best birds in the highest class pens or matings and we have utterly disproved any idea that you cannot get high egg yield from the best exhibition quality.

"No, we do not double mate. Sometimes we expect better cockerels from a certain mating because perhaps of the male at the head of the pen, but we have no separate families with that special object. On the other hand, if one year we get extra-good cockerels or pullets from some particular mating, we retain that same mating, but we do not want to get into double mating, either for males or females, or for exhibition quality and high egg production.''

Example of Intelligent Breeding

Asked about the increased length of back he had mentioned, Mr. Rawnsley continued:

"I'd say we increased the length of the back at least two inches. You can tell this by an examination of the old cuts, as compared with our present prize winners at the Chicago Coliseum, at Madison Square Garden, etc. For example, take the cuts of six or eight years ago—referring to birds that won at our best shows in those days, and the longer back is very noticeable.''

OFFICE OF MORRIS POULTRY FARM, LEBANON, OHIO

To the left at desk is J S. Morris, proprietor; standing is Harold Rawnsley, manager; seated is Mrs. Harold Rawnsley. Cups shown are among those won by Morris Poultry Farm at leading poultry exhibitions of the country on S. C. White Orpingtons bred and exhibited by them. On wall in corner back of desks and letter files are some of the many ribbons they have captured in keen competition.

WHERE EXHIBITION QUALITY AND HIGH EGG PRODUCTION ARE COMBINED

KEEPING TRACK OF THINGS AT MORRIS POULTRY FARM
Type of trap nests used to learn "which hen laid the egg," thus to practice line breeding accurately and to be able to furnish pedigrees with valuable breeding and exhibition specimens. As hen enters nest the lower half or "fold" of trap door drops down, enclosing her. A hen was on the middle nest when this picture was taken.

Question by writer: "Was there any other feature that you sought to change, in the interests of increased egg production?"

Mr. Rawnsley: "Yes, indeed; we got them up off the ground, getting away from the old creepy type, as the Orpingtons originally were in England. Our present birds are closer feathered, with less fluff. And I'll tell you another thing: the old-type, short-bodied birds would bag down behind. Eight to ten years ago prominent breeders would have ten or twelve hens out of every 100 with this bad fault, while now we seldom see one in a hundred. In lengthening the body we of course lengthened the breastbone and gave more room for egg production, so that bagging down, in the case of modern-type White Orpingtons, has now gone out of fashion."

Question: "Do you mean that the legs were lengthened also?"

Mr. Rawnsley: "It was not so much a question of length of shank, but we have bred away that loose fluff. Our birds do stand somewhat higher on the legs, but not as much as some would suppose. What we really have done is to get rid of the excess fluff, doing this by selection through several generations."

Question: "How does our American type of Orpington shape compare with the present English type?"

Mr. Rawnsley: "In England they are still breeding the shorter-backed, more fluffy type. Over there they want them round bodied, while we have, in the American type, a more active bird with smoother, more compact feathering, which in themselves are signs of vigor and greater egg yield."

Used Body Measurements by Finger Plan

Question: "In making your selections during this period of five or six years, you have worked with the Morris-Aldrich birds, also with those original New Zealand birds, and picked accordingly?"

Mr. Rawnsley: "Yes, I at once began to use body measurements by the finger plan to select for our breeding pens. We favored right along, the Hogan recommendations and simply would not use birds that did not measure liberally for egg production. To interest us the pullets had to measure at least four fingers. Another point that may interest your readers in this connection: we never breed from a crooked breastbone. I have found that they reproduce—not every time, but too often to have it profitable or wise to use them. At present, after a Morris Farm White Orpington pullet has been laying a few weeks, she will measure five, six and even seven fingers. Yes, it runs higher with hens—from five fingers up to as high as eight."

Other Points in Careful Selection

This particular interview took place February 16, 1921, and the measurements here given by Mr. Rawnsley were verified by an examination of a dozen or more birds in the breeding pens the day following. Pullets that had been laying six to eight weeks ranged from five to seven fingers, most of them five and six, while several hens that we examined, measured seven and eight fingers. Asked what else he looks for in selecting for a combination or exhibition quality and high egg production in the S. C. White Orpingtons, Mr. Rawnsley said:

"We want a bright, prominent eye, without overhanging eyebrows. Eye must be intelligent and alert—the kind that is looking all the time for more to eat. The experienced poultryman who pays attention can soon tell or 'spot' the naturally good layers—and it is these we should breed from, so far as that value is concerned.

"Also the birds we breed from must be in reasonably good flesh and able, therefore, to offer resistance to the heavy egg-yield drain on the system. We do want a thin pelvic bone, but not one devoid of flesh. Then, also, there is quite a lot of merit in late molting, as compared with early molting; the late molters usually are the best year-round layers. We like to have them molt in September and October, though we do not object if they begin dropping a few feathers in August. However, we like to have the molt all completed by early December.

"Can say, therefore, in the case of White Orpingtons, that our best layers, as a rule, are the last ones to molt, then often they will shed their feathers all at once, almost. On the other hand, a limited number of our hens molt gradually, dropping a few feathers at a time, and these hens hardly stop laying during the process; but in other cases, where hens molt all at once, they stop laying while putting on the new coat of feathers. This keeps them busy four to six weeks and sometimes longer. If it takes them too long we think it indicates lack of vigor and a check mark is put against such hens, with reference to their value for breeding purposes. Low-vigor hens are never permitted in our breeding pens under any condition."

Five Years of Trap Nesting Also

Asked about trap nesting, Mr. Rawnsley replied:

"We started systematic, year-'round trap nesting at the same time we began to Hoganize. The first year we trapped about fifty birds. The egg yield ranged from 150 to as high as 180 eggs. Had numerous birds that did this well. However, only three of them reached 200 eggs and a little better. These birds were mostly of the New Zealand stock, especially those that made the best records in the traps. Then as we lengthened the backs we could see we were averaging to get more eggs, also higher individual records. The highest made on the place was 254 eggs, whereas at Leavenworth 'Princess Pat' reached 303 eggs, the highest record ever made by a White Orpington, either in America or Europe. 'Princess Pat' weighed 7 pounds in September of the year she made the 303-egg record, at which time—September—she was only six months old. She was fully up to weight by November 1 of that year, when she entered the contest.

"No, we do not in all cases trap nest the year 'round—in fact, not in a majority of cases. When hatching time comes we use

ORPINGTON MALE OF FIVE YEARS AGO
Prince I—first cock Madison Square Garden, January, 1916, and again first cock at the Garden, January, 1917. Was prominent among sires of the birds that formed the foundation stock of the Morris exhibition, high-egg-yield White Orpingtons. Note unusual length of body for that purpose, and an absence of excess fluff.

HOLDS WORLD'S RECORD FOR S. C. WHITE ORPINGTONS
Was first-prize pullet, Ohio State Fair, 1918; laid 303 eggs in American Egg-Laying Contest, Leavenworth, Kans., in 365 consecutive days. Length of body is unusual.

trap nests in our breeding pens in order to pedigree the strain; to know the hen as well as the male bird in every case. We find that our customers very often wish to know this, and they are coming more and more to demand it. Also, we are after the good-cycle layers—those that lay three or four eggs in succession and then skip a day, or lay four, five or six eggs in succession, then skip even two days. In the case of such a hen you can count on her being a pretty good 'long-distance' layer. And to know about this we learn by the trap nest. The only other way would be to coop each hen in a separate pen. You can say this, too: That we are just as particular with the males as with the females, in carefully Hoganizing them for use as breeders."

Origin and Prepotency of This Strain

Treating further on the origin of the exhibition-quality, high-egg production, "Proven Leader" strain of S. C. White Orpingtons, Mr. Morris said:

"As before stated, these New Zealand birds soon showed themselves to be unusually good layers. One of the New Zealand males won first cock at Madison Square Garden in 1916 and we gave him the title of Prince I, also we bred him with one of the hens that came with him. As far as I can learn she was a sister. We got 'Prince of the Garden' from that mating. We put 'Prince I' back two years with that same hen, using other hens with him, but it was this particular hen which gave us 'Prince of the Garden' and the remarkable line of males that succeeded him.

"We started another line of extra-white birds from the other New Zealand male. Our finest stay-white birds are from this line or family and the first year it gave us our first young pen at the Chicago Coliseum, 1916, a pen headed by 'White Prince,' as we called him. This bird, 'White Prince,' sired the first cock at Chicago Coliseum, 1918. And this has been a prize-winning line ever since, including second young pen male at Chicago, 1920, and numerous first-prize birds sold to customers, and it has also supplied valuable blood for egg records.

"Yes, we have two other families, which we call the 'Prince Wonder' line and the 'King' family. They aren't specially distinct from the other lines, but we have a notion that our 'Wonder' line has given us extra-well-finished body formation,

including broad backs of the same width throughout, and extra-well-spread tails. The bird 'King' was first cock at the Coliseum. He really traces back to 'Prince I,' while 'Prince Wonder' is a blending of the two lines.

"No, we are not committed, on the basis of experience, to the theory that it is better to start a line or family with a female rather than a male. Sometimes one sex seems to dominate and at another time the other sex. As a matter of fact, these lines of ours branch out more or less from a main trunkline, so to speak. As examples, the 'Prince I' line gave us 'Prince of the Garden,' also 'Prince Marvel,' first cockerel, 1919, but he was a cross between two lines. 'Garden King' also came from this line, while 'Garden King' is a son of 'Prince of the Garden' and a grandson of 'Prince I'. This 'Prince First' line gave us some very choice females. As examples, I recall the first Chicago pullet, 1916, two of the pullets in the first young pen that same year and first pullet at Madison Square Garden in 1918.

"Speaking of the prepotency of males, 'Prince of the Garden' was a wonderful reproducer. He had more or better finish than his father, 'Prince I.' Yes, we keep a written record of matings, results, etc.; also we practice quite a little stud mating, keeping the male separate and testing different females, by which method there can be no doubt as to the ancestry."

General Plan for Developing Combination Strain

Asked for a general plan for starting and developing a combination strain of standard fowl to possess exhibition qualities and high egg production power, Messrs. Morris and Rawnsley joined in outlining the following method, based largely on their experience and success:

Secure from a reliable breeder some well-established, line-bred specimens which conform closely to standard requirements for the breed and variety, house them comfortably and feed them well, making them work for most of what they get to eat, especially in the way of whole or cracked grain, then when laying time comes "put the trap nests on them." Leg band each layer, mark for identification every egg that is to be incubated, and later on toe-mark or leg-band every chick. Keep written records of all breeding operations, rather than trust to mem-

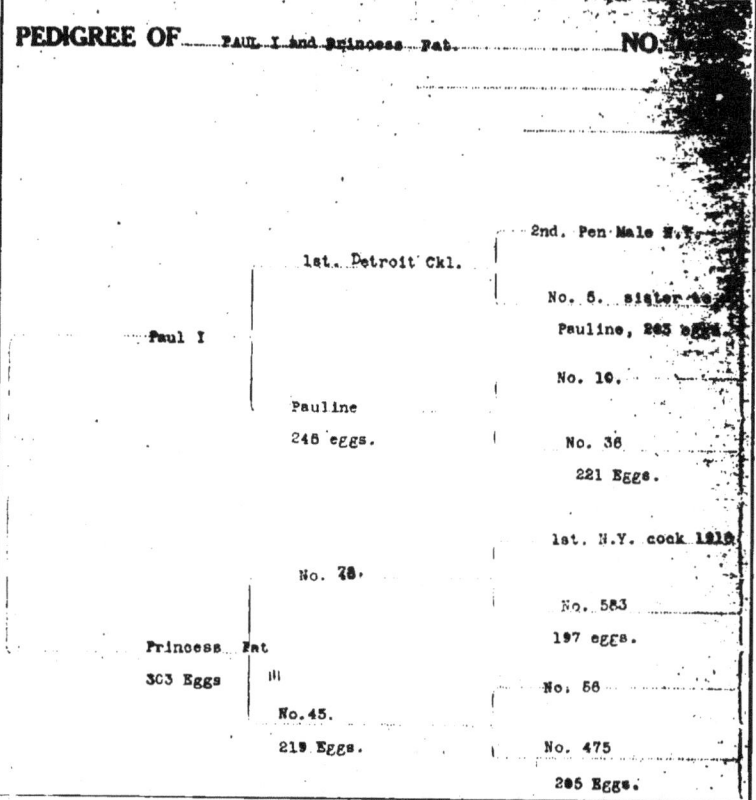

FIGURE 1—PEDIGREE OF PAUL I AND PRINCESS PAT, BRED AND OWNED BY MORRIS POULTRY FARM, LEBANON, OHIO

ory. "You need to keep track of the age and the sire and dam of every good bird, year after year."

First to last, in mating your birds, be governed mainly by two things: their standard qualities, as per descriptions of the breed and variety in the "American Standard of Perfection" and by the trap-nest record of each female admitted to your breeding pens. It is easy to keep track of the male line, and this likewise should be done in written form; also if you have photographs of the exceptionally good birds—showing them in profile, for example—it is a good idea to make these pictures a part of your permanent records.

At the start you probably will not have a trap-nest record of the hens or pullets. This may be unavoidable. In that case, mate them up according to health and standard qualities, then find out as soon as you can how well the different individual females lay—and the one and only sure and economical way to do this is by use of trap nests. After you have found out which are the good layers, including the best layers, you will be ready to proceed intelligently. From the "best" and the "good layers" select your breeders for that generation, from which to produce the next generation, being governed chiefly in this selection by the three big factors: first, health and vigor, including standard size and weight; second, by their trap-nest performance; third, by their standard or exhibition points, per the "Standard of Perfection."

Remember that in poultry breeding we are doubly fortunate: first, it takes only one year to note progress in most respects, from one generation to another; second, we usually have a considerable number of individuals from which to make our selection each year or generation, all being offspring from the one male and the one female, if our records are so kept that they can be relied on. For this favorable reason we can exercise good judgments in the form of individual selections for our future matings and still preserve our blood lines, thus to develop a pedigree strain, using the trap nests every breeding season, year after year. As a matter of course, the male line also must be kept track of, one generation after another, but as before stated, this is comparatively easy to do. Said Mr. Rawnsley in this connection:

"It's no use to split hairs over these problems—as, for example, which came first, the egg or the hen that laid it. We are talking about the breeding and improvement of standardized domestic fowl, not scrubs nor nondescripts. In brief, our practice is this: We use trap nests as our guide, then we do not rely for our breeding advancement on poor layers, on the one hand, nor on inferior specimens as judged by the 'Standard of Perfection,' on the other hand, but we make it a point to retain enough birds every season from each good mating to enable us to preserve the blood lines in both vital respects, namely; good layers, as proved by the trap nests, and high standard qualities, as shown by the 'Standard of Perfection' or as demonstrated in the showroom."

Their System of Breeding Briefly Stated

Asked for their system of line breeding, Mr. Rawnsley replied substantially as follows:

"Mate cock with his granddaughter, and grandson back with grandmother. Mating cock with daughter is not as good, as a rule, as to use the next following generation—cock bird with granddaughter. Keep this up generation after generation. It cannot be done in safety or with large results by the use of one bird; therefore we must use a number of females and select carefully among related individuals to offset defects on both sides of the house and intensify the desirable standard points or values. I never have bred brothers and sisters, following this plan up generation after generation, as some claim they have, but may try it sometime, as a matter of curiosity. I have done that with pigeons during eight straight generations and met with no trouble. These pigeons consequently were of pure blood and they certainly showed remarkable improvement on the lines we are working to establish."

Asked what Morris Farm is now seeking to accomplish further in the way of improvement of their exceptional strain of S. C. White Orpingtons, Mr. Rawnsley said in substance:

"We are now working to reduce the size of combs which often are too large and we are also selecting with special care to enrich the red eyes, as to their attractive color. Have met with fine success in this way and now 95% of our birds come with rich red

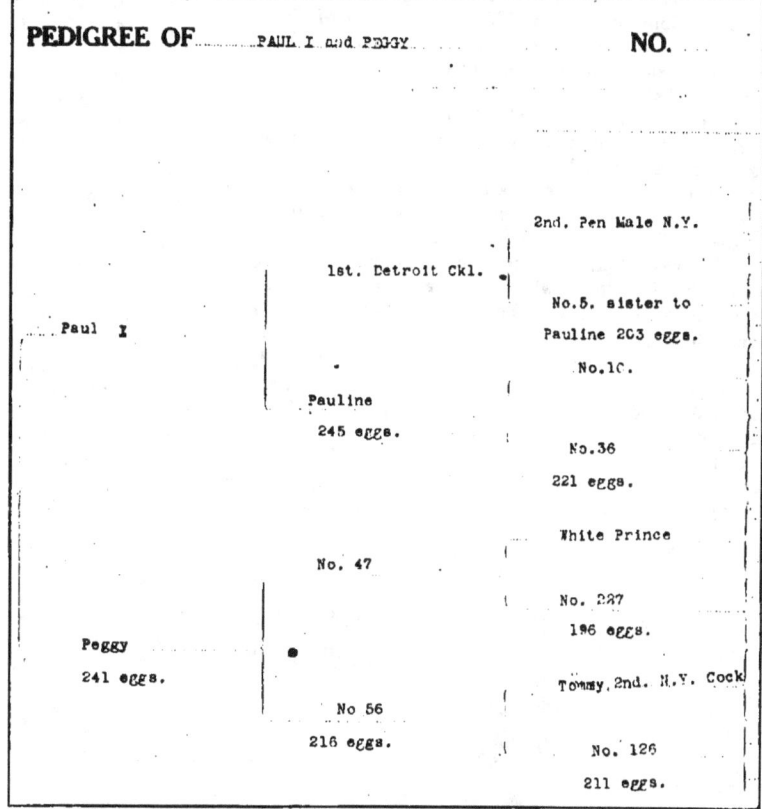

FIGURE II—PEDIGREE OF PAUL I AND PEGGY, BRED AND OWNED BY MORRIS POULTRY FARM, LEBANON, OHIO

eyes. We positively will not breed pale-eyed males, no matter how good they may be in other sections. This work on our part, however, is a little hard sometimes, because judges will not throw down otherwise good birds on account of a pale eye. They never see the eye, or do not seem to, while I think it is an outstanding mark of beauty, if of the right color.

"Another thing: We're strongly of the opinion that red eyes and stay-white plumage come along together. That has been my experience. You can go around this place and cannot find a stay-white bird that does not also have a fiery red eye. Still another peculiarity that these stay-white birds have: as a rule the back of the shank shows more red than is the case with the old-fashioned, unhandsome pale-eyed birds.

Essential for High Egg Production

"Let me mention one more highly important mark of beauty and productiveness that we have established in the 'Proven Leader' strain. I refer to the parallel back formation, without any break in the plumage, near the ends of the wings. Our birds, in order to have body capacity and proper finger measurements for prolific egg yield, should be as broad all the way back to the juncture of the tail, or practically so, as they are at the shoulders, and in the case of such birds the plumage will not break near the end of the wings in front of the tail—which formation is far more attractive, as well as being better for the breed as table fowl, and this also is essential for high egg production. Really this is the biggest advancement up-to-date Orpington breeders have achieved in the last several years and now is a determining factor in the placing of awards by competent judges at our largest shows where competition is keenest, such as Madison Square Garden, the Chicago Coliseum, etc."

At Morris Farm they not only keep pedigrees of their line-bred birds for their own use and benefit, but also furnish them to customers, in the case of valuable specimens. Two examples of such pedigrees are published herewith, for the information of the student-reader. For example, pen No. 7, as described in the 1921 mating list of Morris Poultry Farm, contains Princes Pat—official egg record, 303 eggs—and Peggy—official egg record, 241 eggs—along with three other hens, each of which has an offical record of upwards of 200 eggs. This pen for 1921 is headed by Paul I, a yearling cock that was a Chicago Coliseum winner, 1920, and is a son of Pauline—official egg record, 245 eggs and a member of the first-prize old pen at Madison Square Garden, New York City, January, 1921.

Suppose, as an example, that a customer of Morris Farm wishes to purchase, in season, one or more hatching eggs from Princess Pat or Peggy, as mated with Paul I, or later on wishes to secure a baby chick or a breeder from one or both of these matings, and asks for a pedigree. If the pedigree was to cover the offspring of Paul I and Princess Pat, the one shown herewith as Fig. I, would be furnished, whereas if a pedigree for a descendant of Paul I and Peggy were requested, Fig II would be supplied.

To illustrate, see both pedigrees herewith (Fig. I and Fig. II), and it will be noted that the sire, dam, grandsire and granddam of Paul I are the same in both cases; also that the dam and sire of Pauline (mother of Paul I) are the same, as shown in both pedigrees, but that the sire, dam, grandsire and granddam of Princess Pat and Peggy are not the same. Both were mated to Paul I for the season of 1921, but if hatching eggs or breeding stock were to be supplied from Princess Pat, as distinguished from Peggy, and a pedigree was demanded, then of course the direct ancestry of both Princess Pat and Peggy must be kept separate, as is indicated in Fig. I herewith for Princess Pat and in Fig. II for Peggy.

It involves labor and expense to carry on pedigree breeding, both in the matter of accurate trap nesting and the keeping of reliable records, but it is said by the most successful breeders of standard fowl, whether for exhibition purposes, for high egg production or for a combination of both that this is the only true systematic or scientific method which can be relied on for definite, progressive results. Admittedly the breeding of domestic fowl for any valuable purpose or combination of purposes is still in its infancy, to use an oft-abused phrase, but one that undoubtedly is true when applied to this branch of domestic live-stock breeding. In England and Australia good progress has been made recently along this line, also in Canada and the United States. Various interesting problems are presented, but as yet there is far too little trustworthy information available. To help gather, compile and present to the interested public in serviceable form such worth-while data as do exist, is one of the chief objects of this book and of others similar to it which are being given to the public from time to time by the Reliable Poultry Journal Publishing Company.

LAY FOR CUSTOMERS, TOO

White Orpington in picture here shown was sold by Morris Poultry Farm to A. A. May, Topeka, Kansas. When sending in this picture, Mr. May wrote: "Now, Mr. Rawnsley, this is a REAL layer. Laid her first egg when she was six months and twenty-nine days old and laid 27 eggs her first month, from October 10 to November 10, and laid 41 eggs to date, November 27 (1920). Her pelvis is ⅛ inch thick and she is between seven and eight fingers deep. Had another pullet that laid 25 eggs her first month, then went to sitting."

MEMBERS OF "ROYAL FAMILY" AT MORRIS POULTRY FARM

Right to left:—Moari Queen; Garden King, first cockerel, Madison Square Garden, 1920, and first cock at Garden, 1921; first pullet, Chicago Coliseum, 1920, and Coliseum Prince, head of first young pen, Chicago Coliseum, 1920—all descendants of Moari Queen.

CHAPTER III

A High-Producing Strain of Barred Rocks and How It Was Bred

Methods of Breeding that Have Enabled J. W. Parks of Altoona, Pa., To Develop a Remarkably Productive Strain and To Win Financial Independence—Interesting Examples of Pen and Flock Production—Methods of Line Breeding and "Tracing Back" by Pedigrees

By GRANT M. CURTIS, Editor of Reliable Poultry Journal

TO the best of writer's knowledge, J. W. Parks, proprietor of Wopsey Poultry Farm, located in the heart of the Allegheny Mountains, near Altoona, Pa., has been breeding Barred Plymouth Rocks for increased egg production during a longer period than any other man who has ever undertaken this important service in the interest of mankind. On this point, the following three paragraphs are quoted from Mr. Parks' seventy-two page, illustrated catalogue, 1919-1920 edition:

"This general catalogue marks the thirtieth anniversary of the Parks' strain of Bred-to-Lay-and-Do-Lay Barred Plymouth Rocks. Yes, that is a good many years of breeding for a purpose—increased egg production—but after reading the contents of this catalogue I feel quite sure you will consider that it has paid both the poultry industry and ourselves. Thirty years of what was at first a hobby has with us developed into a trade or profession.

"The Parks' strain of Barred Rocks is now actually being fed by the third generation of Joe's, a record we are proud of and I doubt if it has ever been surpassed with any other breed. Today we are to the Plymouth Rock breed what the seed house is to the farmer, or the nursery to the horticulturist—the source of supply for the practical-utility side of this grand old breed.

"While we always give eggs first consideration, we have not been stupid enough to sacrifice beauty and standard qualities entirely for eggs. On the contrary, our aim has been to bring the egg basket and the 'Standard of Perfection' into closer relationship—and we have succeeded."

Any person who is interested in standard-bred fowl or is a student of poultry culture on up-to-date lines will find a visit to Wopsey Poultry Farm well worth while. From a small beginning this plant has grown steadily until it now consists of thirty-five acres, with numerous buildings and yards, including an unusually large incubator cellar, comfortable, well-equipped offices, an electric lighting and power plant, various kinds of labor-saving machinery, a commodious storage and shipping room, a large root cellar, etc.

J. W. Parks (Joe, everybody calls him), still a young man, is sole proprietor of this plant and has been since his father died a few years ago. Here we have a unique character—a self-educated, self-made man, yet one in whom everyone appears to have full confidence. Locally he is well spoken of by neighbors and business friends, while in poultry circles no one stands higher in public esteem than J. W. Parks. This is especially true on the part of college men—meaning those who are connected with the poultry departments of agricultural colleges and others who are engaged in extension field work, traveling about as lecturers or working as county agents. During one of writer's visits to Mr. Parks' home and poultry plant he displayed a number of letters to that effect which had come to hand in the regular course of correspondence, and they contained expressions of confidence and respect that should cause any man to feel proud, whether or not he has a diploma of his own hanging on the wall.

Began Early to Keep Tab on Egg Yield

The following two paragraphs are quoted from a section of Mr. Parks' latest catalogue entitled, "A Tale of Progress in the History of Parks' Bred-to-Lay-and-Do-Lay Barred Plymouth Rocks":

"From a small flock of sixteen hens first kept tab on in 1889, when the entire flock averaged but 96 eggs, they have been slowly, carefully selected, trap nested and pedigreed for increased egg production, until today they are conceded to be the Daddy Laying Strain of them all, both in years of breeding and in increased egg production.

"They have individual laying records as high as 290 eggs in a year; sworn continuous laying records as high as 148 eggs in 148 days; official laying contest averages as high as 285 eggs, pen average (12 birds) as high as 242 eggs and large flock averages as high as 208 eggs for 126 hens. (It's this latter high flock average that we point to with most pride.)"

J. W. PARKS, ALTOONA, PA.
Intimate picture of the proprietor of Wopsey Poultry Farm, Altoona, Pa. Shows him preparing copy for the thirtieth anniversary (1920) catalogue of the Parks' Bred-to-Lay strain. Catalogue consists of seventy-two pages and tells the story of Mr. Parks' life work.

The foregoing paragraphs were written in the late fall or early winter of 1918. Since then three of Mr. Park's Bred-to-Lay Barred Plymouth Rocks have made new world records, one right after another, in a short period of time—two of them in the hands of customers, one who bought baby chicks and the other eggs for hatching; therefore if Mr. Parks now were to rewrite the foregoing paragraphs (May, 1921) he would be entitled to say: "That have individual laying records as high as 313, 323 and 325 eggs in a year," etc.

The above-mentioned 313-egg record was made by a pullet-hen raised from a baby chick that Mr. Parks sold to E. F. Grundhoeffer of the Engineering Department of the Pennsylvania State College, State College, Pa. Under date November 28, 1919, Mr. Grundhoeffer wrote the Reliable Poultry Journal—independent of any connection with writer's visit to Mr. Parks' place—as follows, and sent us a copy of the complete "laying record of hen No. 9," which record will be found reproduced herewith in the form of a cut.

Letter to R. P. J. from Mr. Grundhoeffer

"State College, Pa., November 28, 1919.
"Editor Reliable Poultry Journal:
"Having read the article on 200 and 300 eggers in your last issue, I assume that the enclosed copy of the laying record of my

Parks' Barred Rock hen might be of interest to you. She is one of a few day-old chicks, purchased from J. W. Parks, Altoona, Pa. In the same lot I had one hen that laid 250 eggs, another that laid 245, and one that reached 202, with two or three others that should average from 220 to 260 each, on the basis of their present rate of production.
"Very truly yours,
"E. F. GRUNDHOEFFER"

Later on, Mr. Parks bought this 313-egg hen and gave her the name of Miss Graduate—believing at the time that she held the world's record as a Barred Plymouth Rock for individual production in 365 consecutive days, which probably was true, as of that date. But Miss Graduate's proud honor was short-lived. Within sixty days Mr. Parks learned that her record had been excelled by an even dozen eggs—and by another Parks' hen. Writing on this subject, Mr. Parks reports:

"Imagine my surprise, after just getting over the remarkable clean-up Miss Graduate had made at State College, Pa., to learn that there was another Parks' strain hen working still harder for a world record. Such was our experience when we learned that James A. Mortensen of Phoenix, Arizona, had a biddy of our strain that had made the wonderful trap-nest record of 325 eggs from January 9, 1919, to January 1, 1920—a record of 325 eggs in nine days less than a year. Having Miss Graduate at home and being anxious to test out these two hens together, I was successful in getting this hen also and a sworn statement of her record, which hen I renamed 'Miss Smarty,' on her arrival at Wopsey Farm."

Nor was that the end of the good news. A few days later, Mr. Parks received another welcome surprise, similar in kind, in the form of the daily record of Lady Martha, a hen hatched from an egg he sold in the regular course of business from one of his pedigree matings, which bird laid 323 eggs in 365 consecutive days. Herewith on page 94 is published her daily record for the year. Also on page 96 are shown photographic reproductions of these three world-record Barred Rock individual layers.

Winners Also at Egg-Laying Contests

Following are several more paragraphs from the "Tale of Progress," as published in Mr. Parks' thirtieth anniversary catalogue, 1919-1920 edition:

"Parks' Bred-to-Lay Barred Rocks are and have been among the topnotchers and winners in nearly every American laying contest. As far back as 1913 they won silver cups in both the Missouri, U. S. A., Contest and the International Laying Contest of Canada. The following year they finished second in the Missouri Utility Laying Contest.
"In 1915 they outlaid every other breed entered in the Missouri contest, as well as outlaying all the Rocks in the North American contest by about 150 eggs, and a customer led the entire Storrs contest for fall and winter laying, and was beaten by only a few eggs by another breed that laid most of its eggs in the summer months at a time when even the sparrows lay.
"The following year they made history by capping the climax of their laying career, by winning first honors and making the best record ever made in any of the five Missouri laying contests, which record also exceeded the best record of the 1,000 birds at Storrs for the same year.
"The five Missouri laying contests were made up of over 2,600 birds. Over forty-five varieties from the best laying strains of thirty-seven states and eight foreign countries were represented, including the famous English and Australian Leghorn and Wyandotte laying strains.
"Their record of three silver cups at Missouri for best winter month laying and 134 eggs in January (five birds) stands without a peer. My pen (10 hens) won first place in the Rock class at Storrs 1918 contest, with a record of 2,087 eggs. 'It was another winning with Parks' Rocks,' writes Richard Allen of Pittsfield, Mass. (Third pen also was Parks' strain.)"

Examples of Pen and Flock Production

As this book you hold in your hands, Reader, bears title, "High Egg Production by Individual Hens, Pens and Flocks," we no doubt should draw on Mr. Parks' files for sample reports of pen and flock production, in our quest for evidence of real progress and success in this field of human endeavor and achievement. Many persons are of the belief that there is no such thing as a two-hundred-egg strain. Possibly they are right about it, but nevertheless it now is a well-known fact that repeatedly small flocks of ten, twelve or fifteen bred-to-lay standard fowls have easily exceeded an average of two hundred eggs per bird in three hundred and sixty-five consecutive days. Let us bear in mind that this is often done in egg-laying contests under official supervision, where a flock or pen consists usually of five to twenty birds.

In May of 1920 writer visited Mr. Parks of Wopsey Poultry Farm, our main object being to secure from him, "on home ground," chiefly for publication in this book, the full story of his twenty or more years of definite work in breeding up the Parks' strain of Barred Plymouth Rock layers. On this errand Mr. Parks showed us every courtesy: allowed us to inspect his records, furnished us copies of record cards and breeding certificates, and also permitted us to select numerous letters, in the original, out of his files, from which letters we publish here the following sample reports and egg records, showing results obtained by Mr. Park's customers, near and far, as they apply to the subject matter of this book. These reports clearly prove that a high level of inherited productiveness has been reached.

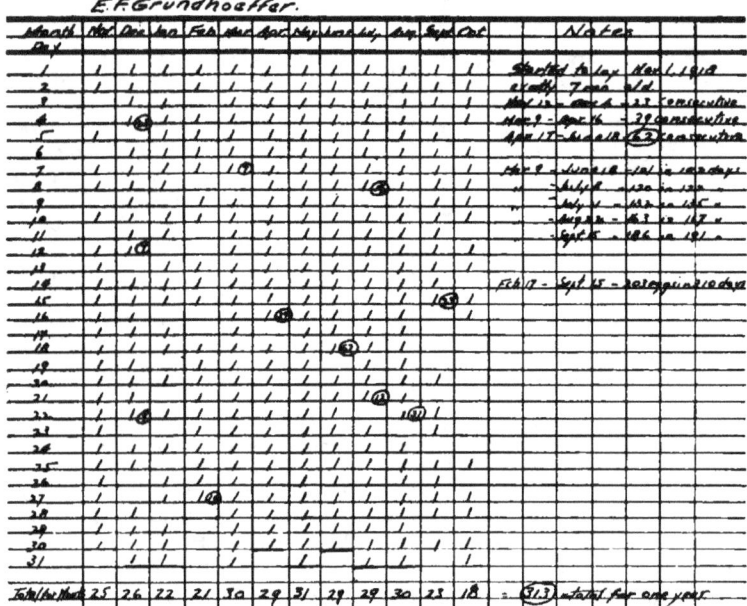

DAILY EGG RECORD FOR ONE YEAR OF A PARKS' BRED-TO-LAY BARRED ROCK THAT PRODUCED 313 EGGS IN ONE YEAR

Above is a facsimile reproduction of "Laying Record of Hen No. 9," a J. W. Parks' Barred Rock, as furnished R. P. J. by E. F. Grundhoeffer, of the Engineering Experiment Station, Pennsylvania State College, as referred to in Mr. Grundhoeffer's letter of November 28, 1919. This is indeed a fine record. Note particularly the long cycles of this hen, as shown at the right under the heading "Notes." These cycles were 23 eggs in 23 days; 39 eggs in 39 days and 62 eggs in an equal number of days. Also note the record just below the foregoing, where this hen laid 101 eggs in 102 days; 120 eggs in 122 days; 132 eggs in 135 days; 163 eggs in 167 days; 186 eggs in 191 days; and 203 eggs in 210 days. That was truly remarkable production, reflecting great credit for J. W. Parks, whose strain produced this "egg machine." Credit is also due Mr. Grundhoeffer for care and feeding.

A HIGH-PRODUCING STRAIN OF BARRED ROCKS AND HOW IT WAS BRED

Writing under date September 3, 1914, James L. Whitbeck, Jermyn, Pa., said:

"Please send me one of your cockerel circulars. I bought fifty hatching eggs from you in March, 1912, and the fourteen pullets I have raised from these eggs averaged 204 eggs in their pullet year. I would like to have a March-hatched cockerel, one that would be a good pullet breeder."

THE PARKS' HEN "VIOLA"
Hen with trap-nest record of 236 eggs from which Mr. Parks, fifteen years ago, started one of the high egg production lines that have made his strain world famous. At present the direct descendants of this bird are decidedly better barred than she was. Note good length of body.

Writing under date January 14, 1915, H. E. Parsons, manager of Parson's Chicken Ranch, El Paso, Texas, said:

"Have been intending to write you since the year's lay has been finished. This past year we have really accomplished something worth while in the way of high records.

"We have one hen, No. 604, whose record was 275 eggs in 354 days; another whose record was 257 eggs, and another whose record was 252.

"We have a pen of hens, thirteen in number, that were bred from a 202-egg Parks' hen, No. 388, that have averaged 208 eggs each. These hens are also from a high-record sire.

"Have six others, full sisters, from a high-record hen, No. 814, whose record was 216 eggs, and these hens have laid an average of 202 eggs each.

"The average record of all hens in our breeding pens this season is 198 eggs per hen.

"We have 126 hens whose average is 208 eggs each.

"We have one pen of ten hens (breeders) whose average is 241 eggs each.

"Have never bred from birds outside of your strain. Had intended getting another male bird from you this year, but managed amongst my own."

John A. Downing, Hyattsville, Md., writing Mr. Parks under date August 21, 1917, said:

"Please ship me five more pedigree pullets, for which enclosed find money order. All eight of my 1916 pullets had laid over 200 eggs, up to date, August 1, which doesn't complete the year, and one has a record to date of 287 eggs. She laid her first egg October 5, 1916. They were all hatched from eggs I got from you on May 5, 1916."

Writing Mr. Parks under date January 21, 1919, J. A. McNary, Fairbault Minn., county agricultural agent, said:

"Will you kindly advise me by return mail what you will charge me for a first-class Barred Rock cockerel? I want a bird that will score 250 eggs according to the Hogan method. I want one that is well barred for a pedigreed bird. I have eleven hens of your breeding that last year averaged 233 eggs and not one of the eleven has offered to set. What I want is a bird good enough to mate with this flock."

Following is a letter and egg record from Byron Miller, of New Kensington, Pa., bearing date January 7, 1920, which Mr. Parks kindly furnished in the original:

A Profit of $9.08 Per Hen

New Kensington, Pa., Jan. 7, 1920.
"Mr. J W. Parks, Altoona, Pa.

"Dear Sir:

"Am sending herewith copy of my egg record for the year 1919. Seven of these hens I bought of you in 1917. Five I raised in 1918 from a male I bought from you as a day-old chick in 1917.

"All these hens are kept in a small up-to-date poultry house 7 by 8 ft. in size, which is always kept in A No. 1. first-class condition. I have a fine bunch of hens and I do not think my record can be beat in this town. I lost two hens during the year. One broke an egg in the egg bag and I killed her. Another, the best layer I had, I lost in the molting season. She dropped most of her feathers in one day and could not stand the strain. Shall have to replenish my stock this year, as I do not have a male for use in getting hatching eggs.

"Yours truly,

"BYRON MILLER"

Mr. Miller's Egg Record for 1919

Month	Hens	Eggs	Month	Hens	Eggs
Jan.	12 hens	88 eggs	July	11 hens	200 eggs
Feb.	12 hens	182 eggs	Aug.	11 hens	190 eggs
March	12 hens	258 eggs	Sept.	11 hens	229 eggs
April	12 hens	251 eggs	Oct.	11 hens	218 eggs
May	12 hens	274 eggs	Nov.	10 hens	136 eggs
June	11 hens	235 eggs	Dec.	10 hens	102 eggs

Total	2,363 eggs
Average per hen	210 eggs
Value of eggs—market price	$186.96
Total expended for feed, straw, etc.	44.75
Profit on the flock	92.21
Average profit per hen above actual expense	9.08

Breeding Records Go Back Twelve to Fifteen Years

It would take several chapters in this book to furnish detailed information about Mr. Parks' poultry career, his houses and equipment, his hatching and brooding methods, how he feeds, yards, cultivates, coops, packs, ships, etc. However, those interesting items are not the subject matter for a book on high egg production. Mr. Parks also has breeding records that go back twelve to fifteen years, which disclose much valuable information—especially to him in his work of mating the birds each season. Following are the items taken care of by his form of breeding chart or record:

Month hatched; pen number and record; band number; hatched from hen number; dam's record; color of eggs (light, dark or medium); size of eggs; total egg yield by year; chicks hatched (light colored or medium); toe marks; remarks.

The "remarks" on the cards have to do with recording the fertility, hatchability, etc., of the eggs as laid by each female. Said Mr. Parks, in substance:

"It has not been my practice to work for the phenomenal hen, as to high egg production, but for a high average—to get all the female progeny, or as many of them as possible above the 200-egg mark. Have been successful to a gratifying extent. Numerous pens and quite a number of moderate-sized flocks in my hands and in the hands of customers have averaged from 200 eggs up to 240 or better. My catalogue gives these reports and I can show you the original correspondence in every case.

"In this work, to be really successful as I view the matter, we must work for a combination, across the years, that embraces several important factors, to-wit: vigor, size, up in weight, type, eggs of good size and the right color, also attractive plumage. Yes, we go by the trap nest and blood lines—by our pedigree records as tested by the trap nest, and we select the best type and color we can, but I give first importance to vigor and to the number and character of eggs, based on pedigree records of actual trap-nest production."

Started with High Egg-Record Females

As is frequently the case in this field of live-stock breeding, Mr. Parks started his present lines of prolific egg yield Barred Rocks from high egg production females. What he calls his standard line, having in mind standard

THE THREE JOES
J. R., J. W., and J. C. Parks, past, present and future breeders and originators of Parks' strain of winter-laying Barred Plymouth Rocks.

color, was started fourteen years ago from a hen named Viola that laid 236 eggs in trap nests within one year. A picture of this hen is shown herewith. Mr. Parks' dark line (cockerel breeding) was started from a hen named Muz, "an excellent dark barred bird that twelve years ago laid 254 eggs in the trap nest." A distinctly pullet line was started ten years ago from hen No. 1014, which laid 219 eggs in trap nests. This bird belongs to the Viola line, but had exceptionally bright and regular plumage. In developing these three lines or families of the Parks' Strain of Bred-to-Lay Barred Plymouth Rocks, Mr. Parks has kept accurate pedigree records of both the males and females. He claims that this practice is indispensable if genuine progress is to be made and if the good results secured are to be maintained year after year.

"Yes," said Mr. Parks, "I DO VALUE body type and plumage color. The present Standard description for both suits me all right. Of course I do not breed birds with the hope that they will win at New York, Boston, Chicago, etc., but several years ago I found among my customers a strong preference for correct standard type and standard color; therefore since then I have paid more attention to these values, with the result that my customers have been able to win at numerous shows in moderate competition. Have improved them a good deal in this respect and I find that my customers like it."

Furnishes Breeding Certificates to Customers

Mr. Parks not only keeps pedigree records of both the male and female sides of the house, but also makes a practice of furnishing what he calls a breeding certificate to purchasers of his line-bred stock. Find herewith a reproduction of one of these certificates—an actual one that not long ago he furnished to Miss See Rice, a poultry club member at Sardis, Miss.

While at Mr. Parks' office on this visit (May, 1920) we looked over several dozen letters which contained reports similar to those here published and can therefore state, for the benefit of readers of this book, that it is a common thing for his customers to get 200 to 260 eggs from individual hens in 365 consecutive days, also to have his strain of Barred Plymouth Rocks average to lay in excess of 200 eggs per hen in pens or small flocks ranging from half a dozen to a dozen and a half birds, and often they do this in limited quarters—frequently on city lots, but we need to remember that stock which does this is raised an range, as a rule, which is the case with all adult birds sold by Mr. Parks. Following is one more sample case—published here because it gives the experience of a back-lotter, also because Mr. Parks made some valuable remarks with reference to it, of a strictly practical nature. T. C. Nicholson, assistant cashier of a bank at Fort Pierre, Florida, wrote Mr. Parks under date March 4, 1920:

"It gives me pleasure to inform you that in the months of January and February, 1920, the fifteen pullets I raised from the twenty-five day-old chicks you sold me in April, 1919 (you sent twenty-eight), laid six hundred and fourteen eggs. They laid two hundred and ninety-one in January and three hundred and twenty-three in February.

"We think this is doing fine, when you take into consideration the fact that they were not culled, selected or fondled by the use of 'lights' or 'dope' of any kind. They just LAY that is all. One day I got fifteen eggs and several times I got fourteen in one day.

"I think this is a good record for a back-yard flock. And one result is that I find that lots of people want eggs for hatching. Here is one practical way to start in the poultry business on a paying basis. The demand is right at hand, when you have something that many other people want."

Referring to this and other similar letters, Mr. Parks said, in substance:

"The surest easy way to pick young good layers is this: Keep for breeding purposes the birds first off the roost in the morning and the last ones to go back on the roost at night. Extra-good layers have to keep at work long hours in order to get food enough for body maintenance and to produce the eggs. Miss Graduate and Miss Smarty are shining examples. They are up with the sun and busy till dark."

Example of "Tracing Back" by Pedigrees

As an example of tracing back the breeding of pedigree stock, we take a pen headed by a strong, typy, quite well-barred male which, as Mr. Parks said, "is one of my latest prepotent descendants from hen Viola, head of one line or family." This cock was from hen No. 307, which hen was from hen No. 1009, with trap-nest record of 223 eggs made in the year 1911-1912 and she in turn was bred from hen No. 899 with a record of 211 eggs in year 1908-1909. Hen No. 1009 was mated with male from dam No. 1059, with record of 216 eggs in 1910. Hen No. 899 was a direct descendant of Viola, whose egg record was 236 eggs, production year of 1907-1908. Viola herself was from a sire whose dam (hen No. 319) had a record of 214 eggs. Mr. Parks does not know the dam of Viola.

In handling this direct line from Viola to the cock bird at the head of this 1919-1920 breeding pen, Mr. Parks, each year or generation, used his best judgment in making selections, as to individuals—that is, each generation he had quite a number of birds to choose from, both males and females; therefore he selected, from each lot of brothers, the most valuable specimens and was governed by trap-nest performance in deciding which female (sisters) to use as breeders in continuing the line and improving the family. But in selecting the females he does not allow himself to be governed solely by the high egg yield, as shown by the trap nests. Far from it, in fact. For example, if one sister of a generation laid as many as 250 or more eggs and was not a good bird otherwise, as to size, body capacity, type, etc., he would not hesitate to pass her by for a 240-egg-record specimen (one of her sisters) which suited him much better in other important respects. Said Mr. Parks:

"It surely is an interesting work! Beats, hands down, any other game of solitaire a man ever tackled. The end of the year for each mating—the one male and the one female, for that is what it is—and the actual trap-nest showing; that is what tells you whether you've won or not, in the way of making further progress. But a good feature of it is that you have a lot of leeway; you're not limited to any one male or female in making the yearly matings, nor even to one line or family, but have a full deck of fifty-two cards to play with—yes, can have several or a dozen decks, as many as you have separate lines or families in your strain."

A bird that lays 323 eggs in 365 consecutive days cannot take many days "off." This egg record of "Lady Martha," a J. W. Parks' strain bird, that in the hands of a customer made this excellent record, is reproduced from a photograph of her score as made day by day. Note how far apart her rest days were after she made up her mind to enter the world's very select class of 300-eggers. Also note that, as examples of intensity in egg production, Lady Martha missed eight days from March 3 to August 5, inclusive, a period of 156 days, and that from February 18, to May 1, inclusive, a period of 74 days, she missed three days only. Three months in succession, March, April and May, she laid 29 eggs per month and during six months in succession, March to August inclusive, she laid 29 eggs each month with the exception of June, in which month (thirty days) she laid 28 eggs.

On Both Sides of the House

Question: Have you concluded, Mr. Parks, on the basis of your experience and records, that for further substantial progress there should be high egg production blood on both sides of the house?

Mr. Parks: "Yes, and also that our best results come from high normal layers, as I call them. My experience to date has been that you cannot take a phenomenal hen and by breeding her keep on piling up the egg yield. Our problem is not as simple as that. Replying directly to your question, will say that on the basis of my experience, confirmed by records of actual performance and the continued breeding results, I rather favor the male as the side of the house that does most to transmit laying power. Am not what you might call 'strong' on this point, but I do want to emphasize the importance of the male in this capacity. To develop a laying strain for what is commonly known as high egg production, I feel that we should start at the bottom and work up, so to state it. What I mean is that we cannot expect to buy a record producer or several such birds and thus start at the top and count on going still higher. We would better begin with what I call high normal layers and build up from there. For example, these 313, 323 and 325-egg-record layers of the Barred Rock variety are my Madison Square Garden Winners, just as the blue-ribbon birds are for breeders who specialize in exhibition qualities. But as is the case with exhibition Barred Rocks that cannot be mated for best results on the basis of prizes won or on their score cards, so it is in breeding for 300-eggers; you'll have to go back to the original source of the good layers and the so-called record layers to get the right blood, then you'll have a fair chance to produce top-notchers, in case you go about it properly and persevere. To do this you should start, as a rule, with birds of normal high production, according to my experience and best advice."

Said Mr. Parks, referring to the hen Viola, a record layer for her day that possessed considerable merit as an exhibition specimen:

"That season of 1906 or 1907 is as far back as our records go for Viola, which is equivalent to fifteen years. I started a new family with her, on account of her combined good qualities—her remarkable trap-nest record for that early day, comparatively speaking, totaling 236 eggs in her pullet-hen year, together with her good standard type and regular and distinct barring. I know too, that her father had egg blood in him. Also Viola won my heart with her good disposition. She was almost human! You could talk to her and she seemed almost to understand. She responded to her name, also to kind treatment. She had great vigor and laid eggs that were dark brown in color. For that period I did not list the size or weight of eggs, but I recall that Viola's eggs were of good size, though not extra large."

Miss Graduate and Miss Smarty

At Wopsey Farm, on this visit, we saw Miss Graduate and Miss Smarty, also the male bird with them. This male was a direct descendant from Viola, whose record, year of 1907-1908, was 236 eggs. Dam of this cockerel, in the year 1918-1919, had a trap-nest record of 220 eggs. Said Mr. Parks:

EXAMPLES OF HIGH EGG PRODUCTION PARKS' STRAIN, BRED-TO-LAY BARRED PLYMOUTH ROCK FEMALES

Laora, the bird on the left, had a trap-nest record of 277 eggs, while the bird at the right, known as "Busy," attained a record of 290 eggs in 365 consecutive days, laid in trap nests. This 290-egg bird also was raised by one of Mr. Parks' customers. Mr. Parks is extra proud of the fine egg records obtained by numerous customers, not only throughout the United States, but in a dozen or more foreign countries.

"No, Miss Smarty was not from the same pen or mating as Miss Graduate. Miss Graduate was from a standard mating and traces back to Viola. Mr. Grundhoeffer bought her as a baby chick and raised her himself by his own methods, therefore there is no mystery about it. I recall that he sent you his original letter, telling what other egg records he got from that same small shipment of baby chicks. As regards Miss Smarty, my records do not show that I sold direct to Mr. Mortensen, but I did sell hatching eggs from one of my record pens to Louis Hanson of Gilbert, Arizona, from which Mr. Mortensen bought Miss Buster, as he called her, but which I have renamed Miss Smarty, because her yield for that same season or year surpassed that of Miss Graduate by exactly twelve eggs, thus making a new world record for this breed and the Parks' strain."

At this point Mr. Parks invited the writer to go out to the yards and join with him in examining these two hens: Miss Graduate, egg yield in one year 313 and Miss Smarty, egg yield 325 eggs in 346 consecutive days. They were found to differ widely as to depth and width, although both had good length of body. Both hens were laying and had been for some weeks, yet Miss Smarty was far deeper and broader by finger measurements. Miss Graduate, in fact, was quite different, comparatively, yet she laid 313 eggs in trap nest in 365 consecutive days. Said Mr. Parks, as he let her go from his hands:

"But she probably would have been thrown out if Hoganized before she made that record. Our problem is no such 'snap' as that. We have to take a lot more into consideration, as I've set forth to you, and when we get through with all these things we can see, then there is that mystery they call 'prepotency,' which I claim exists on both sides of the house, but far more in some individuals—actual brothers and sisters—than in others. To discover these prepotent birds, and work to perpetuate and increase their number, we must use the trap nest—either that or some low-cost and reliable plan of single mating or stud mating.

"Yes, trap nesting and conscientious pedigreeing is costly, both in time and labor, but we must be systematic or scientific in our methods, if we are to 'find out' and make headway toward our goal. Also we need more trap-nesters, more scientific breeders, so greater general progress can be made, and especially with the heavier-weight or dual-purpose breeds—those that can give us both eggs and meat. My experience shows what can be done and I'd welcome more help."

Above chart is the form used by J. W. Parks in supplying pedigree information to customers who buy his Bred-to-Lay Strain of Barred Plymouth Rocks. Reader will note that it gives the band number of the bird purchased; the egg record of the dam that produced her; the size and color of egg; the band number of the father and egg record of his dam, also the color of her egg.

Mr. Parks' Present Plan of Trap-Nest Breeding

Mr. Parks began the use of trap nests in 1904, starting in a moderate way and increasing this part of his poultry plant and operations year by year. Before long he was trap nesting all his breeders and prospective layers, doing this in their first year of production, but later on as the demand increased and he found himself trap nesting several hundred birds each year, trying faithfully to keep pedigree records of all matings, he decided that this was impractical: first, on account of the labor and expense; second, because of difficulties in avoiding confusion of records, among the birds, in the egg racks, etc.

His present plan and basis of operations in this matter of pedigreeing is to trap nest 150 line-bred females each year, for the entire year, thus to get their annual records of individual production. In the early fall, before the new crop of pullets begins to lay—birds of known pedigree, both as to sires and dams—he selects 250 to 300 of the most promising ones. Said Mr. Parks:

"In picking out these birds I select only those that I know the records of, beyond question. During the last six years, however, I have trap nested only every other year for my record birds. This gives more time for a man in my position to study and make sure of using the best birds—and it takes personal study, as well as trap-nest records to determine the proper selections for your breeding pens. What I mean is, we must take into account the physical condition of each bird—its appearance, size, vigor, etc. Each fall or early winter before these birds are placed in pens that are equipped with trap nests, we cull them down to 150—no, to about 165, as I keep each year about fifteen extras as 'reserves,' so I can be sure of having the full number of 150 for the trap-nest pens and to carry through on our records for pedigreeing, etc.

"In keeping our records of the eggs produced, we do not list the number only, but also make a record of the weight and color of the egg. It is not necessary to do this the year 'round, but I do it in the early spring of each year. Half a dozen eggs so recorded from each hen is all that is needed. Have found that if a hen is going to lay an off-colored egg, she will do it in the spring. We prefer a rich, fairly dark-brown egg and for years I have selected our breeders with this object in view—and with a degree of success that brings me approval from many customers.

"Next, before we start to trap these 150 selected pullets, every bird is caught and examined carefully according to physical measurements, as to size of the specimen, thickness of pelvic bones, etc. Will say in this connection, though, that my experience and observation have been that we cannot be too positive or arbitrary on some of these points, but positively we must protect the size or standard weights of the Plymouth Rock breed, otherwise we will sacrifice a good deal, as I look at it. Up to the present we do not know all there is to learn about these physical proportions, therefore it is not wise to be too dogmatic in our methods or opinions. As examples of what I mean, take Miss Graduate and Miss Smarty. They are far from being alike in size, also in physical proportions; and if a hard-and-fast rule were adopted one or the other would be discarded—and I guess it would have been Miss Graduate who laid ONLY 313 eggs in 365 days. Let me say further, though, that Miss Smarty, the larger of the two specimens, with a decidedly greater depth and width of body, lays an egg fully twenty-five per cent larger than those produced by Miss Graduate, evidently because the latter does not have the body capacity.

"Will say also, about pelvic bones, width of the pelvic arch, etc., that I paid little or no attention to these special points previous to the season of 1919-1920, and I am doing it now mainly for my own information. But I do wish to be emphatic on the question of size or standard weight. In breeding Barred Rocks for high egg yield, that perhaps is our worst drawback—to prevent getting what Theo. Wittman once called 'Leghornized' Barred Rocks—birds small in size, too narrow in body and with large combs. If allowed to do so by us breeders, our heavy-laying Rocks would soon become undersize, in which case we could keep them in large flocks like Leghorns, but to do this we would have to sacrifice the popular meat values of the dual-purpose fowl, something I am not in favor of and that I believe would be a bad mistake. Furthermore, the heavier birds can and do lay. As examples, Miss Graduate is somewhat undersize, weighing as a hen a little short of six pounds, while Miss Smarty, who excelled her best year in egg production by just a dozen eggs, weighs in laying condition 6¾ pounds and could easily be fed up to a heavier weight.

Be Considerate of Your Hens

"Returning to an explanation of our present method of trap nesting, we leg-band the hens in the usual manner, for keeping all records, also the chicks later on. An important point here is the matter of releasing the hens from the trap nests. Our rule is to do this four times each day during the winter and five times daily in the period of high egg production. The heavy-laying hen must not be kept penned up in the nest very long. She is likely to become nervous, with danger of breaking her egg. Also, the day at best is not long enough for her. She is a heavy eater and will be anxious to get back to the feed. Fact is, she should not be kept penned any longer than is found necessary by the caretaker. In the spring and summer we release these trapped layers at least three times before noon. To keep a hen penned in the trap too long, especially in the summertime, will result in her getting too warm,

"MISS GRADUATE"—RECORD 313 EGGS   "LADY MARTHA"—RECORD 323 EGGS   "MISS SMARTY"—RECORD 325 EGGS

When a bird makes a specially good record in the hands of her owner, many persons seem inclined to believe that he has some "secret" method of feeding by which he secures the large yield. Here, however, are three birds of Mr. Parks' strain which in the hands of Mr. Parks' customers made the remarkable records of 313, 323 and 325 eggs respectively, in 365 consecutive days—one of them, Miss Smarty, holding the world's record for the breed, having laid her big yield of 325 eggs in 346 consecutive days. In the egg record of Lady Martha (reproduced herewith on page 94), the latest claimant for 300-egg honors, note particularly the long cycles of laying after she settled down to work.

also excited, in which case when released she'll leave the nest with a 'zip' which can do her no good, as a layer and breeder. Our daily practice is to humor our production hens all we can. These trap-birds are doing their best; why shouldn't we help them! Our rule also is to keep them as tame and quiet as we can. No, we do not use specially darkened nests, as I believe that fairly well-lighted nests do not breed vermin."

Question: Taking these 150 trap-nested pullets as selected from the line-bred birds each year, how well do they average in egg production?

Mr. Parks: "Last year (1918-1919) 125 birds that went through the twelve months averaged 177 86/125. Of the remainder, four or five died and the others were marketed during the year, as they were not laying well enough. Of the 125 kept under traps the full year 43 laid upwards of 200 eggs. But last year with us was not a fair test—perhaps not with anybody—as the war seriously affected our records, because we could not always get good reliable meat scrap. This was true also for the year 1917-1918, yet our high individual record last year reached 273 eggs. You will note by the record book here that other high ones were, as examples: 231, 230, 240, 229, 222, 220, 251, 243, 230, 224, 218, 234, 217, 218, 227, 224, 229, etc. Our highest individual record, made here on the place, was in the year 1914-1915 when one hen reached 277 eggs. That year we had 41 among the selected birds that passed the 200 mark.

"Yes, we formerely trapped every year, but as my business grew and labor became hard to get, especially during the war period, we changed to the every-other-year basis. Of course by that plan we can maintain our pedigree, using birds two years old and older. We miss numerous high-egg-record specimens, or a definite knowledge of them, as would be shown by the trap nests, but we know their breeding in every case and have been at it long enough to realize that we can place confidence in them. It simply became impractical to keep track in accurate form of so many egg-record birds. From 1904 to 1912 we trapped every year and there was a steady upward trend in the average annual yield till I coined that phrase "high normal egg production," which is what I am now striving for, not alone in the case of individual birds, but the entire strain, both here on the home place and in the hands of customers.

"My present aim is to build up the flock as a whole and at the same time maintain the vigor of the birds, as they would be in their natural state—in the fields and out in the woods, for example. All this increased production no doubt draws on vitality, therefore we must use every precaution to prevent our breeders and layers from becoming debilitated and run down in size.

Establishes Blood Lines or Families

"Yes, during the nearly twenty years that I have been closely studying this problem, we have followed or aimed to follow along with the same line or family that, year after year, gives the highest average egg yield and in doing this I sometimes pass by or ignore the phenomenal layers, as I call them. To illustrate in a broad way, I would rather go into three egg-laying contests and be third in each of them, than to be first in one and eighth or ninth in the others. The phenomenal layers must have other merits, if I am to keep and use them— if I am to pin my faith to them. Some breeders perhaps—those who seek high individual egg production—would take those two hens, Miss Graduate with her record of 313 eggs and Miss Smarty with her record of 325 eggs, and would head all their breeding pens with males obtained solely from their eggs. Under no circumstances would I do that. I shall use them, of course, but I could not consent to use their 'get' alone, either for myself or ask my customers to do so.

"Frankly, eggs are my first consideration, then I go as far as I consider it safe in the effort to obtain standard type and barring. On the other hand, I do not believe in 'knocking' the Standard, nor standard-bred fowls. Very far from it, because the Standard has done more for the poultry industry than any other one thing, according to my belief.

"Yes indeed, we do practice line breeding, and I believe in it, because our records and my success in the business confirm this view, but I am not convinced that, as some seem to believe, there is no prepotency in the female, for prolific egg yield or otherwise. If there is no prepotency in the female, how could she hand it down to the male bird? My belief is that she cannot transmit something she doesn't possess. I do not and would not depend on the male alone.

New Plan: Pedigreed Pen-Eggs and Pen-Chicks

"No, we do not sell pedigree chicks. If we were to offer to do so I know we could not produce enough of them, and I guess we could get most any price, judging by the correspondence. However, next season, unless I change my mind, we are going to start to sell pedigree pen-eggs and pen-chicks, as I call them. To illustrate: each female in the pen will be with pedigree, also the male, and we shall sell from this pen, but we will not be able to tell the buyer which hen is the mother of each particular egg or chick. As heretofore, shall trap-nest only for my own purposes, using about 150 birds as already described, doing this for males, also to replace pullets and maintain our pedigree records, but then we shall throw the trap nests open.

"Why have I reached this decision? First, it will save trouble and expense in incubation; second, all the average buyer wants to know is that he is getting eggs from high-record layers, headed by high-egg-record males. These customers, as a rule, do not follow up the pedigreeing—they do not need to. What they are most interested in is to buy what they need to enable them to improve the egg yield of their flocks. Increased production of the flock is really what counts in most every case, not an occasional phenomenal layer, which these customers would miss, as most of them do not trap.

"With this new method we shall use also a 'pedigree certificate' and will give the egg record of each female and the production record back of the male. As heretofore, we can and will supply individual hen eggs, but we shall do this only while we are trap nesting and for these eggs we shall charge higher prices, to cover the extra labor."

Is Convinced of Value of Line breeding

Question: You have definitely concluded, as we understand you, Mr. Parks, that line breeding for egg production does count, and quite favorably?

Mr. Parks: "There is no longer any question in my mind about the value of line breeding to increase egg production. If there is not true merit or what our educated friends call 'intrinsic value' in the plan, I surely would not get this mass of testimonials—so many that I cannot keep track of them. It has been so for years. I get enough letters of this kind each season to fill a large catalogue. And, after all, what really counts is what a man's strain does in these little flocks in the hands of ordinary folks—in back yards, on farms, etc. My strain has done very well at official contests, for which I am pleased, but I am still better pleased by what they do—what they have done in many hundreds of cases—in the hands of my customers; what they are doing now and have been doing for years.

"One question I would like to ask: Why don't the laying contests give prizes to birds that earn the most money with the eggs actually laid? That is what we are in business for—is really what we are after, isn't it? What I mean is, maybe the man who wins third prize—as the contests are now managed—sent birds that laid eggs at a time when they brought the most money in the daily market, while the winner of first prize, judged only by the number of eggs laid in twelve months, produced a lot of birds that did their best laying at a time when even the sparrows are laying! Is it simply eggs or is it actual profit we are after? With me the answer is profit.

"Yes, by line breeding our aim is to intensify egg production, or the power and capacity to manufacture a greater average number of eggs in any given length of time by the individual hen, which of course will increase the pen or flock average. In my experience, if I have felt that good enough reasons have existed for doing it, when I came to handle the individual specimens or study their records, I've

EXAMPLES OF PARKS' BRED-TO-LAY BARRED PLYMOUTH ROCK MALES
Author of this chapter saw these males at the time of his visit to Mr. Parks' poultry farm, May, 1920. Photographs were taken later. At the time of forwarding the photographs Mr. Parks sent the accompanying descriptive matter. Photo No. 1—Cock No. 791. Dam record 231 eggs. Granddam record 217 eggs. Bred from my Viola line, record 236 eggs. Photo No. 2—Cockerel sired by No. 791. Dam No. 2311, record 217 eggs. Granddam record 223 eggs. Photo No. 3—Cock No. 676. Dam No. 515, record 215 eggs. Granddam "Mus," record 254 eggs. Photo No. 4—Cockerel sired by cock No. 676, dam No. 499, record 205 eggs. Four of his dam's sisters trap nested at same time she was, made records of 195, 251, 202, 188. All bred from hen 1009, with record of 223 eggs.

line-bred as close as sire on his own pullets, or a cockerel back with his dam, but perhaps through fear of trouble or financial loss I've kept away, as a rule, as far as I could, in this matter of close relationship and still get results. Do not recall that I ever worked as close as brothers and sisters, not on any excuse. I watch size, vigor, etc., every year with extra care, in all matings, then egg yield is the proof, both in the traps and otherwise.

"In conclusion let me repeat, that in order for us poultry breeders to make real progress, as I view the conditions and situation, we need to get away from the one-hen idea. Let our breeders strive to see what average high percentages they can get on their own plants from pens or flocks, in better than 200-eggers and also what higher percentages in 200-eggers their customers can get, year after year. Yes, that is a pretty big order, I'll admit, but you of course get my idea, and so will other persons who care about these things and therefore make a study of them. I have now devoted about thirty years to it and intend to keep at it all the rest of my active life."

CHAPTER IV

Noteworthy Achievement in Production of 300-Eggers

M. E. Atkinson in Ten Years Earns Half Ownership in 10,000-Layer Poultry Farm, Doing an Annual Business of More Than $100,000—How the High-Egg-Yield Hollywood Strain Was Originated and Built Up—Began with Stock of Six or More Strains—From These the Hollywood Strain Was Developed—Rapid Progress in Breeding for High Production Makes Possible the Obtaining of Twelve 300-Eggers or Better in One Season from a Single Small Breeding Pen

By GRANT M. CURTIS

(Introductory Note: In the breeding of fowls on a large scale for increased egg production, no such results ever before have been attained in the history of the poultry industry, either in this or any other country, as have crowned the efforts of M. E. Atkinson of Hollywood Poultry Farm, Hollywood, Washington. In the following pages is given in condensed form the story of Mr. Atkinson's noteworthy achievement, showing how a man who less than a decade ago knew practically nothing about poultry or poultry keeping has in a few short years accomplished truly sensational results in this field. For the COMPLETE STORY of Mr. Atkinson's success the reader is referred to the new book recently issued by Reliable Poultry Journal Publishing Company, entitled, "The Production of 300-Eggers and Better by Line Breeding." This latest addition to R. P. J.'s Library of Poultry Books is by M. E. Atkinson and Grant M. Curtis, Editor of Reliable Poultry Journal. It contains a full and authorized explanation of the breeding principles employed with world-record success on Hollywood Poultry Farm, also a detailed description of the methods used in the housing, yarding and management of adult fowls, in hatching, pedigreeing, brooding and feeding of chicks, etc., etc. Several thousand dollars have been expended in getting together the extensive data and the more than 150 splendid illustrations used. Book contains 350 pages, 6 by 9 inches in size and is published in two standard-size library editions. The cloth-bound edition is printed on enamel book paper, price per copy, postpaid, $3.00. The paper-bound edition, containing exactly the same reading matter and illustrations as the cloth-bound edition, is printed on supercalendered book paper and has a beautiful three-color art cover by Franklane L. Sewell. Price $2.00, postpaid.)

THE record made by M. E. Atkinson in breeding for high egg production may be summarized briefly as follows: He began his career as a poultry keeper in 1911 when, to quote his own words, he "did not know a purebred chicken from a state road." Began trap nesting in the fall of 1914 with 300 pullets, thirty of which made records of 200 to 274 eggs during the following 12 months —an average of ten per cent of 200-eggers. The next year 600 pullets were trapped, 24 5/6 per cent of which produced 200 eggs and better, the highest record being 283. In 1916-17, 900 pullets were trapped and 46 1/3 per cent produced 200 eggs and better. That year two 300-eggers appeared, one laying 304 and the other 307. In 1917-18, 900 pullets were trapped, 52 5/9 per cent of which produced 200 eggs and better. The high record that year was 318, with five birds going over the 300 mark. In 1918-19, 2,700 pullets were trapped, with an average of 57 per cent laying 200 eggs and better. That year twelve birds laid 300 to 324 eggs. In 1919-20 the same number of pullets were trapped again with 60.8 per cent producing 200 eggs and better— and there were twenty-four 300-eggers. In 1920-21 the 2,700 trap-nested pullets produced 60.8 per cent of 200-eggers and better. That year the 300-eggers numbered thirty-two. The general result of the breeding methods here practiced has been to advance the flock average from 154 eggs per bird in the second year, 1914-15, to 199½ each in the laying year, 1920-21, thus bringing entire flocks of hundreds of pullets up to a level of production that, a few years ago, was commonly considered a truly remarkable performance for a single bird or a carefully selected small contest pen.

The foregoing remarkable record is of all the more interest and value to other poultrymen because the results accomplished have been secured under conditions which readily can be duplicated by practical poultry keepers and commercial egg farmers everywhere. As the following description of plant and methods shows, there were no especially favorable conditions, aside from systematic breeding. Contrary to common belief, the climate of the Pacific Northwest is by no means ideal for poultry (see Chapter II, Part III). Methods of feeding and general care afford no outstanding advantages and Hollywood's 200-eggers and all 300-eggers must develop their records in regular laying flocks, numbering from 200 to 400 birds each, kept under ordinary commercial conditions.

How Hollywood Poultry Farm Started

Mr. Atkinson's experience in poultry keeping began in childhood when he took care of his mother's mixed poultry flock. In 1911, after having been in mercantile business for some years, he became interested in commercial poultry keeping and leased five acres of land near Seattle, with the intention of developing there an extensive poultry plant, in hopes of making a living therefrom. He started with twelve Leghorn hens purchased from a woman in the suburbs of Seattle, who informed him that the birds were of the Wyckoff strain. He paid $3 apiece for these and also purchased a trio of Kellerstrass White Orpingtons for $150. Later he bought six more Orpingtons from a local breeder.

Through reading Mr. Kellerstrass' poultry book he became interested in trap nesting and decided to trap both his Orpingtons and Leghorns. This was in the season of 1912. That spring he bought two 70-egg incubators and an oil-heated colony hover, under which he brooded 1,000 day-old Leghorn chicks. From these he reared 350 pullets to maturity. There was no special breeding or egg records back of these birds, so far as he knew. The next spring he increased his incubator capacity and continued to breed Orpingtons, raising 600 chicks of this breed with hens. That spring he bought 1,000 Leghorn chicks from D. Tancred, paying $175 for them, these being from birds having trap-nest records ranging from 170 to 180 eggs—good production for those days.

In the fall of 1913 Mr. Atkinson entered into a partnership with F. S. Stimson, owner of Hollywood Dairy Farm, and moved his equipment and stock at once to its present site on a well-drained plot of cleared land. His stock at that time consisted of 800 Leghorns and 300 Orpingtons. According to the terms of the partnership he was to have a small salary and receive one-half of the profits in the form of a credit to be applied on his prospective half-interest in the plant. Having then plenty of room and enough capital

LADY HOLLYWOOD, OFFICIAL RECORD 275 EGGS
A wonderfully prepotent hen that did much to form the Hollywood strain and create an enviable reputation for Hollywood Farm.

with which to develop, Mr. Atkinson at once laid out a 6,000-bird plant embracing seventeen acres, which later was increased by the addition of 30 acres more. There were no buildings at all on the farm when Mr. Atkinson moved there and he immediately put up a granary and a laying house 18 by 310 feet. This house was of the shed-roof type with curtain front and consisted of two 150-foot wings, with a feed room in the middle. The next year, 1914, all the Orpingtons were sold and attention thenceforth was concentrated exclusively on Leghorns, Mr. Atkinson being led to this step by the fact that practically the entire revenue of the plant was at that time derived from market eggs and the question of getting the most eggs at the lowest cost of production was of vital importance to the success of the enterprise in its initial stages.

Foundation Stock at Hollywood

It was in the fall of this same year (1914) that Mr. Atkinson, feeling that his egg production was not what it should be, began an earnest search for egg-laying stock. His first move was to buy the eight winners, including two reserves, at the Victoria (B. C.) Egg-Laying Contest, year of 1913-14. This was a New Zealand entry and in the eleven months of the contest six of these birds laid 1,330 eggs, an average of 221 and a fraction. Those birds, eight in number, were purchased at a cost of $50 each, and soon after a male of the same strain was imported from the New Zealand breeder who had made the winning entry at Victoria. In the meantime, Mr. Atkinson imported 24 females and 6 males from Australia, the best of these coming from yards of the well-known South Australian breeder, A. H. Padman, whose birds for a decade or more had taken the lion's share of prizes in Australian laying contests. The same year 20 females and 4 males were imported from England. The Australian blood gave best results and the English blood was discarded within two or three years. The Australian and New Zealand birds, systematically line bred and blended with Mr. Atkinson's best birds of American breeding, formed the basis of the three distinct strains or families which he now carries.

Regarding his methods of line breeding he states: "We are able to keep our male lines pretty pure all the way through. These are made up principally of the Padman-New Zealand blood lines, which are back of all three of our present distinct families. With extra-good males of these lines we used pure American-blood females and worked the high-egg blood into the Hollywood strain in that way—taking half American blood as the basis of our female line in building up the flock for better egg production year by year."

The Wonderful Breeding Quality of Lady Hollywood

Speaking in regard to this famous bird (see illustration on page 98), Mr. Atkinson says:

"The bird that made Hollywood Poultry Farm the most money and gave it a wide reputation was the one that laid 275 eggs of a quality marked 'marketable eggs' at Mountain Grove (Mo.) Laying Contest in 1915-16. I know her breeding only on the male side. It went back to the New Zealand family. She was the first direct outcross from our American family, blended with a New Zealand male that had a 274-egg mother, which was the record given at the time I purchased him. I do not know what was back of that record on his sire's side. By this New Zealand male, from a mother with a record of 274 eggs, mated with an American hen without a known record, we got this Mountain Grove hen with her official record of 275 eggs, also a sister that laid 262 eggs in the same contest. I then felt we had made a good start in the right direction.

"We called this 275-egg hen Lady Hollywood. Our strain, to an important extent, is built around her and her dam. Two sons of hers, as an example, sired twelve females in 1919, each of which laid from 300 to 328 eggs inclusive. At Mountain Grove, Mo., November 1, 1915, on entering the contest, Lady Hollywood weighed three and three-fourths pounds. She never afterward took on any extra weight, but all of her female offspring averaged somewhat heavier. At Mountain Grove, Lady Hollywood was first hen, the best in the contest, and her sister—the 262-egg hen—was third best.

"We brought this 275-egg hen back home from her trip to the Mississippi Valley and matched her the first year with an Australian male from a mother that was said to be from a 276-egg hen, and a sire said to have been from a 279-egg hen. There were also other good records back of him, so we were informed. From this mating we did not get many good pullets, the best of which was a 273-egger and one that laid 252 eggs. These two were out of a possible ten or twelve. I was trapping only a few and selected my stuff according to what I wanted and thought was best before I put them into the trap pen. I carefully select all my trap birds, because I want early-maturing fowls; therefore those that lag behind in growth and vigor are not trapped, regardless of the high record breeding back of them.

Twelve 300-Eggers from One Mating

"The next year I mated Lady Hollywood to a Padman line-bred, egg-production male and out of that mating I got some very good pullets, better than from the other mating, also the greatest line of males ever produced on Hollywood Farm. This second breeding year we did not mate Lady Hollywood with a son, because I wanted some of her blood blended with the two different male lines. This would give us something to spread from in case we lost her. Fortunately she kept her stamina and when four years old (the third year we bred her) we mated her in a trap pen with 16 other top-notch females. This pen was headed by two of her cockerels from the second mating. These cockerels were alternated, one with the pen one day, using the other cockerel the next day. From this pen of 17 females, including Lady Hollywood, we secured in one season the greatest lot of high producers we have obtained from one pen to date, consisting of twelve 300-egg hens and better, as follows: 300, 300, 300, 304, 305, 306, 306, 310, 314, 315, 316 and 328.

"Yes, all our trap records are made in flocks ranging from 200 to 400 birds each. We do not trap in small lots, nor give our trapped birds any special care—none whatever. Hen No. 1284 was among the 17 high producers in the pen above described. She laid exactly 300 eggs in 365 days and

FRONT VIEW OF LAYING HOUSE AT HOLLYWOOD FARM
Is 80 feet wide and divided into 50-foot pens, each accommodating 400 layers. Note shaded water troughs in front of pens. Water is piped to each trough, and fowls have access to them through slatted openings in wall.

laid eleven more after she was moved from the trap house to other quarters to make room for the new lot of pullets. From her we produced in one season four daughters that laid 300 or more eggs in 365 days."

Buildings at Hollywood

In addition to the first laying house already mentioned, four brooder houses were built in the fall of 1913 and spring of 1914, and two more 18 by 310-foot laying houses; also 22 movable colony houses 8 by 10 feet in size. For the accommodation of special breeding pens, two 16 by 20-foot houses were built, each of these being divided into two pens, and three more 16 by 30 feet, divided into three pens each. This equipment provided ample room for the 3,300 birds on hand in the fall of 1914. Since then a number of additional brooder houses and special breeding houses have been built and the laying houses also have been increased in number, while the original shed-roof houses were remodeled, making them 30 feet deep. The present laying capacity of Hollywood Farm is 10,000 layers and breeders, plus about 1,250 male birds.

Housing and Management of Layers

The standard Hollywood laying house provides for flocks of 400 layers in pens 30 by 50 feet and is of the semi-monitor type. Ventilation is secured through the open front, excepting in summer. One-sash windows are placed in the rear wall under the droppings boards for light in that space and these also are open in the summertime to increase the circulation of air. A muslin curtain is provided for the front of the house, but is seldom used. To the original houses, which were of the shed-roof style and 18 feet wide, a 12-foot lean-to was later added in front, thus to about double the capacity at small cost. In doing this it was necessary, in order to get headroom, to drop the floor for the front section so that in these houses the floors of the front and rear sections are on different levels—a difference of about two feet. The nests are installed between the two parts, the birds entering from the lower level. These houses face south and are provided with large yards, seeded to a pasture mixture of clover, oats and wheat, and deep-plowed twice each year.

Hollywood birds are fed mixed grains and dry mash in about equal proportions, as on commercial poultry plants generally. The mash hoppers are 5 feet long, 3 feet high, 9 inches wide at the top and 10 inches at the bottom. These are placed in the partitions between pens so that one hopper provides for two pens. The fowls are watered from troughs attached to the outside of the building at the south, to which the birds have access through slatted openings. See illustration on page 99. The troughs are supplied from a pipe line with a tap for each trough.

In the semimonitor-type houses, partitions are installed at the rear of the pens, these reaching from floor to roof and extending eleven feet forward from the rear. Two such partitions are provided in each 50-foot pen, thus dividing the roosting quarters into three draft-proof compartments. Droppings platforms are six feet wide and in each compartment there is an 8 by 16-inch opening in the platform, under which is a drawer sliding on guides and reached through a trap door at the rear of the house. In cleaning the platform the attendant simply removes the cover to the opening and scrapes all droppings down into the drawer. Later a man with a horse and wagon drives along the roadway at the rear of the building, empties the drawers into the wagon and hauls the droppings direct to the concrete dumping pit. Used floor litter is pushed out at the back of the building through the windows located under the platform and thence is hauled out on the fields for fertilization. Dirt floors at first were used in the laying houses but it was found impossible to keep the house dry enough in the rainy season and all houses now are provided with wood floors, using tongue-and-groove flooring.

The laying flocks are kept confined to their houses most of the cold and rainy season, though yards of good size are provided for use in pleasant weather. Mr. Atkinson believes that he gets better production where the pullets are thus confined during inclement weather, and at less expense for feed. During the rainy season, which extends over about two-thirds of the year, an additional reason for keeping the birds in is that their runs then are in a decidedly bad condition, particularly so because the soil is mostly clay. While the winter climate is comparatively

WORLD-RECORD HOLLYWOOD HEN NO. 1284, ONE OF HER SONS AND HER FOUR 300-EGG DAUGHTERS

The photographs from which this illustration was made were taken at Seattle, Wash., February 27, 1922. The mother hen, No. 1284, is five years old this spring and her son and four daughters here shown are three years old—were hatched in spring of 1919. These birds on above date displayed every sign of health, activity and vigor. The cock weighed a little over six and one-half pounds and the hens were above standard weight. Each bird was the picture of health and good type, with well-shaped comb, good-colored eyes, moderate tail carriage, etc.

mild and there are many days when the birds could be outdoors with comfort, it is found that when they are allowed to run out one day and are kept in the next they become dissatisfied and spend a good deal of their time trying to get out. If they are kept in for long periods they soon become accustomed to such conditions, are happy and forget the world outside the confines of their pen walls.

Selection of Laying and Breeding Stock

In the fall, as the pullets begin to lay or are about to do so, they are transferred from their growing quarters to the laying pens, selecting only the most promising, with special preference, of course, for those of the best ancestral breeding, provided always that they show good development. Each brooder house is visited in turn in this way, the most advanced birds being selected in every case and continuing to do this until the round of the brooder houses has been made. These birds thus selected form what Mr. Atkinson calls his "first-choice" pens. Later they go through the brooder flocks again for the "second-choice" birds, invariably leaving the conspicuously slow-developing birds to go to market. As the pullets in a given brooder house are all of the same age, this procedure means that the early-maturing birds of each brood are thus regularly included in the "first-choice" pens, while the pullets that are conspicuously slow to mature can be disposed of without wasting either feed or time in testing them out in the laying pens or trap nests.

During the trap-nesting period the pullets are under constant observation and those that develop any undesirable characters are removed and placed in what are known as cull pens, their bands being removed and trap nesting discontinued. Mr. Atkinson states that he does not want to know anything about the egg-laying ability of a bird which has undesirable characters. As soon as such birds stop laying table eggs at a profit they go to market.

Use of Artificial Light

Artificial lights are used on pullets in the laying pens, being turned on by an automatic time switch at 5 a. m. Lights in the evening are used only on dark, cloudy days, thereby to prolong the workday of the layers to the hour at which they would normally go to roost in clear weather. At Hollywood a working day of 12 to 13 hours is considered best. In lighting a 30 by 50-foot pen, six 40-watt lamps are used, these being provided with reflectors and suspended 3½ to 4 feet from the floor.

Management of Brooder Chicks

Chicks are brooded in lots of 1,000 under colony hovers, each brood having a separate, permanent brooder house. The standard brooder house is 14 by 44 feet long, 4 feet at one end being used as a feed room and stationary oat sprouter, the remainder being divided into two rooms, each 14 by 20 feet. One of these rooms, known as the "warm" room, is carefully ceiled to conserve heat and has an oil-burning brooder stove in the center. A temperature of 90 degrees is carried 6 inches from the top of the litter and 27 inches from base of stove. As the temperature nearer the stove is still higher, the chicks have an opportunity to select any degree of heat they require.

To secure ventilation without drafts, four intake air shafts 12 inches square are provided—one in each corner of this "warm" room. In providing these, a hole twelve inches square is cut in the wall about four feet from the floor, and on the outside of the building a 12-inch square shaft is placed which goes down to within about 12 inches of the ground. The fresh air thus enters the brooder room at about four feet from the floor. The outlet is in the gable of the building, through an opening in the ceiling of the room, near the center, that is covered with a slide that can be opened or closed as desired.

When the chicks are 5 to 6 days old they are given access to the second chamber, known as the "cool" room and used for exercising and hardening the chicks, especially those early hatched. This room has single walls, with a three-foot opening along one side, practically the full length of the room. This is covered with inch-mesh netting and also has a muslin curtain for use in stormy weather. When the chicks are 5 to 6 days old they are fed here during the warm part of the day and are herded in and out from one room to the other until they learn the ropes, after which all feeding is done in the cool room. This system permits of hardening the chicks, without driving them out into the rain—a particularly important detail when they must get their start during the rainy season, which often includes the entire spring in that climate. As soon

"FIRST-CHOICE" PULLETS AT HOLLYWOOD

As the pullets mature they are culled out of the growing flock and placed in laying quarters. "First-choice" birds, or those that mature first in each brood, are found almost invariably to be the best layers of the lot.

as the chicks become accustomed to the cool room they are encouraged to remain outdoors in nice weather and when they once get well acquainted with their quarters and know how to find their way readily back to the warm room, the exit into the yard is left open all the time, rain or shine. There are 15 of these brooder houses at Hollywood, each having a yard the size of which varies from 22,500 square feet to nearly double this area for the largest broods.

In the summertime the birds are out every day and it is considered especially desirable to have them out early in the morning—so much so, in fact, that Mr. Atkinson provides automatic doors which the fowls can open themselves, thus allowing them to get out at daybreak without necessitating an early trip all around on the part of the attendant. A special door is provided with a screen-door spring at the top. A lever is attached to the door to hold it shut but the first bird stepping on the lever at daybreak releases the door which flies open and remains so until closed by the attendant in the evening.

When the cockerels no longer need heat they are removed from the brooder houses, thus to give the pullets undisputed possession of the brooder houses, where they remain until they are ready to go into the laying pens in the early fall. The cockerels are sorted at the time they are removed, the best ones being given more liberty and reared for use as breeders, while the poor ones are placed in fattening pens and sold as broilers.

Hollywood's Greatest Breeding Achievement

Space is not here available for going into detail, either in regard to the breeding methods practiced at Hollywood or the splendid results achieved. For these the reader is referred to our book, "The Production of 300-Eggers and Better by Line Breeding," which is briefly described on page 98. By way of illustrating the possibilities of systematic line breeding as practiced at Hollywood this merely general sketch is concluded with the following statement of the remarkable results secured from a single, small, special breeding pen of this truly extraordinary production-bred strain. The greatest achievement at Hollywood Poultry Farm thus far (August 27, 1921) in high and consistent egg production was in the results from Pen 5, as mated in the fall of 1918 for 1919 production, which pen was made up of the following:

Headed by cockerels 544 and 545, full brothers, bred from Lady Hollywood and sired by a Padman male. In this pen were hens Nos. 1027, 1202, 1088, 1157, 1276, 1284, 1057, 1219, 253 (Lady Hollywood), 681, 1212, 1115, 1278, 1179, 1014, 1178 and 1071.

No. 1027 had laid 270 eggs. She was bred from a pen of American hens with records of 200-229 eggs, headed by a New Zealand cockerel bred from the Victoria (B. C. 1913-14) Contest pen.

No. 1202 had laid 284 eggs. She was bred from a pen of pen-sisters that laid 230-289 eggs, sired by a male from a 254-egg hen.

No. 1088, with record of 273 eggs, was bred from a mating of five sisters with trap records of 260-269 eggs and her sire was a full brother to sire of 1027.

No. 1157, with record of 287 eggs, was a pen sister of 1088. By "pen sister" is meant from the same pen, carrying the same blood lines.

No. 1276, with record of 272 eggs, was from the same pen as 1157 and 1088.

No. 1284, laid exactly 300 eggs; was from the same pen as the three just given.

BY THEIR FRUITS WE MAY KNOW THEM

One day's gathering of new-laid eggs at Hollywood Poultry Farm—5,895 eggs from 9,322 hens. On this plant they breed systematically for weight, shape and color of shell, as well as for numbers.

No. 1057, with record of 268 eggs; from same pen as foregoing.
No. 1219, with record of 252 eggs, was a daughter of Lady Hollywood with record of 275 eggs in official contest.
No. 253 was Lady Hollywood.
No. 681, with record of 284 eggs; from an American hen and a New Zealand male, bred from the Victoria Contest pen.
No. 1212, with record of 272 eggs; a pen sister of 1202.
No. 1115, with record of 271 eggs, a pen sister of 1212 and 1202.
No. 1278, with record of 273 eggs, was a daughter of Lady Hollywood and was a full sister to 1219.
No. 1179, with record of 270 eggs; a pen sister of 1088.
No. 1014, with record of 276 eggs; from pen of hens with records of 270-279 eggs (pen sisters) and a male from No. 283.
No. 1178, with record of 282 eggs; a pen sister of 1014.
No. 1071, with record of 254 eggs; a pen sister of 1202.

From this mating seventy-four pullets were trap nested and made 365-day records as follows:

Five birds made records of 251-259 eggs, as follows:
No. 257, from hen 1088, laid 258 eggs.
No. 1419, from hen 1284, laid 259 eggs.
No. 1562, from unidentified mother, laid 252 eggs.
No. 1601, from unidentified mother, laid 251 eggs.
No. 1641, from unidentified mother, laid 252 eggs.

Six hens laid from 260-269 eggs, as follows:
No. 601, from hen 1219, laid 261 eggs.
No. 699, from hen 1202, laid 269 eggs.
No. 732, from unidentified mother, laid 260 eggs.
No. 1497, from hen 1202, laid 262 eggs.
No. 1502, from hen 1115, laid 261 eggs.
No. 686, from unidentified mother, laid 261 eggs.

Nine hens made records of 270-279 eggs as follows:
No. 637, from unidentified mother, laid 273 eggs.
No. 774, from hen 1027, laid 277 eggs.
No. 241, from unidentified mother, laid 276 eggs.
No. 1441, from hen 1179, laid 273 eggs.
No. 1568, from hen 1027, laid 276 eggs.
No. 1572, from hen 1115, laid 279 eggs.
No. 1668, from hen 1088, laid 272 eggs.
No. 1577, from hen 1179, laid 279 eggs.
No. 1702, from unidentified mother, laid 274 eggs.

Six hens laid 280-289 eggs, as follows:
No. 643, from hen 1276, laid 287 eggs.
No. 677, from hen 631, laid 286 eggs.
No. 673, from hen 1202, laid 280 eggs.
No. 715, from hen 1212, laid 280 eggs.
No. 279 from hen 1088, laid 288 eggs.
No. 1447, from hen 1276, laid 286 eggs.

Also this mating produced two that laid as follows:
No. 256, from hen 1202, laid 291 eggs.
No. 766 from hen 1178, laid 292 eggs.

This mating produced thirteen contest birds that laid from 206-315 eggs, from November 1, 1919, to October 31, 1920, under official care and supervision, as follows:
No. 119, from unidentified mother, laid 206 eggs at Western Washington Contest.

No. 5, from hen 1088, laid 210 eggs at Storrs, Conn., Contest.
No. 2, from hen 1202, laid 222 eggs at the Vineland, N. J., Contest.
No. 132, from hen 631, laid 241 eggs at Western Washington Contest.
No. 3, from 1179, laid 241 eggs at Philadelphia, North American Contest.
No. 120, from unidentified mother, laid 247 eggs at Western Washington Contest.
No. 131, from hen 1027, laid 249 eggs at Western Washington Contest.
No. 18, from hen 1284, laid 249 eggs at the Vineland, N. J., Contest.
No. 79, from hen 1276, laid 259 eggs at the Western Washington Contest.
No. 115, from unidentified mother, laid 279 eggs at the Western Washington Contest.
No. 129, from hen 115, laid 279 eggs at the Western Washington Contest.
No. 118, from unidentified mother, laid 298 eggs at the Western Washington Contest.
No. 116, from unidentified mother, laid 312 eggs at the same contest as the four preceding birds, and the management held her until November 4, up to which time she laid 315 eggs, completing her 365th day as certified to by the management as an official record.

This Pen 5 mating produced no less than twelve birds that laid 300 or more eggs in 365 consecutive days. Additional to No. 116 above mentioned, there were eleven others as follows:
No. 640, from hen 1284, laid 316 eggs.
No. 655, from hen 1284, laid 300 eggs.
No. 1471, from hen 1284, laid 305 eggs.
No. 1430, from hen 1284, laid 306 eggs.
No. 311, from hen 1088, laid 310 eggs.
No. 1409, from hen 1088, laid 328 eggs.
No. 704, from hen 1027, laid 314 eggs.
No. 1457, from hen 1276, laid 300 eggs.
No. 1536, from hen 1278, laid 306 eggs.
No. 1542, from hen 1219, laid 300 eggs.
No. 1667, from hen 1179, laid 304 eggs.

Of the progeny of Pen 5, as above described, we trapped only seventy-four pullets and additional to the foregoing records in excess of 250 eggs and the list of contest winners we obtained the following:
No. 1646, from hen 1157, laid 245 eggs.
No. 1513, from hen 1212, laid 249 eggs.
No. 718, from hen 1212, laid 230 eggs.
No. 1415, from hen 1284, laid 231 eggs.
No. 284, from hen 1088, laid 234 eggs.
No. 1402, from hen 1202, laid 236 eggs.
No. 1467, from hen 1219, laid 228 eggs.
No. 1680, from unidentified mother, laid 228 eggs.
No. 1627, from hen 1115, laid 211 eggs.
No. 791, from hen 1202, laid 214 eggs.
No. 1730, from hen 1057, laid 217 eggs.
No. 1698, from hen 1027, laid 219 eggs.
No. 888, from hen 1027, laid 202 eggs.
No. 790, from hen 1276, laid 207 eggs.
No. 670, from hen 1027, laid 197 eggs.
No. 1757, from hen 1057, laid 186 eggs.
No. 854, from hen 1027, laid 188 eggs.
No. 688, from hen 1202, laid 156 eggs to May 1st and died.
No. 691, from hen 1027, laid 208 eggs to May 1st and was sold.
No. 698, from hen 1027, laid 160 eggs to May 1st and was sold.
No. 621, from hen 1027, laid 196 eggs to July and was sold.
No. 635, from hen 1027, laid 198 eggs to May 15th and was sold.

Commenting on this remarkable production, Mr. Atkinson said:

"This accounts for all birds we trapped from that pen and shows the high average breeding results on a production basis, which truly are extraordinary, as represented by the figures given you, which speak for themselves. To the best of my knowledge, results to equal these were never before obtained, referring to the twelve 300-eggers and better and an average of 255 46/74 for each one of the pullets we were so fortunate as to trap. If I could have known soon enough how wonderfully the pullets saved from this mating were going to lay, you can be sure that I would have entered every one in some official contest, if I could have found room for them. The bulk of the eggs from that mating, Pen No. 5, 1918-1919, were sold by us to customers throughout the hatching season, February to May inclusive, at the rate of $30 per fifteen. I will admit two things: first, if we had known how remarkably those birds were going to 'nick' for high egg yield, we would not have parted with but very few of the eggs for hatching at any price within reason; second, you may be certain that we would have trapped every well-matured, good-quality, vigorous pullet possible from that mating."

AMERICA'S DEAN OF LARGE-SCALE TRAP NESTING

D. Tancred, Kent, Washington, Began in 1905 with a Few Dozen Wyckoff-Strain Birds of Standard Quality—To Date Over 19,000 Line-Bred Leghorn Layers Have Passed Under His Close Observation and Severe Test—Has Produced Numerous 300-Eggers in Last Few Years—Tancred Birds Now Hold World's Contest Record for Pen of Five Leghorns—Flock Average First Year's Production Now Above 200 Eggs per Bird

By Grant M. Curtis

FOR ten years or more we had heard occasional reports, which found their way eastward across the Great Divide, about the extraordinary egg-yield and beauty-strain of S. C. White Leghorns created by D. Tancred, of Kent, Washington, as the result of years of accurate trap nesting on the basis of systematic line breeding, but it was not until this fall that we had the good fortune to become acquainted with Mr. Tancred, and the opportunity to learn the facts at first hand, for the benefit of those who are genuine students of poultry culture.*

Really it is to ill health that poultrydom owes the work and astonishing results attained by Mr. Tancred in this important field of live-stock breeding. He had followed the sea from boyhood, filling difficult and responsible positions until at the age of only 36 years an ailment of the spine forced him to retire.

It was in the fall of 1905 that Mr. Tancred found it necessary to give up his position as a sea-faring man. We quote here briefly from a stenographic interview of recent date with Mrs. Tancred:

"I think Mr. Tancred has told you about Mr. Moses Locke, an old man who lived across the street from us in Kent. He was seventy years old at that time and had 300 Brown Leghorns which he kept in one long continuous house. He fairly lived with those fowls! His practice was to raise one hundred pullets each year and retain a few good ones as breeders, which plan we ourselves adopted from the start. It was Mr. Locke's success that interested us in keeping poultry and we also began with Brown Leghorns. There were very few White Leghorns out here at that time, but owing partly to the fact that this variety has no black pin-feathers and because we were admirers of C. H. Wyckoff and his methods as a poultryman, we soon decided to keep only White Leghorns. Our start was made before Mr. Tancred had to give up his position and it wasn't long before we were making a good living from our hens, about 500 in number, selling market eggs alone."

In those early days Mr. and Mrs. Tancred lived on a two and one-third acre place in Kent; now they occupy a twenty-five acre farm (two farms in fact—ten acres in the home place and fifteen acres directly across the road),

high up in the partly wooded hills about one mile out of Kent, on well-drained soil, overlooking Kent and the White River valley. Following, we quote from one of many interviews with Mr. Tancred, taken stenographically:

"We made our start with stock from Mr. Locke and that first year we raised seventy pullets. Soon I got to thinking the matter over seriously, as to the best breed or variety to keep and concluded it would be difficult to breed Brown Leghorns to good advantage, so I decided that the White Leghorn must be the coming bird. Was influenced in this conclusion by the fact that wherever egg production was the determining factor, the White Leghorns seemed to be 'It,' especially in a moderate climate such as we have here in the Puget Sound country. Was finally able to talk Mrs. Tancred into changing over, so we disposed of the Brown Leghorn pullets at 60 cents apiece. She was disappointed, because she then had a nice modern poultry house and nothing in it, till we could make a start the coming spring with White Leghorns.

First Knowledge of Trap Nesting

"We did not become acquainted with trap nests until the following year, when breeding the Whites. Then old Mr. Locke discovered it through Prof. Gowells bulletins of the Maine Agricultural College at Orono, in which pictures were shown of the open-front house and his style of trap nest. Mr. Locke had built one of these trap nests for himself and later we got him and a neighbor to take a contract to build us an improved one, according to Prof. Gowell's and Mr. Locke's ideas. The next bulletin Prof. Gowell got out—the following year—dealt solely with trapping experiments and it had pictures and diagrams of the nest itself. When I arrived home one day, over came Mr. Locke, bringing a trap nest. He had gone down town and got an orange box and by cutting a hole through the middle partition of the box, all he had to do was to make a gate and work the trap. We set it on the floor of the kitchen, got one of our Leghorns and we poked her in there and pulled her out again till she couldn't stand on a leg. After we had tired her out, we got another one and practiced on her awhile. Then I got an orange box so I, too, could have a trap nest.

"Before I went away on my next voyage, Mrs. Tancred and I decided we would put in a bunch of trap nests and try the plan out. We had thirty nests made by a carpenter and put them in that house of ours, which was built for 100 breeders, and we began trapping about ninety birds. When I came home for the season, about the 1st of November, I found that Mrs. Tancred had been trapping since midsummer. Some of the pullets were hatched in March and a few had begun to lay as early as July, which then was remarkable. Our records showed they were hatched the first day of March and laid before the end of July. Mrs.

MR. TANCRED'S HIGHEST RECORD HEN TO DATE, NO. 340
Laid 330 eggs in 365 consecutive days, from September 17, 1918, to September 16, 1919, inclusive, as sworn to by the men who were in charge of trapping her and her mates at Tancred Farms.

---

*Note:—The foregoing was written in the fall of 1921. The complete account of this master-breeder's remarkable advancement in breeding high-production, standard-quality S. C. W. Leghorns is to be presented in a book by Grant M. Curtis bearing title of, "D. Tancred, Dean of Large-Scale Trap Nesting in America," which book is to be issued in 1923. Like "The Production of 300-Eggers and Better by Line Breeding" (see page 98), it will be issued in both paper and cloth binding, size 6 by 9 inches, to contain 350 pages and splendidly illustrated with a large number of half-tone photoengravings, the photos for which were taken expressly for this use. The paper-covered edition will carry a beautiful art cover by Sewell and will sell at $2.00. The cloth-bound edition will be printed on heavyweight enamel paper; price $3.00.

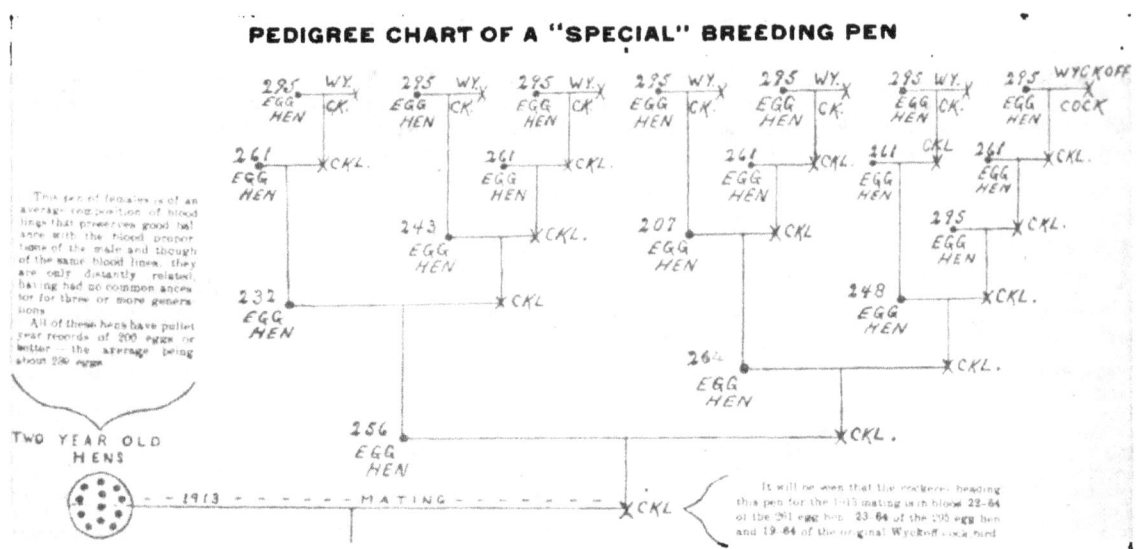

AN ILLUSTRATION OF THE THOROUGH-GOING METHODS OF PEDIGREE BREEDING PRACTICED BY D. TANCRED
The pedigree chart illustrated above shows the breeding of one of Mr. Tancred's special breeding pens and is reproduced from a full-page ad published by him in 1912 in "The Pacific Poultryman." It illustrates not only the remarkable breeding back of the pen but shows that at this early date the Tancred strain had already been bred in line for high egg production for seven years.

Tancred had carried the work along until I got home in November, then I started to make records. Soon I found that our 1905 records were not exactly dependable, on account of our not knowing better than to leave eggs in the nests, the result being that they accumulated during the day and we would take a hen out and credit her with an egg, but did not discover for some time that some hens had the habit of getting on a nest and loafing, without laying at all—were just useless drones, which later we got rid of systematically. So we considered our trapping for 1905 to be half a failure, but from 1906 on I have every confidence in our records, as they have been kept absolutely as nearly correct as is humanly possible."

Origin of the Tancred Utility-Beauty Strain

Asked about the origin of the Tancred strain, which combines prolific egg-yield and standard qualities to a noteworthy degree, Mr. Tancred said:

"We started with three lines, so it may be stated, but as a matter of fact all three lines originated with the Wyckoff strain. We bought eggs and males from Thomas Robinson, then of Spanaway, Washington; some eggs also from C. S. Gorline, of Utah, and later we bought male birds direct from C. H. Wyckoff, of Aurora, N. Y. Bought 300 eggs of Mr. Robinson, from which we raised less than 100 pullets—about 90, in fact. Got our 295-egg hen of 1906-1907 from the Robinson stock, which originally was all Wyckoff, but Mr. Robinson had bred this stock more than half his lifetime, starting with it early in the Wyckoff career. This 295-egg hen, known as No. 22, came the following year (after purchase of the 300 eggs), out of the choicest of the Robinson pullets, and in the fall of 1907 was mated with the Wyckoff male. From the Gorline eggs, bought a year later, we got our hen that laid 261 eggs, and that was only part of her year's record. She began that record on the seventh day of January, but had been laying no one knows how long before this date. She had been laying out in the yard and laying frequently, which caused me to put her in the trap-nest house in January. At the end of her year she had made the record of 261 eggs, laid in trap nests.

"I had eleven or twelve other 200-eggers from the Robinson eggs, otherwise I would not have had such confidence in his stock. That first year the Robinson pullets averaged 144 eggs, which was way ahead of any other stock I had. These two outstanding birds—that from the Robinson eggs with trap-nest record of 295 eggs and the Gorline-Wyckoff-strain hen that laid 261 eggs in the trap, starting January 7 of her production year, these two hens were mainly the start of the Tancred strain, on their side of the house. No, there were no individual egg records back of the Robinson strain, nor of the Gorline stock either, so far as I know. Mr. Robinson had pen records and one pen produced extra well. In selling me eggs he gave us some eggs out of that pen, as well as from others.

Started with Standard-Bred Stock

"Mr. Robinson in his day was a leading exhibitor of S. C. White Leghorns in this section, winning the bulk of the prizes wherever his birds were placed in competition. In other words, he was a breeder of exhibition stock who gave special attention to egg production, following quite closely in those early days the methods employed by Mr. Wyckoff, as I understood them at the time. This meant that Mrs. Tancred and I started with good-quality, standard-bred White Leghorns and we never have departed from that course. During all these years Judges Collier and Coats—Mr. Collier for eleven years and Mr. Coats during the past four years—have visited our plant repeatedly every season to go over our prospective breeders for the purpose of helping us maintain and improve their standard qualities by the rigid elimination of disqualified birds and inferior specimens. Not once during the entire period have we used a disqualified bird for breeding purposes, nor a bird with brassy plumage, the result being that our birds now come pure white and remain so, and they also lay white-shelled eggs without exception, which is a very important fact in their favor now that Leghorn eggs are closely graded by color of shell in the Pacific Coast and New York markets."

Tancred Farms now carry about thirty-five hundred birds each season, more than two thousand of which are trapped and many are pedigreed as individuals. Speaking broadly of the laying pedigree represented at present by his combination utility-beauty strain, Mr. Tancred said, in a recent interview, after referring to his records:

"More than 9,000 of the female ancestors of all chicks hatched on this farm in 1921 were fine layers of known records, every one of them trap-nested. That is what I mean by an unbroken pedigree, reaching back fifteen generations on the male side and eight on the female side—back to 1906, when I began accurate trap nesting and systematic line breeding. A Tancred pullet of the present generation, therefore, would have to jump back over the heads of her nearly nine thousand foremothers in order to inherit the characteristics of a poor layer. It is hardly probable that she would do that—and very few of them do.

Strain Reached 300-Egg Mark Six Years Ago

"Yes, we reached and passed the 300-egg individual-hen mark some time ago, doing this in 1917—one bird at home and another that won in a contest; but my customers got them a year earlier, and one customer trapped out a 300-egger from my strain two years earlier—in 1915-1916. This 300-egg Tancred-strain hen was widely advertised in Oregon, after she made her record of 304 eggs. Last year, here on the home place, we trapped out eleven 300-eggers —birds that in each case laid 300 eggs or better in 365 consecutive days. Our highest record bird to date is No. 340, a standard-weight, line-bred hen that two years ago reached 330 eggs, and I am positive she would have laid two eggs more, during the last two days of the 365-day period, if a local photographer had not got after her without my knowledge and frightened the notion out of her, by rough handling. According to her cycles, which appeared to be about as regular as clock work, she should have laid on her 364th and 365th days, but she stopped on the 363rd day, after laying that afternoon.

"No, I do not use artificial lighting and never have. This alone makes a considerable difference, no doubt, in the pullet-year production. In our case I am depending solely on breeding for prolific egg yield, along with sensible handling and proper feeding. I do not use artificial light, in any of my poultry houses at any time, on birds of any age, for increased egg production or any other purpose. Every egg record of my hens, therefore, is made naturally and without forcing, either in lighting or feeding. And I go farther than this. We weaken our pullet ration every season during the early months of their laying, in order to accomplish two things in particular: first, to delay egg production until they are of a proper age; second, to conserve their strength and vitality, which I prefer to keep undiminished, even at the cost of a few eggs at the high-price season for the daily market."

Record Winnings in Egg-Laying Contests

The Tancred strain during the last several years, has made exceptional winnings at egg-laying contests, in competition with all comers, the most notable being as follows:

Western Washington Experiment Station, Puyallup, November 1st, 1920, to October 31st, 1921, 1384 eggs laid by five pullets (original birds entered, no substitutes), an average of 276 4/5 eggs per fowl.

Western Washington Experiment Station, November 1st, 1919, to October 31st, 1920, 1306 eggs laid by five pullets, an average of 261 eggs per fowl.

At Leavenworth, Kansas, November 1st, 1918, to October 31st, 1919, 1319 eggs laid by five pullets, an average of 264 eggs per fowl.

Said Mr. Tancred, in this connection:

"Tancred Farms have many other laying contest winnings to their credit, some of them scarcely less notable than the three I have just given you. Please observe that they occur one year after another, without a break. This strain does not have 'off' years, because it has been bred right and very rigidly for too long a period. We can now tell pretty nearly what our blood lines will do, under proper conditions. Yet it is not high individual records that we are after, as our main consideration, but high flock averages. That is where the real profit is for our customers —for commercial egg plants, for the ordinary farmer, for the back-lotter, for everybody who wishes to keep poultry in order to make money, in place of paying feed bills without a profitable return on the investment."

Breeding To Perpetuate Egg-Laying Capacity

When Mr. Tancred, by trap nest, discovered hen No. 22, probably the world's record layer for that time, doubtless he would, if he had understood line breeding in 1907 as he does now, or the special value of the male in poultry culture, have gone back to Mr. Robinson in quest of a closely related male—perhaps for the father of this 295-egg hen, or a full brother of her sire. Mr. Robinson was not trapping his stock, yet it is quite likely that he knew the sire of the pullet that was the mother of the 295-egg hen, hence if Mr. Tancred could have bought this bird it would have been practical to mate him with his granddaughter, a strictly allowable mating, even in the line of present-day knowledge, especially for the purpose of concentrating and perpetuating the high-production blood of this extraordinary hen—a hen that, not satisfied with breaking the then world record (so far as our knowledge goes) by laying 295 eggs in her first year of production, proceeded to lay 202 eggs in her second year, starting August 31, 1907, and finishing August 30, 1908, and that laid 152 eggs her third year, 109 eggs her fourth year and 47 eggs her fifth year, as sworn to by Mr. Tancred before a notary public (H. B. Madison, Kent, Wash.), at the end of each production year, in the form of an affidavit attached to the tabulated egg charts of her annual yield.

Facts are that, instead of going back to Mr. Robinson, Mr. Tancred scanned the poultry horizon and, being an admirer of the work and methods of C. H. Wyckoff, Aurora, N. Y., as a long-time breeder of S. C. White Leghorns who, for twenty years or more, had combined standard qualities with prolific egg yield, he sent to Mr. Wyckoff for a cock bird, specifying in particular that he wanted a male from a good layer. Evidently he wished to keep within the Wyckoff strain, even if he did not at that time fully appreciate the added value of a closer blood relationship to this 295-egg hen.

GENERAL VIEW OF ONE OF THE TANCRED FARMS
Shows view of one section of ten-acre farm on which the Tancred home is located. Long building in foreground is one in which hundreds of line-bred pullets are trapped each year.

Best Daughter of the 295-Egger

The best layer obtained directly from the 295-egg hen was No. 375, which laid 243 eggs in 365 days, beginning November 10, 1907, and ending November 8, 1908. This bird in turn was mated with a cockerel bred from the Gorline 261-egg Wyckoff-strain hen, which had been mated with a cockerel direct from the 295-egg hen and the Wyckoff cock.

A study of the pedigree chart on page 104 will be of real help to the interested reader. It illustrates in simple form the systematic method of line breeding practiced, shows how the 261-egg hen was introduced, merging her blood with that of the 295-egg-Robinson hen (Wyckoff strain originally) and that of the Wyckoff cock which Mr. Tancred purchased from the Wyckoffs; gives egg records of the individuals that were used in this line (as low as 207, 232, etc.); tells the year (1911) when a hen was obtained that passed the trap record of the Gorline hen by laying 264 eggs; shows also at the right how hen No. 22, with her 295-egg record, was again used, mated with a cockerel bred

LOOK INTERESTING, DON'T THEY! AND ORDERLY, TOO

Exterior view of row of four special breeding houses, each 8 by 10 feet, on Tancred Farms, in which small matings of 300-eggers and better are made each season. In these small-sized, low-cost houses Mr. Tancred each season carries "Private Matings," made up of high-record and prepotent birds which "are not for sale at any price."

from the Gorline hen which the season before had been mated to a cockerel produced by this 295-egg hen and the original Wyckoff cock, etc., etc.

It was the blood of these great breeders, carried down carefully from year to year, that gave to the Tancred strain its present-day preeminence. Just as Mr. Tancred might have gone back to Mr. Robinson for a near-relation male to mate with the 295-egg hen; so he could have tried his luck with Mr. Gorline for a male closely related to the 261-egg hen, but as before explained he did not do this. The general result, as he now knows from experience, was to have been expected. While he did keep to the same strain, (Wyckoff) the relationship was so remote that it could have but little bearing on these two highly exceptional females, so far as the immediate benefits from established line breeding were concerned; hence, in fact, it required seven years for Mr. Tancred to produce a hen that, as proved out by trap nests, was the equal of No. 22 in prolific egg yield. In 1913 he passed the incomplete trap record of the 261-egg hen with hens Nos. 06 and 933, which laid respectively 278 and 282 eggs in trap nests—and a year later his trap nests disclosed hen No. 809—one that laid 299 eggs, thus eclipsing the proud record of hen No. 22 and coming within one egg of the 300-egg mark.

Then the Topnotchers Came Faster

After that the "toppers," as they are called, came along more rapidly—birds laying in the 270's, 280's and 290's; also the 300-eggers in the Tancred strain were then close at hand. Notable examples of the latter include "Brewster's Oregon Girl," of the year 1916-1917, bred by Mr. Tancred and sold to Frank Brewster, Portland, Oregon, that laid 304 eggs in 365 days; also "Lady Kinmont," bred by Mr. Tancred and sold to Mrs. Jean A. Patterson, Richmond Highlands, Washington, which laid 313 eggs in 365 days. Then, a year later, hen No. 251, bred by Mr. Tancred and entered by him in the Washington State College Egg-Laying Contest, Pullman, made an official record of 311 eggs in 365 days, her year ending October 31st, 1918. Next, in 1918-1919, came Mr. Tancred's highest record bird to date—No. 340 that laid 330 eggs in 365 days, which record was attested in affidavit form before a notary public by the Tancred Farm trap-nest attendants.

High Flock Averages the Main Goal

So much for the almost incredible increase in individual records, as accomplished by Mr. Tancred during practically fifteen years of systematic and persistent efforts, except let us add, as further evidence of his cumulative success, that during 1920 the trap nests on the home place disclosed eleven hens that laid 300 eggs or better, also the fact that at the time of our last interview on this phase of the subject he expected to obtain more than that number for the year to end in October and November, 1921, as the result of trapping about 2,000 pullets—the usual number now trapped annually by him.

But it is not high individual records that Mr. Tancred most desires. They are truly welcome and of course contribute directly to the main object for which he is working—HIGH FLOCK AVERAGES, both in his own behalf and for the financial benefit of his customers, which now number a good many hundreds, in fact several thousand. On this important question, the following instructive quotation is made from Mr. Tancred's 1917 catalogues:

"The first generation of pullets that I trap-nested averaged only 144 eggs in their pullet year. That was a fine flock record for those days, however. * * * Four years after starting, the average annual production for the whole flock had increased to 166 1/10 eggs, a gain of twenty-two eggs over the production of the best birds I started with. The main cause for this heavy gain at the beginning was the elimination of the 'drones.'

"With the drones gone, the flock settled down to a slower but steady and certain gain in production. Two years later the total flock average was 170½ eggs. This rate of gain has been maintained to the present time and the yearly average for the whole pullet flock is now 180 eggs (year 1916-1917), as closely as may be determined."

Has Since Reached and Passed the 200-Egg Flock Average

During a recent interview (late October, 1921), Mr. Tancred made the following statement regarding the present flock average:

"It is impossible to know, with absolute accuracy and in full justice to the birds, but at present it is not far from 200 eggs—above that rather than below. However, my flock average doesn't begin to equal that obtained by a majority of my customers. As a rule, they go in strong for eggs and still more eggs, while I hold back the production. With most of my customers, eggs are the main objective, but here we are conducting a breeding establishment. In the early fall I cut down the birds' ration about fifty per cent, meaning those that are being trapped, because I find them going too fast. On the other hand, many of my customers get flock averages of 215, 224, etc., while I am well satisfied with around 200 eggs per bird, because it would be unwise for me to push our birds; therefore my own pullet-year record is no real guide of their attainable performance under average conditions. Will state further that customers obtained 300-egg birds out of my stock before I did, because we slow down their pullet-year production, for the reason that we want to keep them in condition for breeding. Also, as previously stated, we do not use artificial lighting at all, either on pullets or breeders, nor any form of forced feeding for egg production."

CHAPTER V

High Egg Production and Superior Table Quality Combined

How R. A. Richardson of Haverhill, Mass., Has Been Able Successfully To Develop a Strain of White Wyandottes in Which Fifty per cent of the Pullets Regularly Exceed 200 Eggs in Their First Year—Strain Is also Distinguished for Exceptional Table Quality

By MR. JOHN H. ROBINSON, Associate Editor of Reliable Poultry Journal

THE strain of pedigree-bred, heavy-laying White Wyandottes developed by Mr. R. A. Richardson, of Haverhill, Mass., is dated by him from 1904-1905—that being the time when he established the first of the pure lines that have been bred by him continuously ever since. The history of the stock in his hands, however, goes back to 1902 and, knowing the sources of his foundation stock, I can take one ancestral line some five years, and another ten years back of that. Hence, while his pedigree stock has at this time (1921) a history of sixteen years, records in its line can be carried back to 1892—thirty-one years.

Mr. Richardson states that he became interested in the matter of individual high egg production in 1902. He had for some time been interested in breeding standard poultry on a small scale, without paying particular attention to high egg production. Being a thoroughly practical man, his interest turned more and more to the product of his hens, and to methods of determining absolutely what each hen was doing and how each hen he kept was bred. There was at that time a great deal of interest in "200-egg hens" in New England, and this interest centered largely on the White Wyandotte, which was then probably the favorite breed among amateur breeders in this section.

Two Historic Strains Combined

Mr. Richardson bought a few birds of each of the two heavy-laying strains of White Wyandottes that were then best known and in highest repute—the Pulsifer and Bricault strains. He bred these strains separately, and also made some cross matings, keeping individual laying records of all the hens that he reserved. He kept at this time only a small stock, for his regular employment as a shoe-cutter left him only spare hours for this work, and he was then—as always since—more concerned about doing thoroughly whatever he undertook than about operating on a large scale. He consistently limited his stock to what he could manage while keeping complete records and accurate pedigrees, and to what Mrs. Richardson could give careful attention as necessary in his absence, without interfering too much with her household work.

The Pulsifer strain of heavy-laying White Wyandottes was established by F. P. Pulsifer, of Natick, Mass. In a symposium on "200-Egg Hens," in "Farm Poultry," January 15, 1901, Mr. Pulsifer gave this account of the origin of his strain:

"The first pen of White Wyandottes (12 pullets) I owned averaged 213 eggs each in one year. I did not use a trap nest, so cannot give individual records. It is needless to say that this heavy yield was obtained by great care and devotion, and the most favorable circumstances. Up to this time, 1892, I had bred Barred Plymouth Rocks, but the following season disposed of all of them, and have since raised White Wyandottes only from this valuable pen and its descendants. I have not kept a pen record since, and I doubt very much if I have ever equaled this record, and can simply say that my egg yield has been and is very satisfactory."

Mr. F. P. Pulsifer advertised this stock for some years as the "213-Egg Strain," but never made any other claims for definite records. A Mr. F. E. Woods, also of Natick, for at least two years before Pulsifer began to advertise his birds as a "213-Egg Strain," advertised a "210-Egg Strain," stating in the advertisement that he gave the strain this name because he had a pullet of it that had laid 210 eggs in a year. The Pulsifer strain was of good standard quality, not top-notch show birds but good enough to come in for an occasional C in Boston when White Wyandottes were at the zenith of their popularity in New England.

For some years before and after 1900 there were two other Pulsifers, one at Natick and one at Gloucester, Mass., advertising heavy-laying strains of White Wyandottes. These men did not identify their stock as of the F. P. Pulsifer "213-Egg Strain," but it was commonly supposed to be all the same, and with three breeders of the same name advertising heavy-laying White Wyandottes this strain soon had a good reputation as reliable layers. As a matter of fact, practically all the well-known strains and stocks of standard-bred White Wyandottes, up to near 1910, were good layers. There were, besides, in the early history of the variety, a great many stocks as bred by egg farmers that were generally good layers, but very indifferent in type and color—even by the standards that prevailed when White Wyandottes and White Rocks might differ only in comb, and pure white birds were rare.

Development of the Bricault Strain

The Bricault strain, bred by Dr. C. Bricault, of Lawrence and Andover, Mass., was started about 1897 and, as far as my information goes, was first brought before the public in 1899. Dr. Bricault, a veterinary surgeon, was one of the first—if not the first—to advertise "bred-to-lay" stock. His advertisements proclaimed that he had White Wyandottes and Barred Plymouth Rocks—"carefully bred from dams of standard weight, whose egg-laying record for each succeeding generation was individually known to have increased from year to year. Males are all from 200-egg hen."

As far as I know, Dr. Bricault never published nor gave any information as to the origin of his "bred-to-lay" stock of either breed. There were a good many breeders of Barred Rocks in this section at that time who had good laying stocks, and J. W. Parks, of Altoona, Pa., and H. F. Cox, of Sabbath Rest, Pa., were breeding and advertising heavy-laying strains of Barred Rocks at the time Dr. Bricault began. His connection with Barred Rocks, however, did not last long. He concentrated his attention on the White Wyandottes and in a few years his strain became widely known. According to reports of those who bought it, the Bricault strain was a good-laying strain, but did not often produce hens that would go over 200 eggs. I personally did not take much stock in the records—as complete and accurate records—because both from poultrymen in his vicinity, and from Dr. Bricault himself, I knew that he was much away from home, and that the conditions were not favorable to accurate record keeping.

Late in 1903 an occasional contributor to "Farm Poultry," who had visited the Bricault Farm, sent me an account of his visit and interview with Dr. Bricault which was so frank an admission of shortcomings in his records that before publishing it I sent it to Dr Bricault for approval, which was promptly given. Said this contributor:

HEAVY-LAYING WHITE WYANDOTTE
Bred and owned by R. A. Richardson of Haverhill, Mass. This pullet made a winter record of 132 eggs in 175 days.

"Dr. Bricault has had and still has 200-egg hens, he has bred from them, he has made something—if not the most—of them. It is early yet to speak of positive results in breeding for eggs; suffice it to say that Dr. Bricault has achieved enough in that direction to encourage him to persevere.

"It is interesting to know just how much Dr. Bricault pretends to devote himself to the trap nests. At this time of year (October) all his nests are frankly and shamelessly fastened open. The record keeping ceased in July. He will base his opinion of the hens on their work for the nine months preceding. From November to July he was a slave to the trap nest—not absolutely, but more or less, and within reason. He makes no bones of taking a day off occasionally. He is not a scientist but a business man, and attempts only the practical. His records then lay no claim to scientific accuracy, but they are thought to serve the purpose if they discriminate roughly among the good, bad and indifferent layers. * * * *

"Dr. Bricault has this to say about his position: 'I do not insist too much on the 200-egg hens. Call them heavy layers. I took their measures when somewhat more enthusiastic than I have been this year or intend to be henceforth. Strictly speaking, I shall have no 200-egg hens this year. My best record is 160 eggs in a period of one year less three months. What the three months would have brought forth we shall not quarrel about. I am content to rest on what I am sure of. I do not wish, and never did wish, to be identified with the 200-egg interest. I wish to be known for what I am doing more than for what I possess. What if I have a few 200-egg hens? So have others. I am distinguished from most of these by what I am doing with my 200-egg hens. I am of those who believe that the 200-egg hens indicate an upward tendency that may be hastened by artificial selection; that they may be made to uplift all hendom in productiveness, and this without detriment to the stamina of the stock. I breed accordingly.'

Systematic Trap Nesting Begun

Such were the stocks with which Mr. Richardson started. The Pulsifer stock had not even a pen record for ten years before he bought it. The Bricault stock had more or less deficient individual records for nine months of the year for some four or five generations before he acquired it. The Pulsifer stock had been bred by ordinary selection by appearance—keeping within the blood lines of the pen that had made the 213-egg average in 1892. The Bricault stock had been bred by selection according to performance as shown by the partial records kept. While defective from the standpoint of completeness and accuracy, the Bricault method did without doubt show the relative breeding value of the birds in breeding for high egg production. Both were unquestionably good-laying stocks. That was shown by their general performance in the hands of buyers and by the records which Mr. Richardson made with them.

To the hens he obtained from Pulsifer and Bricault, Mr. Richardson applied what the methods of those breeders lacked. His hens had the most careful and unremitting attention, and his trap-nest records were kept with absolute fidelity—purely for his own information in the early years of his work, and always primarily for that purpose. When Mr. Richardson was away at work, Mrs. Richardson—who took the same interest in the work in those days that her husband did, and who after he began to devote all his time to poultry still kept informed on what the birds were doing—looked after the trap nests. Any deficiencies in the records of this stock since they began to keep records count against the hens. They have never recorded for any hen an egg that was in doubt, either through two hens getting into a nest, and only one laying, or being laid outside the nest. Mr. Richardson even goes so far in this that he will not make records at all for the occasional pullet that refuses to use the trap nest and regularly lays outside. Even though he might be able to identify those eggs with certainty, he will not band the hen nor record them, because he wants to maintain the condition that enables him to vouch absolutely for every one of his records as a record of the eggs which it is positively known that hen has laid in the trap nest. He is never away from the place long enough to interfere with his keeping close tab on the records.

High Breeding Standards Applied

From the time he started with them, Mr. Richardson made each year some matings of each of the two strains pure, and some crossing the strains. He was simply trying out and observing the stock until in 1904 he bred three pullets which in 1904-5 laid, respectively, 218, 230 and 242 eggs. The hen that laid the 230 eggs appears in his records as No. 3, and her blood lines predominate in his

BREEDING HOUSE AND TRAP-NESTED STOCK ON FARM OF R. A. RICHARDSON
Each pen is given free range for a few hours daily while the others are confined. Pen No. 1 was out at the time photo was taken.

stock today. In fact, no new blood has been introduced into his stock since the mating that produced these three hens was made. He has occasionally bought eggs or stock of other laying strains for observation, with the intention of making new lines, combining his strain with any that would improve it, but he has never found anything he has wanted to use in this way.

On this point he says, in substance:

"There are many good-laying strains. I have had stock of some of the others that I think might perhaps make bigger records, in number of eggs, than I have made with my strain. But I never worked for number of eggs at the sacrifice of other desirable qualities. I want good-sized eggs. The heavy layers that I started with laid rather small eggs. The Bricault line furnished the best layers, but they did not lay as good eggs as the others, nor were they as good from the Standard point of view. Now the general run of my hens' eggs, just as they are gathered, will weigh 26 to 28 ounces to the dozen. I can't afford to put the blood of strains that lay small eggs or light-colored eggs into mine. Neither can I see any advantage in using stock that is poorer than mine from the Standard viewpoint. I want all the quality I can get without letting go anything of what I have already gained.

"I never intend to breed from a bird that is not a pretty good typical Wyandotte. Of course, you know even the Wyandotte breeders who are breeding for ex-

from Pulsifer's own statement which has been quoted it is plain that there had been no intensive breeding for egg production in this stock in the ten years before Mr. Richardson began with it. Mr. Richardson says that this Black Line has always been the best looking, giving many birds of good standard type and quality.

From a combination of the Pulsifer and Bricault stocks he made what he calls his Yellow Line, which is intermediate between the others both in egg production and in show quality. His highest record in the Yellow Line is 265 eggs.

These three lines take their color designations from the leg bands used on the males. The females are not given colored leg bands unless for some special reason it is desired to have a bird's identification conspicuous as the colored leg band makes it.

Different Lines Readily Identified

Mr. Richardson uses a very simple system of marking his different breeding lines for identification. Chicks of the Blue Line are not toe punched at all, except when bred from his best mating, when they are toe punched in the outside web of the right foot. Those of the Black Line are toe punched in the inside web of the right foot;

PRACTICAL OPEN-FRONT LAYING HOUSE ON FARM OF R. A. RICHARDSON

hibition don't always hold to the best type. They have had the extremes and everything between in my time. Several times I have found that in breeding for eggs and for size, shape and color of eggs, my birds of a certain line were beginning to get away from the type. At one time I got them too rangy, but it took only a few years of selection to avoid that fault to breed it out.

"Egg type—laying type? I've had good layers in them all. They come of all shapes and sizes, the good layers do, unless you select for the type you want as well as for heavy laying. The only type I've found in Wyandottes that will not make a heavy layer of good eggs is the very compact, short, round-bodied bird. If hens of this kind lay many eggs, they are small, round eggs—like marbles, because the hen, hasn't room to make big eggs."

From the No. 3 hen Mr. Richardson had three sisters that laid 272, 267 and 238 eggs. The first two he used as the foundation of what he calls his Blue Line, which is pure Bricault stock, and has always given him his heaviest layers. His best record in this line was 280 eggs.

The pure Pulsifer stock constitutes his Black Line. The highest record he has ever had in this line is 214 eggs, which is only one egg better than the average Pulsifer made with his first 12 Wyandotte pullets. But

those of the Yellow Line in the inside web of the left foot. When he crosses the Black Line with one of the others he gives the chicks the toe mark for the Black Line, and punches the outside web of the right foot. When he crosses the Yellow Line with one of the others, he punches the outside as well as the inside web of the left foot.

These three lines take their color designations from that a pure line is falling off a little in egg production, as shown by the average trap-nest records. The cross then made is in the nature of an outcross, the progeny of it being bred back to the line it is desired to stimulate until it is again practically a pure line, having only one-eighth or less of the blood of the other line, and the mark to indicate the cross is no longer used.

Mr. Richardson says that in his experience a pure line that has the vigor to reproduce its laying quality, and perhaps to increase it, will run from five to eight years and then the records will begin to go down. That if then a cross is made with another line, vigor will be fully restored and the line will again run on for five to eight years before showing any falling off in production. It is interesting to note that this observation accords very closely with the experience of breeders of line-bred exhibition stock as to the duration of lines of superior quality. Considering the highest records in his strain, before

R. A. RICHARDSON'S COLONY BROODING HOUSE

and after it came into his possession, Mr. Richardson would not appear to have made striking progress. But when we take account of other things very substantial progress is shown and—in addition—he has produced a strain of heavy layers that are hard to excel as table poultry, which of itself is something of an accomplishment.

Average Over 200 Eggs in Pullet Year

While 280 eggs is his highest record, and in the Black Line he has not gone over 214, MORE THAN FIFTY PER CENT OF THE PULLETS HE RAISES GO OVER 200 EGGS IN THEIR PULLET YEAR. The Bricault hens, among which he found his best layers when he began pedigree breeding, gave a large proportion of broodies—more than half. The 272-egg hen of this line, however, went broody in the five or six years she lived. Mr. Richardson has now practically eliminated broodiness, and in doing so has so increased the average production that among about three hundred pullets trap nested each year he will have not more than one or two go below 150 eggs, and most of his low-record hens are around 180.

The average weight in his flock is approximately standard weight, and his birds are quick growers, mature early, and are very meaty and plump. He has developed the table qualities of his birds because he has an excellent market for table poultry, and fully appreciates the value of that market. The meaty character of his strain and its readiness to fatten, does not—apparently—have any tendency to reduce egg production in the pullet year. The pullets begin to lay at maturity and are easily kept in good laying condition until they molt a year later. Then it requires judicious feeding and management to keep the hens from becoming too fat before the breeding season. This of course is what has to be done with the hens of all stock and breeds that make good table poultry, so it can hardly be considered an objection. It can be avoided only by breeding a spare type that has little tendency to put on fat.

Mr. Richardson considers the meat qualities of his strain just as valuable as the laying qualities. He gets a large proportion of cockerels that will make two-pound broilers at eight to nine weeks, and will dress six pounds at about five months. Such birds as small roasters will bring him three dollars or more apiece in September, and anything that will not bring ten dollars for breeding he prefers to sell for three dollars at that time rather than carry it over for the trade in low-priced cockerels for breeding.

He has been constantly making more severe the requirements both for his own breeding pens and for stock sold. Except in the Black Line, which has not yet reached that mark, he will not breed from a male whose dam has a record of less than 250 eggs. In all lines a hen must have a record of 200 or more eggs in her first year of laying to be either used in his breeding pens or sold for a breeder. He sells males for breeding only from the No. 1 Pen of his Blue Line. For these he has a minimum price for a bird that is simply a utility bird, and an ascending scale of prices for those of various degrees of standard quality.

From the commercial point of view it does not seem "good business" to put the standards of performance so high. There is a market for heavy-laying stock, both males and females, below his standards, at profitable prices. But Mr. Richardson says:

"I don't want anything that does not meet the standards I have set, to go out as representative of my stock." From which it appears that in his line he is very much of a fancier. He says: "I am just as enthusiastic about trap-nest breeding now as when I began nearly twenty years ago. Not that I believe in following the trap nest blindly, as some seem to do. That seems to me all wrong. I want the trap nest to tell me exactly what each hen does. What she is in appearance I can see with my eyes. As for pelvic measurements, loss of yellow pigment and such things, they are interesting, and perhaps they apply sometimes, but the only thing you are absolutely sure of about laying qualities is what you learn from the individual record of the hen, and the only practical way to do that with large numbers of hens is with the trap nest. I find all kinds of pelvic measurements from barely one finger to three fingers in males that sire great layers, and that are themselves from great layers. In my experience the way hens are kept and fed has so much effect on the color of the leg that it is not at all reliable for judging how a hen has laid."

Mr. Richardson breeds only from hens approaching their second year at the beginning of the breeding season. He does not trap nest his breeders. All the pullets that suit him in appearance are trap nested for a full year (excepting, as previously noted, those that persist in laying outside the trap nests). It is on these records that he then selects his breeders. What surplus he may have of hens with a record of 200 eggs or more in their pullet year is then sold.

Mr. Richardson regards good feeding and good care as of equal importance with proper methods of breeding in making a flock of consistently good performers. He does not attach as much importance to accurate balancing of rations as to good quality in the feeds used, and feeding with judgment to keep appetites keen.

DEVELOPING A HEAVY-LAYING STRAIN OF S. C. R. I. REDS

The "200-Egg Strain" of Reds at West Mansfield Poultry Farm, Mr. L. J. Moss, Proprietor, Originated with a 309-Egg Hen—This Was the Highest Individual Record Claimed at the Time It Was Made—Special Helpful Hints on Management of Brooder Chicks and Laying Hens

By John H. Robinson

THIS strain was built up by H. W. Sanborn, who established the farm where it is still bred, and operated the plant for some eight or nine years, during which period the farm became nationally known for the heavy-laying qualities of its Rhode Island Reds. As the farm passed into the hands of a purchaser who bought for investment, and was subsequently acquired from him by the present owner, Mr. L. J. Moss, details of the origin and early history of the strain are not available. Mr. Moss knows the history of the stock only for the four years it has been under his management, two years as lessee of the plant, and two years as owner.

This plant had for a long time the best equipment for trap nesting on a large scale in the country. It is still one of the largest plants trap-nesting all its pullets. With its facilities it can keep individual records for fifteen hundred hens, and during most of the time it is operated well up to capacity. The stock was somewhat reduced during the war, but is now (1921) about 1,300 laying-breeding hens. Its sales of day-old chicks up to the middle of May in this year were over 40,000.

The Sanborn strain as it came into the hands of Mr. Moss was a strictly utility strain—big, rugged birds with the whole range of shades of color found in Rhode Island Reds, just as they happened to come in mating on the plan adopted in breeding for egg production. Mr. Moss began at once to pay attention to mating for color as well as egg production and size. Marked improvement in color has been made by selection within the strain, and to a less extent—in numbers of birds affected—by introducing males of good exhibition quality when he could obtain them line bred for high egg production. In this he considers that he is only making a beginning. While unwilling to take long chances on results of combinations with new blood, by allowing it to be extensively distributed through his stock before he has thoroughly tested a combination, he proposes to go just as far in the production of new lines of superior exhibition quality as the performance of the hens produced in any case will warrant. Still he relies more at present on steady improvement of color by selection within his own stock than on outcrossing with stock better than his in color.

The First 309-Egg Record

The great reputation of the Sanborn strain was based on some very high private records, and especially on the performance of one hen which laid 309 eggs in 365 days, the highest individual record claimed at the time. This record has not again been equaled here, but the lines bred from this hen are consistently extra-good layers.

The system of breeding followed by Mr. Moss aims to use all grades of heavy-laying females with males in nearly all cases from higher producers than the females mated with them. The exceptions are the pens containing the highest record females that are themselves the daughters of some of the highest producing hens.

In mating according to production, with as much consideration for other points as is practical, Mr. Moss' policy is to grade the hens according to performance, and mate with each pen males from hens of a higher productive grade than the females in that pen. Except as to the very highest producing females, the males in his matings are always better bred from the production standpoint than the females—representing higher average production.* While he wants to get all other good points possible with high production, he will not in any case, in a regular mating from which he sells stock, relax the severity of selection for laying quality to introduce improvement in color or other superficial quality.

His prices are graded according to quality, ranging from $1.00 each for eggs, and $1.50 each for baby chicks from his "Championship Mating" (which in 1921 contained hens with records of 252 to 283 eggs, mated with a male from a 283-egg hen that was a granddaughter of the 309-egg hen mentioned) to 14 cts. each for hatching eggs, and 28 cts. each for baby chicks from pullets on test, that are laying at a rate indicating that they will at least qualify for some place in a breeding pen next year.

The lowest grade breeding pen of hens contains those with records of 170 to 190 eggs. The prices for stock

LAYING HOUSE AT WEST MANSFIELD POULTRY FARM

from these pens is 15 cts. each for hatching eggs, and 30 cts. each for baby chicks. Hence the buyer of pullets' eggs and of baby chicks from them is getting a proportion of chicks from birds of a higher producing grade than the hens in the lowest grade pens, at a lower price than from any hens that have completed a year's test.

Cultivating Trade in High-Producing Stock

This plan of catering to all grades of interest in heavy layers, and to all grades of ability to buy, has had much to do with building up the large demand for stock from this farm, and extending the interest in breeding for heavy egg production. Mr. Moss states his attitude in substance as follows:

"We all know that pullets do not always produce according to the records and pedigrees of their parents. Some birds produce offspring that are better layers than themselves. Some produce none as good as themselves. But as a rule, the average of the pullets corresponds

ONE OF TEN PORTABLE BROODER HOUSES ON WEST MANSFIELD POULTRY FARM

pretty well with the expectation based on the records of their dams and of their sires' dams. That justifies the grading of prices according to records and pedigrees. The time, labor and money required to build up a laying strain, keeping accurate records and pedigrees, justifies the high prices we and other breeders who have high record stock ask for eggs and stock from our best matings; and we consider that the fact that we have no trouble in disposing of all eggs through the breeding season at the prices asked for eggs and chicks shows that our estimate of the values of the different grades is as near right as possible, and is so considered by customers.

"Indeed, I have always considered as to all our grades of stock that we are giving the customer who makes the effort necessary to get the highest record his hens can make a little better than he is paying for; because such customers regularly get bigger records from our stock than we do. We are not in a position to work for the highest possible records except by cutting down our stock, and that would seem poor policy to me when we have all we can do to supply the demand with all the stock we carry by our present method, in which the production is really just what our hens will do with ordinary good care. I think that, in general, a pullet from our stock is capable of making a higher record than her dam made with us. From a selling point of view I like to have a lot of customers telling their friends and neighbors that the stock they bought from us laid better than the stock it came from. If we crowd our stock to the limit to make good records here and then sell on those records, not nearly so many of our customers will get results that satisfy them, and make them want to buy from our higher-priced matings, if they first bought from the cheaper ones."

Management of Brooder Chicks and Layers

While regarding the trap-nest record as the key to progress in breeding for egg production, Mr. Moss considers that the foundation of a strain is in the general physical quality of the stock. The two things that he regards as of most importance in growing and handling stock for good production are: the care of the chicks when in brooders, and the proper ventilation of the poultry houses in summer. He puts emphasis on these not as really more important than other things, but because they are the points at which it seems most people fail. He uses stove brooders in colony houses 10 feet square, putting 300 chicks in each house. Of the handling of chicks under these conditions he says:

"All things considered we think this is the best way to grow chickens to get uniformly good growth and development. But, frankly, in our experience, chicks cannot be left to themselves in these brooder houses as much as most people leave them, and do their best. When the house is warm in the early evening many chicks will settle down near the outside wall, especially in corners. You can wire or board so that there is no corner, but the chicks settle down near the wall anyway, and when they begin to feel cold crowd together next the wall instead of moving toword the heat. The only way we have ever found to prevent this is to watch the chickens and make positively sure that they are all in the circle around the stove after dark. It is the usual custom to do this for a few nights, then take it for granted that the chicks are 'hover broke' and will go where they will be comfortable all night, of their own accord. Ever once in a while a big bunch of chicks fails to do this, and one uncomfortable night injures them permanently. It is tiresome watching chicks as closely as necessary in the brooders, but it pays."

In regard to the management of laying hens in summer Mr. Moss says: "The one thing that I am absolutely sure about in regard to making hens lay is that if the houses are properly ventilated when an extremely hot day comes in the late spring or early summer, hens will not go off laying, as so many people complain that theirs do at this season. Hens feel the heat of these first hot days more than they do the same degree of heat later in the season after they become accustomed to the heat. Most poultrymen do not ventilate for summer until they think settled summer weather has arrived. We open the ventilators in the back of our laying houses whenever it becomes uncomfortable with them closed. We take out half the back of the house if it is necessary, but we keep the house cool. If you can keep a cool house, that is half the battle for good egg production in summer.

"The other point is in feeding to keep up the appetite. We find that hens that are fed constantly on dry mash and have been on range through the spring are apt to lose their appetite for the mash at this time. Then we confine them more closely and give the same mash wet instead of dry for about two weeks. This change breaks the monotony of their diet. They eat more of the wet mash than they would of the dry. But after about two weeks of the wet mash, hens that have been accustomed to dry mash will go back to the dry mash with as good appetite as ever".

CHAPTER VI

How Some Well-Known Breeders Have Developed High Egg Producing Strains

In This Chapter Are Described the Methods and Successes of Tom Barron, England; Professor W. R. Graham, Canada; O. F. Mittendorff, Illinois; John S. Martin, Canada; Pennsylvania Farms, Pennsylvania; M. K. Bohlander, Illinois and C. L. Manwaring, Indiana—Success in High Egg Production Is Not Conditioned Upon Any Particular Breed, Location or Method

AMONG the many breeders who have been successful in developing a high degree of productiveness in their strains, the selection of a few for representation in this chapter has been a somewhat embarrassing task. Obviously it was possible to include only a small number, and details in regard to methods and results have necessarily been limited even in the few reports presented. Such as have been selected, however, serve to supplement the successful experiences given in greater detail in preceding chapters, and to illustrate many of the points brought out in the discussion of special subjects treated in Part I.

At the time that egg contests were first being developed in this country and interest in increased production was becoming widespread, Tom Barron, of England, was the best known breeder of record layers, and his White Leghorns and White Wyandottes were successful in carrying off a majority of the prizes at the first contests. Mr. Barron is no longer the easy winner that he was some years ago, but unquestionably he has exerted a marked influence on the "utility" branch of the industry, and it is fitting that this chapter should begin with his own statement regarding his method of breeding and selection.

BREEDING AND SELECTION FOR HIGH EGG PRODUCTION

Extracts from a Paper Read by Mr. Barron at International Poultry Conference, London, March 11-15, 1919, Describing the Methods by Which He Developed His Famous High-Producing Strains of S. C. W. Leghorns and White Wyandottes

By Tom Barron, Catforth, Preston, England

THE first and most important point in breeding laying hens is health. If you have not healthy and hardy foundation stock, then all your trouble and labor will be in vain. Anything that is conducive to good health and hardiness in breeding laying stock must not be overlooked. Good housing, management, feeding and cleanliness are most essential if success is to be attained.

I breed from different families of fowls on my farm and have quite a number of these families. It is surprising how long one can go on, especially on a large farm, without seriously inbreeding. I do not believe in inbreeding laying hens more than is absolutely essential.

Sometimes I go out for a change of blood, and the way that suits me best is to buy good hatching eggs, rearing the chickens and putting them when reared on the trap nest, always testing them myself. If I only obtain a few good ones, this serves my purpose. I can breed a good cockerel, which is all that is required for my purpose. In this way I can continue for quite a long time.

After the question of health, as mentioned before, the stock intended to be bred must not be forced in any way, but should be kept as naturally as possible in good-laying condition and, above all, the birds must not be too fat.

All pullets intended for breeders on my farm are trap nested the first year. Those that lay 180 to 230 eggs in a year are selected for stock to be used in the second season's breeding—that is, as hens. Those that lay over 250, provided they lay good-sized eggs, are the ones my cockerels are bred from to mate to the above hens. My best laying pullets have always been bred from cockerels whose dam laid a large number of eggs.

Tells of His Ideal Pen

I will here give what is considered an ideal pen for breeding first-class layers. Use hens coming on to lay in February that have laid 200 to 230 eggs each during the previous years—good-sized eggs. These hens are mated to a cockerel twelve months old bred from a hen that laid over 250 large eggs in her pullet year.

My experience is that the above will breed far better layers than hens with higher records during that season.

The system of management found to be the best is as follows: After the pullets have completed their first laying year and you have finished trap nesting them, remove them to the stubbles or free range for the winter (English climate), feeding them very sparingly and in no way inducing them to lay. Try to keep them from laying if possible. Then when you wish to begin breeding, mate a cockerel to them and begin to increase their feed. In a few weeks you will have an abundance of good, sound, fertile eggs.

At this time the birds ought to be very carefully examined for vermin. A good dusting with insect powder every fourteen days, especially the cockerels, will be found to exert a great influence on the fertility of the eggs.

Another thing ought to be mentioned: eggs for hatching should never become chilled. Eggs ought to be brought in from the nest as often as possible during the early months of the year, be kept in a room with temperature about 50 to 60 degrees and be set as fresh as possible.

Tells of Two Experiments

Some people say it is not possible to breed good females from a high-record hen. Here is an experiment that may be worth noting. I had a hen which laid 289 eggs in her pullet year. She was recorded with four sisters. The average of the pen was 261 eggs. An unusual occurrence happened. Why I say unusual is because it is not often that the hens that lay the largest number in the pen also lay the largest eggs. But in this case the 289-hen laid the bigger eggs. All her eggs were hatched, and the pullets bred from her were carefully recorded. These laid good-sized eggs, but the average of the pullets was only 150 eggs each during the year. Several pullets laid from 180 to 200. These were the best, and were bred from and mated to an unrelated cockerel. When they were in their second season's laying the progeny produced from these hens averaged 250 eggs each in the year.

Another test I once carried out was that brothers and sisters were mated together. The result was, as one would expect, very poor laying of the progeny. These were carefully trapped and the best also bred from, mated to an unrelated male. The result was wonderfully good layers of over 250 eggs from each bird during the year. I give these two instances for what they are worth.

Selecting the Layers

My ideal of a breeding pen is eight hens of the general-purpose breeds and ten of the lighter breeds. I will now give my method of selecting these layers, and for the man who does not trap nest, equally with those who do, this selection is useful for the purpose of choosing birds for a laying competition, or to use as breeders. Get together in a house about twenty to thirty pullets that have been bred on the lines already given and select for the following points:

Large, bright eye with as large a black core in center as you can find. It is from this source that the quickness of the eye is derived for procuring a great amount of food which a bird with a dull, sleepy eye would not see.

A CONTEST PEN OF BARRON S. C. W. LEGHORNS
Good representatives of this English heavy-laying strain. This pen won first place in the 1912-13 Storrs Laying Contest, with a record of 1,190 eggs, or an average of 238 for each bird.

Rather narrow head with eyes projecting well from the skull. The skull should project low down the neck.

Fine texture of comb, which should not be too small nor too beefy. The comb and wattles should feel like silk.

Stout, short beak and short face. I do not like a gypsy-faced bird; these are not good layers.

The neck ought to be thin and medium in length; the back as long and broad as possible and very wide across the wings.

A short breastbone of about three and three-quarters to four inches I find in the best layers.

The male bird ought to have a broad breast and be very wide behind. He should possess a wide cushion and, when handled, this ought to feel like a sponge, with no hard substance whatever in the abdomen. when standing upright, his appearance ought to be wedge-shaped; not deep in front, but very deep behind. Also I prefer a thick-set bird, with rather short legs and good carriage. The legs ought to be thin. In the females these lose their color after a very short period of laying.

The pelvic bones of females ought to be very thin and straight, if possible, without having any fat on them. I discard those that are fatty or thick at the end of these bones. And the greater the distance from the breast or keel bone to the pelvic bones, the better layer the bird will be, as a rule. The measurement of a good layer (when in lay) should be about four inches in the "general-purpose" breeds and three to three and a half in light breeds. The tail of a good layer is nearly always carried rather high.

Small birds for the breed or variety are the best layers. But I do not think it wise to get laying stock too small. My belief is that it is the small birds that generally breed the small-egg layers. (By this I do hope we shall not get our layers too large.) The size of egg ought, I admit, to be reasonably considered.

The cockerels must always be bred from a hen that has a good egg record and that is sound and healthy. These points can be taken also for the selection of cockerels, only a cockerel will not measure the same as a hen from the breastbone to the pelvic bones. I should say a cockerel that would measure two inches from breast to pelvic would be a first-class breeder. I like a vicious male bird—one that will take care of his hens.

HOW A HIGH-PRODUCING STRAIN OF BARRED ROCKS WAS ESTABLISHED

Report of a Recent Interview With Professor W. R. Graham of Ontario, Who Has Developed a Strain, the Pullets of Which Mature in Less Than Six Months and Average to Lay Over 160 Eggs Each, and 190 with a Little Culling

PROFESSOR W. R. Graham of Ontario Agricultural College, is one of the pioneers in breeding for egg production. For over 12 years he has been studying this problem and, following a definite scheme of breeding, has fixed this character in his strain to an unusual degree. Now, after a dozen years of scientific effort, his pullets, taken as they run, will average better than 160 eggs each, or with such culling at the beginning of the laying season as a reasonably skillful breeder can give, will average 190. They are characterized also by exceptionally early maturity, regularly beginning to lay when around 165 days old. This character is so definitely fixed that whole flocks reach 50 per cent production by the time the birds are six months old and they do this not now and then but every year. As a matter of fact, this strain of Rocks reaches maturity at almost exactly the same age as the strain of Leghorns bred by Professor Graham, at which time the pullets average to weigh 5½ pounds, or 2 pounds more than Leghorns of the same age. The pullet-year production is almost the same for each strain, the average for the Leghorns being 166.4 taking the pullet flocks as they run, without culling. Such results as these constitute a distinct achievement in poultry breeding and the methods by which they were secured are worthy of close study.

Professor Graham has been breeding along somewhat the same lines as were followed by Dr. Pearl at the Maine Station, and in more or less close cooperation with him for a time, so that the results secured must be regarded as to some extent confirming Dr. Pearl's Mendelian viewpoint. However, Professor Graham is a practical breeder, as well as a scientist, and it would be difficult to find any important difference between his applied methods and those of progeny-testing breeders generally. As a matter of fact, he does not follow Dr. Pearl in some important details. For example, in discussing the latter's procedure in basing his estimate of the laying value of the hen on her winter production, he gave the accompanying table for a pen of birds taken as they run, to show that there is no necessary correlation between winter production and the total for the year.

It will be noted that, aside from the extremes, winter production cannot be accepted as an indication of the probable total record of any of the birds represented. Hen No. 231, with a winter production of 22, laid only

(*The information on which this article is based was secured in an interview which the writer had with Professor Graham, fall of 1920, a special trip having been made to Guelph for this particular purpose.—H. W. J.)

93 eggs for the entire year, while the highest hen, No. 239, laying 91 eggs in the winter, laid 263 for the year. But the highest record for the year (264 eggs) was made by a hen that laid only 63 in the winter while the lowest record aside from Hen No. 231, was made by Hen No. 234, whose winter record was 64. These individual records also show that there was only one poor layer in the flock and only two that laid fewer than 150 eggs—an excellent illustration of the uniformity with which the characters of high production have been established in this strain.

Professor Graham's comments on this table were to the effect that winter production must be confirmed by July and August records, to afford a reasonably good clue

Winter Production at Ontario Agricultural College Contrasted With Total Production for Year

No. of Bird	Winter Eggs	Total Eggs for Year	Eggs Set	Hatch	Mortality
231	22	93	11	8	0
232	63	264	21	21	5
233	75	234	17	14	4
234	64	134	2	0	0
235	68	188	16	10	5
236	71	243	24	11	8
237	57	193	14	11	5
239	91	263	27	12	8
241	45	167	24	17	7
243	41	156	15	8	6
244	41	154	10	5	2
245	45	194	14	8	7
246	48	236	18	8	3
247	17	164	14	12	8
248	80	207	27	15	5
249	56	150	18	5	3
250	55	218	13	9	3

This table, which gives laying records for a number of Barred Plymouth Rocks bred at Ontario Agricultural College, not only contrasts winter production with the total for the year, but also gives interesting data in regard to hatching and mortality, which show marked differences in the value of individual birds as breeders.

to laying ability. Dr. Pearl's rule certainly enables the poultry breeder to hold onto early maturity, which is important in securing high records, but that alone is not sufficient.

Professor Graham regards high production as quite largely a question of nutrition. In other words, the birds that are skillfully fed and that have the ability to digest and assimilate large quantities of feed are pretty apt to be good producers, and he is of the opinion that any bird that laid heavily in July and August was capable of being a good producer in the early winter preceding if it had had a fair chance.

Efforts To Improve Barring

His records are full of most interesting facts similar to those presented in the foregoing table relating to breeding, and everyone of them appears to be in his memory as well as on his books. The writer managed to catch a few on the run, but missed a great many that, lacking a stenographer to take full notes of the conversation, could not be recorded. One striking illustration, however, of the cumulative effect of consistent breeding for production may be mentioned. This was the case of Hen No. 262 that had a record of 271 eggs. She was hatched April 8, 1918; her mother, hatched April 1, 1915, laid 179 eggs; her grandmother, hatched April 27, 1912, laid 140 eggs and her great-grandmother, hatched April 10, 1910, laid 129. Some other records are referred to elsewhere in this article.

Professor Graham does not regard himself as having "arrived." On the contrary, he is frankly a student of the subject and is constantly experimenting with a view to further improvement in his strain, not only with reference to production but also size of egg and standard qualities. As a matter of fact, he appears to have been more or less generally misunderstood with reference to the last-named subject. There is among many a belief that he has been indifferent to standard qualities, or at least has made no effort to breed for them, which is far from the truth. As a matter of fact, while he has placed productiveness first he has constantly striven to improve the color of his birds, and if they still fall short of satisfying the breeder of exhibition stock it is not for lack of effort to make them do so. The fact appears to be that he had the misfortune to start with a family of Rocks that is distinctly lacking in certain standard characters and in spite of his best efforts he has not been able entirely to remedy the defect, though he has made a marked advance in this direction.

The foundation stock for this strain was secured from

TYPE OF HEAVY-LAYING STRAIN OF BARRED PLYMOUTH ROCKS DEVELOPED AT ONTARIO AGRICULTURAL COLLEGE
Hen on the left has a record of 242 eggs, one in the center 264 and one on the right 244. This is one of the first heavy-laying strains founded in America, and has been bred to a good degree of uniformity in productiveness, also in early maturity. Birds of this strain average 50 per cent production by the time they are six months old, and flocks taken as they run will average 160 eggs per bird in their first laying year.

Professor Gowell, while he was in charge of the Poultry Department at the Maine Experiment Station, and it has been carefully line bred ever since. For some reason which has never been satisfactorily explained, it seems to be a peculiarly difficult matter to introduce standard barring into this family. At Guelph, last October, were pullets that after all these years of careful breeding very closely resembled the color of the original birds secured from Professor Gowell, these having the wide, bluish-white and bluish-black bars that are to be found so generally in inferior Barred flocks and quite different from the narrow, snappy black and white of the modern standard Rock. Professor Graham has made repeated efforts to correct this, both by selection and by introducing new blood but he reports regretfully that whenever he has secured the popular narrow, clear-cut barring he has always got, along with it, slower maturity and lower production. He does NOT say that narrow barring and high production cannot be combined. On the other hand, he believes it can be, but frankly admits that, so far, he has not been able to accomplish the feat with his strain. By way of proving that he does not look upon high production in exhibition Barred Rocks as at all impossible we may mention one test with an exhibition male without a known record back of him, but selected on egg type, which when mated with Hen No. 271 produced pullets with records of 228, 219, 171 and 152. The same hen mated with a male from a 277-egg hen produced pullets with records of 213, 188 and 182.

Recognizing the fact that his strain did not have standard qualities and that he must go outside to get them, Professor Graham has resorted to experimental outbreeding practically through the entire period in which his strain has been in process of formation. While holding onto high production so skillfully that he has made it a definite character of his strain, he has introduced blood from the Thompson, Cosh and Buck strains, none of which, however, appears to have satisfactorily blended with his own strain. His method of introducing new blood without risking loss of ground in production has been to inject it through the female line, and never in greater strength than 25 per cent. This means that when he has selected a bird which is to be used in introducing new blood it is bred to one of his own birds and the offspring again bred to his own strain, and not until the foreign blood is thus reduced to one-fourth does he risk using it in any of his regular matings.

Even with this care in the introduction of new blood there will be misfits and disappointments, and Professor Graham called my attention to some curious outcroppings of undesirable characters in the last year's flock, years after they presumably had been bred out. It was with these in mind that he said with great earnestness: "When you put anything bad into your strain please remember that it is there—NEVER FORGET IT!"

To Get High Production, Pullets Must Be Hatched At Exactly the Right Time

Professor Graham is extremely careful with regard to date of hatching. With his early-maturing strain he finds that if the pullets are hatched a little too soon they are almost as apt to molt in the fall after a short laying period as are Leghorns; and if hatched too late, winter production (and therefore the entire year's production) is cut down. In his climate and under his methods he regards the period from March 25 to April 15 as the best date. Pullets hatched at this time are almost certain to escape the fall molt, but can be expected to come into 50 per cent laying by October 15 to November 1, and with proper handling they will hold this rate of production right through the winter.

In discussing the effect of time of hatching upon production Professor Graham gave the following illustration: The son of a 310-egg bird was bred to a hen with a record of 271 eggs, the mating being continued through April, May and June. Three pullets hatched in April laid 229, 218 and 207 eggs; five May pullets laid 194, 177, 167, 162 and 150, and three June pullets, 122, 119 and 75.

Professor Graham had no explanation to offer as to how he had succeeded in developing such an early-maturing strain of Rocks aside from the fact that he has consistently selected his breeders with that in mind. It will be remembered, however, by those who have followed the work at Maine, from which flock Professor Graham's stock was originally secured, that it also was early maturing. In one of his bulletins Dr. Pearl mentions the fact that the pullets are apt to come into laying as early as September, before they are moved to winter quarters.

Professor Graham's insistence upon early maturity must be considered in the light of his northern latitude and the shortness of the growing season. Obviously no character that results in delaying maturity can be tolerated there, but even if it is conceded that narrow barring does mean slower maturity, that is not necessarily a handicap further south where the growing season is longer. The reader should note that it is not claimed that a Ply-

HEAVY-LAYING, EXHIBITION-TYPE WHITE LEGHORNS ON CYPHERS POULTRY FARM
The Cyphers Poultry Farm of Elma Center, N. Y., was one of the first large plants to take up systematic breeding of trap-nested stock to secure increased egg production. The D. W. Young hen in the middle had a pullet-year record of 288 eggs and was of good show type, being twice a member of first pen at Madison Square Garden Show. On the Cyphers Farm she became the dam of a long line of a daughter and granddaughters which, like the two shown on her left and right, were of excellent exhibition quality and made records of 202 to 251 eggs, each.

mouth Rock pullet that begins laying in six months is any more likely to be a heavy producer than one that matures in seven or seven and one-half months, provided the latter is hatched correspondingly early in the season. What is claimed is that early maturity is imperative in securing high production in that climate. The pullets of this strain average to weigh 5½ pounds by the time they begin laying, and reach fair size as hens. The cockerels also are exceptionally quick growers. We handled one that at 6½ months weighed 11 pounds.

Breeding Out Broodiness

Professor Graham believes that broodiness is a character that can be bred out and also that it is regularly associated with a bulging skull, or one that overhangs the eye. He states that in all his experience he has never found a persistently broody Rock that did not have that external character. If general observation should confirm this theory it will greatly facilitate the effort to breed out broodiness or reduce it to a point where it would no longer be a serious objection, doing this simply by eliminating from the breeding pen birds that have this peculiar skull formation.

How the Layers Are Fed

Without a doubt, a great deal of Professor Graham's success in securing high production among his birds is due to his skill in practical management. There are few station workers, or practical poultry keepers either for that matter, who are so expert in their handling of young stock and adults. In many respects his methods differ from those that are commonly employed on poultry plants. For example, he feeds his young chicks quite largely on soft feed, including relatively large quantities of boiled, mashed turnips and other roots. He is convinced that young chicks need a great deal of "filler," and regards boiled roots as exceptionally desirable for this use. Doubtless the free use of soft feed, milk after the chicks are a week old, and the comparatively cool summers which are characteristic of this section of the country, have much to do with the rapid growth and development that he regularly secures.

Pullets are carefully culled in the fall when they go into the laying pen, and the ration thereafter consists of the usual dry mash and scratch grain. The formula for the laying mash is 500 pounds of shorts, 500 pounds of low-grade flour, 750 pounds of corn meal, 300 pounds of bran, 500 pounds of oat chop, and 100 pounds of tankage. The grain mixture consists of 40 pounds of wheat, 40 pounds of corn, 20 pounds of oats, though prepared scratch feed is also used. It will be noticed that the proportion of tankage is extremely low, amounting to only 4 per cent of the total weight of the mash. Where this formula is used however, the hens have plenty of milk to drink, which supplies the need for animal food to a great extent. Once a week milk is omitted from the ration and the birds are given Epsom salts dissolved in the drinking water.

As contrasted with the feeding at Vineland where an abundance of mash is regarded as vital to high production and where it is consumed about in the proportion of 1.4 parts of mash to 1 part of grain, Professor Graham's ration carries only 20 to 25 per cent of mash, the total feed consumption at Guelph averaging about 72 pounds, of which 16½ pounds only are dry mash. For green feed the average hen's ration carries 12 pounds of dry oats sprouted. Cabbage also is fed quite extensively and he regards it as by all odds the best green feed he can supply—better even than sprouted oats, but necessarily limited in quantity owing to the difficulty of storing it for winter and spring use. He also feeds rape to some extent in the fall but limits the quantity, as he finds that too much rape will color the yolks. He believes that this trouble is most likely to occur after frost.

LINE BREEDING AND TRAP NESTING FOR INCREASED EGG YIELD

General Statement of Remarkable Results Obtained in Prolific Egg Production by a Long-Time Breeder of Utility Barred Plymouth Rocks—For Fourteen Years He Has Been a Daily Student of the Actual Trap-Nest Yield of a Line-Bred Strain

By Grant M. Curtis

AT Lincoln, Illinois—population about 12,000—in a choice residence district within two hundred feet of a popular street-car line and where the mail man twice a day brings the inquiries and orders to the front door, lives O. F. Mittendorff, proprietor of Mapleside Poultry Plant, and originator of the Mapleside bred-to-lay strain of Barred Plymouth Rocks. After Mr. Mittendorff had made a financial success of his poultry business he paid $5,000 for two acres of land on which his present home and poultry plant are located, paying this price for the land bare of any buildings. The residence and poultry buildings cost $16,000 and the entire plant has been paid for out of the profits of the poultry business, conducted on lines worked out by Mr. Mittendorff during the last twelve years. Additional to this he has made a good living for a family of four, has educated two daughters and has paid for a farm of eighty-two acres, price $250 per acre. And the poultry did it. Fifteen years ago last summer Mr. Mittendorff had to retire from the grocery business, conducted in Lincoln, as a result of ill health—a physical breakdown from overwork. He says that the poultry business, in addition to paying him handsomely, has saved his life, as he believes. About the time he was forced to give up the grocery business and other hard physical labor, he read in R. P. J. about an egg-laying contest conducted in Australia. It was this that put the idea into his head to test out some Barred Plymouth Rocks he then had, doing so with a view to learning what their egg-production ability might be. Becoming more and more interested, he sent to J. W. Parks, Altoona, Pa., for some eggs for hatching; also to the Maine State Experiment Station, Orono, for fifty hatching eggs from the Gowell strain of Barred Plymouth Rocks, which then was mentioned frequently in the poultry press.

A TYPICAL HIGH-VITALITY MITTENDORFF MALE

Next, and without loss of time, he put in trap nests and during the fifteen years they have been rigidly faithful to trap nests—"from the first day we installed them to the present hour," as Mr. Mittendorff expressed it. This was not his first experience with poultry. He is now 58 years old and kept his first chickens 40 years ago. Was taken out of school at 16 years of age and put to work in his father's grocery store. However, during those early years, when living with his parents and also after his marriage, they merely "kept chickens" according to the general ideas of that day in Lincoln and vicinity, although those early chickens were Barred Plymouth Rocks, so called. But no effort was made to breed them to standard requirements, nor to obtain prolific egg yield. Let us now report our interview with Mr. Mittendorff in the order that it took place in his pleasant home.

WELL-MARKED EARLY-MATURING PULLET

This "Mapleside" pullet was 5 months old at time photo was taken and shows early development and good type.

The Interview with Mr. Mittendorff

"We aim to raise, including those raised for us, about 3,000 head each year. As a rule, we keep about 450 to 500 on the home place. Raise here on these two acres about 1,000 each year, but we begin to sell them off at eight weeks old. Yes, we sell baby chicks, as conditions permit. The eight-weeks-old birds are sold as breeders. After that we let them go at any age.

"We have line-bred at least thirteen years and in every generation we get 200-eggers. If a trap-nest bird does not produce 200 eggs or better, she does not look good to me and if we have a family that failed to produce a good share of 200-eggers we would discontinue that line.

"The highest trap-nest record made by a bird of the Mapleside bred-to-lay line was a pullet I sold to Edgar Weber, Williamston, Mich., in October, 1913. There were three pullets in the lot and they laid as follows, as per his report: No. 329 laid 221 eggs from November 20, 1913, to November 8, 1914; No. 353 laid 219 eggs from December 14, 1913, to November 16, 1914, and No. 352 laid 290 eggs from November 25, 1913, to November 23, 1914.

"But it was 'Liberty Lass,' with a record of 268 eggs in 365 days that put me on the map. I discovered her by trap nest after I came near selling her for $2.00. This was back in the season of 1911-1912. It was her blood that had the prepotency, giving us, in surprising degree, the heavy-laying trait now possessed by our strain. Here are some of the facts, as taken from our carefully kept records:

"Twenty daughters of 'Liberty Lass' laid 4,064 eggs in one year, an average of 203 eggs each, and one of these daughters, No. 403, laid her 250th egg by the time she was eighteen months old. There were only the twenty birds in the this flock at time of test.

"Ten of the first granddaughters of 'Liberty Lass' laid 2,123 eggs in one year, an average of 212 each. There were twenty-two birds in the flock at time of this test.

"One of the great-granddaughters of 'Liberty Lass' laid 274 eggs in one year.

"Six sisters, that were sired by sons of 'Liberty Lass,' laid 1,457 eggs in one year, an average of 248 eggs each.

Females Also Transmit Laying Powers

"In my opinion there is no question but that a high-production male will greatly improve a low-production female, but the reverse also is true—a high-production female mated with a low-production male will transmit increased egg production power into the offspring, both males and females. In the last thirteen years here at Mapleside, we have tried all sorts of combinations, but we are not prepared to say that prepotency is carried more by the male than by the female.

"No, we do not find that type governs. 'Liberty Lass' was not a long-bodied nor a long-backed hen. We have a limited number of birds each season among the high producers that have long backs and frequently a high producer with a reasonably long body and back, but we have equally high producers with what would be called short bodies and short backs, comparatively speaking. Our observation and records show that good layers come in all types, in all shapes and in all colors, generally speaking. In other words, we have not yet found any distinct type that stands out as a better layer than the average of their mates, generation after generation.

"The Hogan system, as we understand it, has not worked out well with us. Have studied it and tested it but did not find it reliable. With us that plan failed badly, and both ways. Birds that by the test should have done well, did not; while others that ought not to have laid well, proved to be good layers. One hen that laid 249 eggs did not show up well by the Hogan test, while another that did show up extra well laid only 179 eggs.

"When a man uses trap nests fifteen years—uses them every day in the year, keeps careful records and studies these records—he has a lot of theories knocked out of him. His day of judgment is the 365th day.

Firm Believer in Blood Lines

"Yes, I am a firm believer in blood lines. I know the value of line breeding as well as I know my name. I do not believe in breeding them too close, not after we get our lines established. We do use the same blood, but it is somewhat distantly related. From what I have read, we do not breed as close as the feather-line breeders do. I tried mating brothers and sisters, doing this two or three times, but I quit partly on suspicion and partly on account of visible results.

"THE FOUNDATION OF HIGH EGG PRODUCTION IS VITALITY. Without a strong constitution you are lost. Among our male birds you will note a number with frozen combs. That came about because of my strong belief in plenty of fresh air at all times of the year. Am also trying to establish or produce a strain that will stand up well under average farm conditions. You have got to 'breed the lay into them and feed the eggs out of them,' as Mr. Quisenberry has put it. That statement agrees with my fifteen years of close study of the subject, also with our practices here at Mapleside. April is our favorite month for hatching. We never breed from the earliest layers—meaning pullets that begin to lay at five months or five and one-half months old. The April-1 hatched pullets begin laying in October, at the end of seven months, and we do not want them to begin earlier than that. Blood lines have produced the goods with us and our records furnish the proof. Here is a resume of our records, dating back several years:

"Eight years ago we produced 20 per cent more hens laying 200 eggs each in one year than we produced the previous year.

"Next we produced 30 per cent more hens laying 220 to 240 eggs in one year than were produced in the pre-

LIBERTY LASS—RECORD 268 EGGS

This bird not only made a splendid trap-nest record herself but has proved exceptional in her ability to transmit productiveness to her offspring. One of her daughters made a record of 250 and in the third generation a record of 261 was developed. Lass 4th made a record of 249; Lass 5th, 274; Lass 6th, 267; Lass 7th, 254; Lass 8th, 270.

vious year.

"Later, 54 per cent of all our 1914 hatched pullets on hand January 7, 1916, averaged 202 eggs each.

"Another achievement was that 80 per cent of all hens on hand January 1, 1917, averaged 201 eggs each.

"Our latest and greatest achievement was that more hens during 1918 made trap-nest records of 200 to 267 eggs than during any previous year, and 1920 records show a further advance to 270—our best individual record to date.

Some Hens Lay Through Molt

"Yes, we have had hens lay right through the molt, including the time when they were putting on new plumage, doing this right along when covered with new feathers one-half inch long and not enough old plumage left to cover the palm of my hand.

"Naturally we like the long-cycle birds, those that lay several eggs at a stretch and we have had them lay forty or more eggs in a cycle, but we do not work for this nor recognize any special value in it; what we are after are yearly records and we want our birds to lay well during the fall and winter months when eggs are of most value—that is, bring the highest prices when sold on the market.

"No, the birds when molting do not lay every day, but they continue to do well. I believe they can be bred so that they will lay during the molt, and it isn't necessary for them to be forced into full molt. Molting can be regulated by feeding so that it will be gradual and our experience shows that if the 'lay' is bred into them, they will keep it up in some cases during the molt even when it is a full molt.

"We practice the plan of 'resting' our male birds. In our pedigree pens they are kept up during the breeding season till 4 p. m. At that time most of the eggs have been laid and are out of the way. We think this practice helps matters. Our idea is that we should do all we can think of that will conserve and promote vigor and stamina. We notice the special activity of males before roosting time and have decided that Nature must have some good reason for this.

"If I wish to start a new family or when we do start a new family, we use a high-record, vigorous hen with desirable qualities as to shape and color, mating her with a blood-line male, some distance removed, and with this pair we go to work. Have done it repeatedly. Sometimes the progeny go beyond the high record made by the hen at the head of this family, but often they do not; but, year after year, there has been a distinct gain in the flock average and repeatedly a new individual record has been made that was higher than any previous record we had reached.

"Back there, fifteen years ago, we first tried about 100 hens, mostly our own, and only seven laid above 200 eggs. There was exhibition blood in some of those birds. Next we sent to Professor Gowell for 50 eggs and I got a few from J. W. Parks; then we dipped back into our own flock, using the 200-eggers, as proved by trap-nest performance. After this first cross or 'mixing' they went back a little, then gradually forged ahead. Since then we have introduced no outside blood—that is, during recent years.

Favors Standard Type and Color

"We favor standard type and a bright, attractive color. We appreciate good looks and, to be frank about it, we notice that other people, judging by our correspondence, like the attractive fowls better and we know they take better care of them, as a rule.

"Yes, we practice single matings, generally speaking, based almost solely on trap-nest records and blood lines. We breed only from pedigree hens in carrying on our pedigree work. We never hatch pullet eggs. Tried that for a number of years and feel sure that we can keep up vigor and stamina with greater ease by the sole use of hatching eggs produced by well-matured birds a year old and older."

On the two acres, with the help of his wife and daughter (Mrs. Mittendorff looking after the trap-nest records, helping Mr. Mittendorff in this work and with the baby chicks, and the daughter doing most of the stenographic work), Mr. Mittendorff has built up an annual business close to $16,000 per year, "more than one-third of which is returns for our labor," as he expressed it. About 60 per cent of this amount is received for breeding stock and surplus cockerels and the remaining 40 per cent from hatching eggs.

ESTABLISHING A COMBINATION EXHIBITION HEAVY-LAYING STRAIN

Interesting Proof that It Is not Necessary To Sacrifice Standard Exhibition Quality in Order To Secure High Egg Production—Regal-Dorcas Strain of White Wyandottes Makes General Flock Average of 160 Eggs per Bird and Individual Records up to 287 Eggs, also Produces Winners at Largest Fall Fairs and Winter Poultry Shows

POULTRY breeders, in the past decade or so, have seen many practical, clear-cut demonstrations of the truth of the old saying that "The longest way around is often the shortest way home." There were breeders (and their name was Legion) who, having discovered the great possibilities in the way of securing increased egg production by trap nesting and pedigree breeding, succumbed to the temptation to cut across lots by breeding for productiveness with almost complete indifference to every other consideration. A high egg record admitted almost any bird to their breeding pens regardless of how inferior it might be in the characters associated with breed quality.

They got egg yields that way—no doubt about it. But they also got birds of so inferior quality that they were salable only at prices approximating those quoted for market fowls. It came as a distinct surprise to many of these "utility" breeders to learn that, while the average American buyer demands stock of high-laying ability, he just as certainly demands birds of first-class standard quality. And many a breeder who has made the attempt to meet this additional requirement has found it a long, uphill struggle to get back the standard quality that he threw away in his corner-cutting pursuit of high egg records. There were breeders, however, who while quick to see the advantage of developing egg-laying ability in their strain never lost sight of the fact that permanent

A PEN OF REGAL-DORCAS-STRAIN WHITE WYANDOTTES
The hen with the "X" over her head is Dorcas (record 241 eggs), on which was founded the Regal-Dorcas line of high producers. Illustrated on next page are descendants from this bird.

This article is based upon information secured by the writer during an interview with Mr. Martin at his farm, fall of 1921.—H. W. J.

success in this line is dependent upon maintaining a high standard of breed excellence. That meant slower progress of course, but it meant permanent progress.

John S. Martin, originator of the Regal-Dorcas strain, of White Wyandottes, a foremost Canadian poultry breeder, was one of the first to see the importance of effecting a practical combination of exhibition quality with high egg production and he has been exceptionally successful in doing this. Fifteen years ago—years before the laying contests at Storrs and Missouri had fanned public interest in heavy production to white heat—Mr. Martin had started his laying strain with "Dorcas," an exhibition-bred White Wyandotte with a record of 241 eggs in twelve consecutive months of laying. From this bird, mated to a male from a 205-egg dam, has descended the Regal-Dorcas strain that has won high honors in laying contests and that at home and in the flocks of hundreds of customers has demonstrated beyond a doubt that it is an established heavy-laying strain and this result has been attained without compromising on standard quality.

In other words, Mr. Martin, in developing this strain, did not breed just layers, he bred WHITE WYANDOTTE LAYERS—birds that had to pass his exacting requirements as worthy representatives of the breed and variety, as well as to make high records. To this day, if any individuals are not typical of the breed in all points "out they go"—regardless of what their records or ancestry may be. Fifteen years of such constructive breeding for high egg production has given Mr. Martin a strain in which the general run of pullets average to lay 160 eggs in 12 months, with official individual records as high as 262 eggs and sworn private records up to 287.

Regal-Dorcas birds have made exceptional records in laying contests, not only in Canada but also in the United States, where they have necessarily to travel long distances and must become acclimated. For example, at the 1919-1920 American Laying Contest at Leavenworth, Kan., Mr. Martin entered three pens of five birds each, these winning first, third and fourth in White Wyandottes, the best pen leading all heavy and medium-weight varieties with a record of 1,038 eggs. Eight pullets out of the fifteen laid a total of 1,761 eggs, or an average of 220.

Writing us in regard to these birds Mr. Martin says: "Had three pens of five birds each in contest; won first, third and fourth in White Wyandottes, against five other breeders. Records, 1,038, 947 and 866. Best pens led all

A PEN OF HEAVY-LAYING, EXHIBITION-QUALITY REGAL-DORCAS WHITE WYANDOTTES

These birds, with records of 217, 255, 223, 185, 209 and 213 eggs at American Egg-Laying Contest, are good illustrations of John S. Martin's success in combining high egg production and standard quality. Pullets of this strain have made many other excellent official records running as high as 262 eggs in one year, and sworn private records up to 287.

HOW SOME WELL-KNOWN BREEDERS HAVE DEVELOPED HIGH EGG-PRODUCING STRAINS

heavy and middle-weight varieties; were beaten only by White Leghorns. Best pullet record Dorcas Wyandottes, 262 eggs in 365 consecutive days and this pullet did not begin to lay until Dec. 1st—30 days after the contest started. Eight of my pullets, out of the fifteen entered, laid a total of 1,761 eggs, or an average of 220. One pullet laid 62 eggs in 62 days without missing. These pullets were typical Wyandottes with fairly good show qualities. My birds arrived at contest ten days late. Above record therefore really shows 355 days' work, or ten days less than a full year and they had traveled more than a thousand miles to get to the contest pens."

INTERIOR OF A LAYING HOUSE ON POULTRY FARM OF JOHN S. MARTIN

Pedigrees His Male Line

Aside from the fact that the males must come from dams with records of over 200 eggs, Mr. Martin does not attach a great deal of importance to extremely high records, and in his trap nesting he has pedigreed only his male line. In other words, selection through all these years has been based on pedigreed males, to which exacting requirements as to standard quality have been regularly applied, while in selecting females the qualifications demanded are: "1st, vigor and stamina; 2nd, a good egg record; 3rd, exhibition quality." Following this method of selection, exhibition quality has become so definitely established that Mr. Martin does not hesitate to exhibit birds from his laying strain at the largest fairs and shows.

As is well known, Mr. Martin breeds a special exhibition strain—the "Regal." These birds also are excellent layers and there does not appear to be any doubt that many high records would be secured in this line if it were practical for him to trap nest so many fowls. Since he cannot do this he wisely limits his efforts to his Dorcas strain.

In our tour of inspection we noted with special interest a Regal-Dorcas cockerel that had just won first at the great New York State Fair at Syracuse, 1921. There was also in the yards a splendid lot of Regal-Dorcas pullets, some of which had taken high rank at Syracuse and elsewhere, and while a close observer could note that the general run of females in this strain appeared to be somewhat longer in body and with less finish than those of the Regal strain, there were numbers of them which we doubt if any observer, no matter how keen his eye, would be able in any way to distinguish from the best Regals.

Feed and Care Given the Birds

Mr. Martin's method of feeding and general care does not differ materially from that with which poultry keepers everywhere are familiar. His birds are kept, for the most part, in comparatively small pens, flock matings consisting of 25 to 30 females with three males, the pen matings being of various sizes, depending upon quality and individuality. Houses are kept scrupulously clean and littered with straw, though planer shavings also are employed in warm weather. Mr. Martin clings to the moist mash as a regular feature of the daily feeding in the case of his best birds, but most of the pens have dry mash in hoppers as a matter of convenience.

Both the moist mash and dry mash pens have two feeds of grain a day. In order to keep the birds active and prevent their getting logy or fat, the feed hoppers are open only in the afternoon. This is regarded as a better plan with Wyandottes than giving them access to hoppers all day long. Mr. Martin wants his birds to spend the forenoon digging for grain in the litter, with no inducement to fill up at the dry mash hopper and then loaf. He feels that if he can keep the birds busy until noon they can be left to their own devices thereafter.

He depends mainly upon fish for animal feed, using nonsalable varieties caught in the nets of Port Dover fishermen. He is able to purchase these at a very low price. The fish are cooked and fed twice a week, giving the birds all they will eat—"and they eat a lot." Inquiry as to whether feeding fish in this way did not affect the quality of eggs produced brought out the statement that all danger of this is avoided by removing the oil before feeding. Not only is the oil skimmed off after the fish are cooked, but the meat is afterwards placed on a screen and bucketfuls of water poured over it, thus to wash out every particle of oil, which is then buried. In the wintertime Mr. Martin buys quarters of horse meat from farmers who are disposing of old or disabled animals. He also is able to use milk freely, as he maintains a good-sized dairy on one of his farms.

Gets Out Early Chicks

Mr. Martin has always specialized in early hatches, getting the first chicks out in January and continuing to hatch right along until the end of the season. Much of his phenomenal success in winning prizes at fall fairs and early shows is due to this early hatching and to exceptional skill in rearing the birds. In winter brooding he uses a Candee hot-water system, which has underground pipes with flues discharging warmed air under circular hovers. At a time when pipe-heated brooder systems are popularly supposed to be out of date it is interesting to note that Mr. Martin finds this method thoroughly satisfactory and for cold-weather brooding preferable to any other with which he has had experience. He attributes his success with these early chicks quite largely to the fact that he keeps his brooder house at a high temperature—around 80 degrees for newly hatched chicks—as a result of which they spend but little time under the hover.

He insists that the chicks must spend some time outdoors each day from the time they are a week old, regardless of what the temperature may be. Handled in this way he has few losses and the chicks are thrifty and grow rapidly.

Another factor which he regards as of special importance in his success in raising his early-hatched chicks is that nearly one-half of their feed consists of sprouted oats. These he feeds with the sprouts very short, in which condition he finds that they are most digestible and also have the maximum of feed value. In addition to oats the chicks get all the buttermilk they will drink.

The acreage devoted to fowls and general farming is considerably in excess of 100 acres. There are four large laying-breeding houses and (to the casual visitor) a bewildering

RED ROSE—RECORD 254 EGGS
This Pennsylvania Poultry Farm Rhode Island Red made the above record at the North American Contest, 1919-20.

PRODUCE RECORD LAYERS IN THREE BREEDS

The Pennsylvania Poultry Farm Has Been Producing Birds that Have Won in Laying Competitions in Three Breeds—Leghorns and Wyandotte Strains Were Founded on Direct Importations of Heavy-Laying English Stock

MOST breeders are content if they are able to establish a high-producing strain in a single breed, but from the Pennsylvania Poultry Farm at Lancaster, Pa., excellent results have been reported, not with one breed but with three—S. C. W. Leghorn, White Wyandotte and S. C. Rhode Island Red. There are splendid records from numerous official and semiofficial laying contests to prove the claims of superior productiveness for each breed, as well as the trap-nest records made on the farm.

LIBERTY BELL—RECORD 294 EGGS
Another Pennsylvania Poultry Farm heavy layer, with a record of 294 eggs in 1916-17, at the North American Contest.

The winning pen at the National Laying Contest back in 1912-13 consisted of 10 White Leghorns from this farm, these making, at that time, remarkable record of 2,073 eggs. Every year since then pens from this farm have been among the leaders at one or more annual contests up to 1918-19, when a pen of 5 Leghorns at the American (Leavenworth) Contest made the extraordinary record of 1,301 eggs, an average of 260 per bird. In this pen was Keystone Maid, the leading bird in the contest, with a record of 306 eggs. During the years 1912-19, 50 Penna. Farm Leghorns, Wyandottes and Reds, entered in various contests, produced a total of 10,869 eggs, an average of 217 each.

The Leghorn strain bred on this farm was founded originally on importations of stock direct from Tom Barron. That the strain has been systematically and successfully bred for increased productiveness is shown by the fact that representatives of it have defeated Mr. Barron's own birds in laying contests in late years.

The Wyandotte strain was also founded on importations of heavy-laying English stock, and has made an excellent showing. The highest individual record (semi-official) to date was made by Liberty Bell, in the North American Contest in 1916-17, who has 294 eggs to her credit, the entire pen of five laying 1,165.

We do not have exact information regarding the breeding of the strain of Rhode Island Reds on this farm, but that they are excellent layers is proved by the fact that a pen of five entered in the North American Contest in 1913-14 laid 1,043—an average of 209 each, the highest individual reaching 251. Private records up to 264 have been secured.

It is impossible here to go into full details regarding the special breeding methods and the management of the stock on this farm. Trap nesting is constantly practiced and high producers are bred to males from very high-record dams and the best conditions for growth and development are sought. The farm is located five miles from Lancaster and is of good size, thus to afford sufficient range for the breeding stock and for rearing the large numbers of birds annually produced in each of the three breeds.

While efforts have been concentrated upon high production, standard quality also is sought and the management of the farm feels that it has been quite successful in developing birds that more closely conform to American ideals than is usually the case with imported birds. Herewith are presented some half-tone reproductions from photos, showing the type of heavy layers produced on this farm.

SOME EXTRAORDINARY BARRED ROCK RECORDS

How M. K. Bohlander of El Paso, Illinois, Bred and Cared for the Birds That, in Their Pullet Year, Made Sworn Trap-Nest Records of 338 and 336 Eggs—Believes that It Is Readily Possible to Combine Beauty and High Egg Production

By M. K. Bohlander, El Paso, Illinois

(The extraordinary results with Barred Rocks, reported by M. K. Bohlander of El Paso, Illinois, have set a new record of achievement with this popular breed. The following brief statement regarding his breeding and feeding methods was prepared at our request and is here given substantially in Mr. Bohlander's own words.—Ed.)

I WILL endeavor to give you all the information I can in answer to your questions regarding my Barred Rocks, their breeding and records. I can say for one thing that it is no easy task to get a high-record flock. It takes both time and patience to be on the job every day and visit trap nests at stated intervals, and keep the records correctly for each one, but it is the only way to know your valuable birds and to be able to weed out the loafers.

We are rapidly approaching the time when the demand will be for both eggs and fine feathers and the poultryman will find that in order to satisfy his customers he will have to bring the egg basket and the "Standard" together—which is by no means an impossible feat. Our greatest records come from well-marked birds. Lady Bluebird finished her second laying year with the grand total of 622 eggs. She produced Lady Superior and Gypsy Queen, who have sworn trap-nest records of 336 and 338 eggs respectively in one year, thus setting a new worlds record both as an individual for two years, and as a trio—mother and daughters. And these birds were not mongrels either, with merely eggs to their credit but a pleasing sight to the eye as well. The male used with Lady Bluebird in producing these two daughters was one of our finest exhibition males.

As to the management of these record layers, would say that there were but three hens in the pen from which I obtained these high records. They were kept in a comfortable paper-lined house with plenty of windows for light and I always had plenty of clean straw for them to exercise in. I use trap nests of my own invention and have never used artificial light.

Lady Bluebird was hatched toward the latter part of May, began laying December 5, 1918, and laid 147 eggs in 150 consecutive days, and a total of 315 up to December 5, 1919. She laid two eggs in one day three times during the year. At the close of her first laying year she took a rest of 30 days and then laid 307 in the remaining

KEYSTONE MAID—RECORD 306 EGGS
Was one of a pen of five entered by the Pennsylvania Poultry Farm in the American Laying Contest in 1918-19, where she made her record of 306, being the leading bird.

eleven months, making a total of 622 in two years, or from December 5, 1918 to December 5, 1920.

Gypsy Queen and Lady Superior were hatched sometime in May, 1919, but I cannot give the exact day. The first named began laying October 11 and up to October 11, 1920, laid 335 eggs. Lady Superior also began to lay on October 11, and up to the same date, 1920, laid 338 eggs. Her production by months was as follows: Oct. 11 to 31, 19; Nov., 27; Dec., 28; Jan., 30; Feb., 29; Mar. 29; Apr., 30; May, 29; June, 29; July, 30; Aug., 27; Sept., 26; Oct. 1 to 11, 5.

All three birds were late in molting, starting about the latter part of October. Lady Bluebird laid steadily through the molt but molted so slowly and gradually that it was almost unnoticeable. She did not stop her laying even while getting her wing feathers. Gypsy Queen was directly opposite. She began to molt in November and dropped her feathers rapidly, at one time being almost bare. Lady Superior was much the same as Gypsy Queen but did not lose her feathers as fast.

In the way of feeding, we have so far been unable to find anything that will take the place of plenty of milk, sweet or sour, for making hens lay. We give all the grain, such as cracked corn, wheat, oats and bran, that they will clean up, feeding it in deep litter, and a warm mash is fed occasionally in cold weather. I always keep plenty of water before them at all times.

My birds run at range in the summer and I keep them housed in good comfortable quarters in the winter, letting them out on pleasant days in yards about 60 by 60 feet. In summer they have grass for green feed, and in winter, vegetables such as cabbage, beets, sprouted oats, etc.

DEVELOPMENT OF THE "BELLE OF JERSEY" HEAVY-LAYING STRAIN OF WHITE LEGHORNS

Origin and Development of This Famous Strain Here Described by Professor Harry R. Lewis of New Jersey Experiment Station

(NOTE: The "Belle of Jersey" strain of White Leghorns, which was originated and developed by Professor H. R. Lewis, has a most interesting history. At our request Professor Lewis furnished the following statement in regard to the strain, which statement was first published in Reliable Poultry Journal, August, 1919.)

BELLE of Jersey, a Single Comb White Leghorn hen (No. 531), bred and owned by the New Jersey Agricultural Experiment Station, proved herself to be a wonderful transformer of food material into the finished product, eggs. She was a small hen, weighing only three and eight-tenths pounds and during her first year's production she laid 246 eggs, or twenty and one-half dozen, weighing twenty-nine and one-half pounds, which is an average of 1.918 ounces each. This record means that for every pound of her body weight this little three and eight-tenths-pound hen in one year produced 7.76 pounds of eggs. It is estimated that in her body there was 1.68 pounds of dry matter and that during her first year's production she consumed food containing 92 pounds of dry matter and produced from this eggs containing 9.82 of dry matter, or 5.84 pounds of dry matter for each pound of dry matter in her body. This is a remarkable performance, both from the commercial as well as from the physiological standpoint.

The eggs produced that first year sold on the local retail market for $7.18 and it is calculated that she ate 118.5 pounds of food, costing $1.79, which showed for her first year's production the following financial balance:

Income—
    Market value of eggs.....................$7.18
    Value of manure...............................33

        Total income...........................$7.51

Expense—
    Food cost........................$1.79
    Labor cost (estimate)..........75
    Interest on investment..........25

        Total expense..................$2.79
        Total net profit................$4.72

Of still greater interest is the wonderful record which this hen made during her second year's laying. From November 1, 1912, till September 1, 1913, or in ten months, Belle of Jersey laid 205 eggs, weighing 26.9 pounds, or a

PREPOTENT S. C. WHITE LEGHORN MALE, NO. 1286
Used several years in succession by Professor H. R. Lewis, to transmit, intensify and establish the high-egg producing ability of "Belle of Jersey" and her line-bred descendants. Photo taken at end of breeding season at a time when plumage of male was considerably broken. Birds with him are two line-bred daughters.

total of 451 eggs in 669 days, or one egg every one and one-half days for this period.

Belle of Jersey was one of twenty birds which were the product of definite breeding on the New Jersey State Agricultural Experiment Station farm for high fecundity (high production) and whose yearly average for the first season's laying was over 190 eggs in 365 days. She was a direct descendant, two generations removed, from an exceptional, heavy-producing Leghorn hen who was first noticed on the College Farm because of her exceptionally high winter egg yield. She was known on the records as Number 70-C, and it is from her, inbred with her sons, that the present high-fecundity birds were originated.

Following is the pedigree of Belle of Jersey, as far back as the records go:

Belle of Jersey   70-C Female
  (No. 531)   70-C Female
Record, 246 eggs in   Record, 80 eggs
  365 days   Nov. to Feb., inclusive
         721, male of unknown breeding
        721—Male

Records of daughters:
  No. 531..............................246 eggs
  No. 15...............................221 eggs
  No. 77...............................196 eggs

No. 581, full brother of 531, was the father of a great number of heavy-producing daughters.

The breeding policy pursued has been as follows: to get as much of the blood of 70-C into both the male and female line as was consistent with high vitality and good constitution, to accomplish which very close selection has been practiced, except that all birds, regardless of their actual pedigree, which did not exemplify the highest in stamina and vigor, have been discarded, so that only six of the progeny of the second mating were kept for breeding. Not only should high-producing females be used to breed for egg production, but males which by actual test have begot daughters capable of heavy production are of even greater consideration, hence the great mistake often made of killing the males too young and without regard for their gametic constitution.

Belle of Jersey has been used to begin a strain of line-bred Leghorns which to date have made a wonderful record. Belle of Jersey was hatched in 1911. In 1913 she was mated to her father, No. 721, and produced seven pullets and four cockerels. All these pullets laid over 200 eggs each during their first laying season. (In 1914 she was mated to one of her sons, No. 1286, and produced nine pullets and five cockerels. All but two of these pullets succeeded in laying over 200 eggs during their pullet year.)

All the progeny from this particular mating were of uniform size and remarkably typical in respect to characters and conformation, and were, without question, a very valuable family of heavy-producing birds. (This male, No. 1286, has been used each succeeding year since 1914 and has been bred back to his daughters of the previous year, using each year only a very few of the earliest-hatched, quickest-maturing daughters for this line-breeding work.) To date (May, 1919), I have a flock of sixty-four pullets, daughters from this No. 1286 in his fifth year. The mothers of these sixty-four pullets are all daughters of 1286, being one, two and three years old. This strain of birds is now approaching close to an average of 220 eggs per year and we rarely find an individual that is healthy and in normal physical condition which falls below the 200-egg point.

This strain of birds is peculiar in that it is possessed of an exceptional amount of pep and vigor, the birds standing up under sustained production remarkably well, with practically no mortality. The pullets as a lot are representative of typical, standard-bred Leghorns, weighing from three and one-half to four pounds each. The eggs laid run uniformly two ounces or a little better and are exceptionally well shaped and entirely free from tint. In my case this line-breeding work is being carried on largely as a hobby, yet with the idea of working out and supplementing our present knowledge regarding the effect of close inbreeding and line breeding through successive generations.

I shall continue to breed No. 1286 (see photographic reproduction on page 123) to his daughters just as long as he remains in fit breeding condition and just as long as the birds show their present degree of vitality, stamina and productiveness. Dozens of cockerels from this particular male have been used in our experimental pedigree breeding work at the College Farm, and this past year nothing but his sons were used in our commercial Leghorn matings. I expect by following this close line of breeding to establish in the near future a large commercial flock, the breeding of which will trace directly back to this Belle of Jersey line. The highest egg production we have had from any one pullet in this line-breeding work to date has been 263 eggs. The average production last year was 215 eggs per hen for a flock of thirty-seven birds—all I had room for on my back-yard plant. This year we have about 300 pedigree chicks from this one male, No. 1286, and we are anticipating a wonderful lot of pullets.

From time to time the writer has made certain outcrosses in separate matings, using a few selected Belle of Jersey pullets and males from outside flocks. Some of these outcrosses have shown remarkable producing ability, while others have not. There is no question in my mind but that the ideal way to begin a strain of uniform, healthy, high-producing birds that will breed true, with great regularity, is to isolate one individual hen, peculiarly qualified to head such a strain—a bird that has the right body conformation and that by trap-nest records has demonstrated a satisfactory egg production. To this bird should be mated, if possible, one of her own sons, and from the line thus established, if the progeny continues to breed true in the characteristics desired, it will not take many generations to build up a large flock, ALL OF WHICH may contain ninety-nine and a fraction per cent of the blood of the original hen selected for the work. In my judgment where a great many of our breeders fail, is in trying to start pedigree lines with too many individuals. One superior hen at the end of from three to five years, if closely line bred and helped by intelligent selection, will get one farther than by starting out with a dozen at the beginning.

(NOTE:—The word gametes (from which the adjective form gametic is derived) is used to designate germ cells as distinguished from ordinary body (somatic) cells. These germ cells carry with them the characteristics of each parent, determining in turn the characteristics of the offspring. Naturally germ cells vary in their character in individual fowls, wherefore it is said that they differ in their "gametic constitution".—Ed.)

INCREASED EGG YIELDS SECURED BY SIMPLE METHODS OF SELECTION

Interesting Record of Achievement in Securing Gradually Increasing Flock Averages by Simple Methods of Selection, that Anyone Can Adopt—Striking Results Secured Without Resorting to Use of Trap Nests

IN practically all instances of success in developing high-producing strains it will be found that trap nesting and careful pedigree breeding have been persistently followed in securing the results reported. One practical objection to this method, however, which makes it impossible for the average poultry keeper to adopt it, is the high labor cost and the close, unremitting attention required. Probably nothing will supplant trap nesting and pedigree breeding, where it is practicable to carry it on, but simple methods are greatly needed that will enable the average man also to effect gradual (if slower and less certain) improvement in his flock. In a recent letter to the authors, Professor J. Holmes Martin of the Poultry Department of University of Kentucky called attention to a notable instance of success with an easily practiced method of selective flock breeding at the White City Egg Farm, Mentone, Indiana. Professor Martin had just returned from a visit to this farm and wrote: "Mr. Manwaring, the proprietor, has 3,500 Leghorns and is able to secure averages above 150. His breeding programme is very simple and practical and I believe it can readily be adapted to the farm flock." We immediately wrote Mr. Manwaring and received from him the following statement of his methods and of results secured:

1,000 Pullets Average 161.5 Eggs Each

"Professor Martin mentioned to you our last year's 150-egg flock average. It is with great pleasure that we inform you that we have reached 161.5 eggs per bird this year. This record for a flock of 1,000 pullets in one house was computed by Professor A. G. Philips of Purdue University, who states that he thinks it is as high an egg production as he ever heard of in a flock of 1,000 birds. Professor Philips' computations from the daily records are as follows:

Egg Production of 1,000 Hens on White City Egg Farm
Per cent of Egg Production per Month

Month	%	Month	%
October	26.4	April	62.6
November	38.1	May	63.6
December	32.5	June	56.4
January	32.	July	43.8
February	44.9	August	42.5
March	57.4	September	30.1

Average egg production per bird per year, 161.5
Per cent mortality, 6.2
Average number of birds, 940.5
Total number of eggs, 151,918

"We think this record is remarkable in view of the fact that we had considerable molt during December and January. Furthermore, the birds received a very simple and extremely economical ration, i. e.—corn and oats as a grain, bran, middlings and tankage as a mash and dry alfalfa hay, of our own raising, as green stuff. These birds received absolutely nothing else. We dare say that with a little pampering they could have been made to produce 170 eggs per bird. By pampering we mean smaller flocks, sprouted oats, moist mashes, plenty of wheat, etc. We believe you will agree with us however, that the average we attained and produced in an economical wholesale manner gives a larger net profit than a somewhat larger average produced at greater expense in both feed and labor.

"Personally, we think flock production in a real business-sized flock to be the true criterion of valuable egg production. The individual hen with the unusually large egg record is valuable only in so far as she is able to produce offspring with the same tendency for heavy laying, and in the majority of cases she does not do it, no matter how she is mated.

"We are keeping, at the present time, 2,000 laying pullets and 1,200 select breeders. This number of breeders is made necessary due to the fact that we have an unusual demand for 'White City' baby chicks. We might add that we produce these chicks entirely from our own selected hens. We breed from no pullets whatsoever, nor do we buy eggs elsewhere. We hatch for our own use about 10,000 chicks annually. This number enables us to dispose of about 2,000 pullets each fall. As a matter of possible interest we held an auction sale on 1,800 breeding hens this fall and sold the entire lot in less than 25 minutes.

Practical Flock Breeding Method

"Our breeding programme is, as Professor Martin says, 'very simple.' Our highest averaging pullets at the end of their pullet year are carefully selected and culled and mated to their nephews. For instance, starting with any season, we have:

P—pullet
H—breeding hen
C—cockerel

1920—H   C   P
H and C are mated, giving P1 and C1 as offspring
P becomes H1 breeders in 1921
1921—H1   C1   P1
H1 and C1 are mated giving P2 and C2 as offspring
P1 becomes H2 breeders in 1922
1922—H2   C2   P2

"We breed in this manner in large units—500 to 1,000, and the results both in vitality and egg yield have been exceedingly good. Ten years ago our egg average was about 100 eggs. We reached an average production of 152½ in 1920, while this year will see our average at 160 eggs per bird (pullets). This average is based on flocks as large as 1,000 birds in one flock and not on a few individuals.

"We consider vitality to be the cornerstone of heavy egg production and our large flock breeding units of aunts to nephews seem to be conducive to this. In photo sent you (see illustration on this page) please note the vitality of these four-week-old youngsters as indicated by their evenness in size and their closely fitting wings—not a weak, listless one in the entire 5,000. Professor Jones or Professor Philips of Purdue can verify this statement. In this brood our mortality did not exceed 4 per cent."

Another way of stating the methods practiced on this farm is given by Professor Martin in the letter previously referred to, and is here quoted as it gives some additional details and may assist readers in understanding and applying Mr. Manwaring's breeding formula:

GROWING PULLETS ON RANGE AT WHITE CITY EGG FARM

"The earliest maturing 1,000 pullets from the range are placed in pen 5. The following fall the 500 latest molters from this group are placed in pen 1. Consequently pen 1 contains hens which were the earliest of all pullets raised to come into laying and the last to stop laying and go into a molt. In other words, they were early layers and late layers and 'burned the candle at both ends.' We find by our trap-nest records that it is only hens of this class that make records over 200 eggs. From pen 1 are saved all the cockerels that will be used during the following breeding season. These cockerels are sired by cockerels that were from the previous year's best layers. Each year after the breeding season the cockerels are sold on the market. Mr. Manwaring believes that the following year's cockerels, or rather that the cockerels raised that year, will be better than the old birds he has, since they will be from dams that made higher averages. You will see that this method involves a minimum amount of labor as there is no trap nesting or handling of the entire flock. The pullets would have to be brought up from the range in the fall anyway and at this time they are selected. The following summer the birds are culled from the pens and sold as fast as they come into a molt, hence the best layers are left in the flock and do not need to be handled until taken to the permanent breeding house. He keeps at the rate of one cockerel to every 20 to 25 hens.

"This programme may be adapted to the farm flock by banding with a distinctive color band the first pullets to come into laying. The following summer the latest layers from this banded group should be selected for the breeding pen. Hence all cockerels raised are out of the best layers in the flock and are sired by cockerels that were out of the previous year's best layers."

GENERAL VIEW OF A PORTION OF PLANT AT WHITE CITY EGG FARM, INDIANA

In the foreground are the colony brooder houses in which 5,000 Leghorn chicks were raised during the season of 1921, with a total loss of not more than 4 per cent—a record which speaks well for the vigor of the breeding stock. Pullets on this farm in flocks of 1,000 averaged to lay 161.5 eggs each in 1920-21.

PART III

CHAPTER I

Egg-Laying Contests and Their Lessons

No Estimate of the Influences that Have Brought About Recent Extraordinary Developments in the Poultry Industry Is Complete Which Fails To Take into Consideration the Part Played by Public Egg-Laying Contests—Brief History of Important Contests in Many Countries and Summary of Results Secured at Each

WITHIN the past 12 years or so, public egg-laying contests have had a tremendous influence upon the poultry industry. Not only have they aroused great public interest in the possibilities in market egg production, but they have also directed the attention of an army of practical breeders to the almost unsuspected capacity of fowls for increased egg yields under skillful breeding and management.

It would be unreasonable, however, to expect that, in the comparatively short time that has elapsed since the first contest was inaugurated in this country, the awakened interest in commercial poultry keeping and the better knowledge of breeding for productiveness should have effected a noticeable improvement in the average egg yield of fowls the country over. Even at some of the contests, the general average from year to year appears to be almost stationary, while others show only such increases as can readily be accounted for by greater skill and improved methods of care. A few contests, however, show gradually mounting averages that clearly must be attributed to improved breeding in the flocks from which the contest pens are recruited. The following tables will serve to show what increases, if any, in general averages, have been secured at a few widely separated contests.

Contest Averages at Harper Adams Agricultural College (Eng.)

Year	Average per Hen
1912-13	231.5
1913-14	226.5
1914-15	212.0
1915-16	194.66
1916-17	249.0
1917-18	233.0
1918-19	109.8

In studying the above table it should be noted that it does not represent the entire life of the contest but takes up the record after the first (and experimental) stage of the contest was over.

Contest at Hawkesbury Agricultural College, New South Wales

No. of Competition	Average per Hen
1st	130
2nd	163
3rd	152
4th	166
5th	171
6th	173
7th	180
8th	181
9th	168
10th	184
11th	178
12th	177
13th	181
14th	192
15th	205.7
16th	206
17th	194.7

The above averages are quoted from the "Agricultural Gazette of New South Wales." In the last three contests the birds were divided into "light" and "heavy" classes and the contest averages for these years have been computed from the class averages. This table shows a fairly distinct upward trend. Omitting the first year, which at all contests is more or less experimental and the value of its data affected by the inferiority of many of the birds entered by unskilled contestants, and reducing the yearly records to four-year averages, the gradual increase is more readily perceived: First four-year period, 163; second period, 175.5; third period, 180; fourth period, 199.6.

International Contest Averages, Storrs, Conn.

Year	Average per Hen
1st	152
2nd	156
3rd	145
4th	151.8
5th	162
6th	163.4
7th	159
8th	145
9th	161.5

The averages at this contest fluctuated too widely for them to have any bearing upon the question at issue. It must be remembered however, in considering results at this and all other contests, that the war greatly disorganized government experimental work, and Storrs suffered unusually in this respect.

North American Contest (Semiofficial)

Year	Average per Hen
1st	152
2nd	156
3rd	170
4th	165
5th	177
6th	167
7th	141
8th	152
9th	155.4

The North American Contest shows the expected increase in the first few years followed, however, by a decrease, so that even with the first year's record included no definite advance in the general average is shown.

National Egg-Laying Contest at Mountain Grove, Mo.

Year	Average per Hen
1911-12	135.6
1912-13	142.6
1913-14	157.8
1914-15	152.
1915-16	164.7
1916-17	175.2
1917-18	171.8
1918-19	184.
1919-20	187.7

Among American contests the Missouri National is the only one that shows a noticeable and fairly regular increase in average production from year to year.

The First Laying Contests

To the best of our knowledge the first "laying contest" was conducted in this country by the National Stockman and Farmer in 1889-90. This contest was of the class now known as "farm" contests, the birds remaining in the custody of the owners, records for entire flocks being reported weekly or month by month. A second contest of the same sort, on a much larger scale, was conducted by the same paper in 1894-95, which brought out some exceptionally good records. Details in regard to these contests will be found elsewhere in this chapter.

The first contest in which small pens of selected birds were brought together at a common center in order that they might be kept under uniform conditions and records kept under more or less official supervision was conducted at the Harper Adams Agricultural College in England, beginning in 1897. This contest was controlled by the National Utility Poultry Society. During the first few years the contests were limited to 16 winter weeks.

In 1907 the contest period was increased to 12 months and has so continued ever since.

To Australia goes the credit for having established the first 12-month contest, this having been started April 1, 1902, at the Hawkesbury Agricultural College, New South Wales, with the support of a local daily paper. This contest has been conducted regularly since that date.

In this country the first contests were organized in November, 1911, at which time the International Contest at Storrs, Conn., and the National Contest at Mountain Grove, Mo., were started. Both of these contests have run continuously to date. In the meantime, many others have been organized in various parts of the country. It is impossible, in the space available, to go into details in regard to these numerous contests and the results secured, but brief mention of the more important ones is here given for comparison and as a matter of history.

Egg-Laying Contests in England

The following review of English contests is condensed from an article by Edward Brown, London, England, which article appeared in Reliable Poultry Journal, March, 1921.

"The declared object (of the first contest at Harper Adams College) was to indicate that heavy laying is influenced by strain—i. e., family, rather than by breed, and to show by actual demonstration where the best laying stocks were to be found. For ten years these sixteen-week competitions continued annually. It was not until 1907 that a twelve months' test was made.

"At the outset number of eggs laid per pen was the determining factor. In the sixth of these competitions two important changes were made, namely, introduction of the trap nest, so that individual records could be made, and, second, that the weight of eggs in accordance with market standards was taken into consideration, undersized eggs being penalized.

"It is of interest to note that, in the first competition, Black Minorcas took first and second places and Langshans were third. In the second, Buff Leghorns were first, Barred Rocks second and Golden Wyandottes third. Not until the fifth year did White Leghorns enter into the winning group, and in the sixth year White Wyandottes stepped into the front rank.

"In the first year, during sixteen weeks, the highest pen, Minorcas, laid 161 eggs (35.93 per cent production). What appeared phenomenal at that period was in 1902-03 when the four White Wyandottes recorded 276 eggs, or a 61.6 per cent production—a remarkable feat from October to February at that time, though exceeded since, for at the Burnley Trials of 1907-08 a pen of four Buff Orpingtons laid in 112 days 301 eggs—that is, a 66.98 per cent production.

"As already stated, in 1907 a twelve months' competition was inaugurated and that period is now the general rule for these trials. Unfortunately for comparisons, some of these do not extend over the entire year. A two-year trial has also been in operation.

"In 1905 what is known as the Northern Utility Poultry Society, whose headquarters are at Burnley, Lancashire, one of our great cotton-manufacturing areas, where poultry keeping has always been most popular as a supplementary pursuit, began laying trials, which for eight years were on the sixteen-winter-weeks' basis. Since 1913 these have also been extended over an entire year. As might be anticipated, so long as the shorter period was maintained, the heavier bodied races scored most heavily. For example, at Burnley from 1905 to 1913 out of 25 prize pens 20 were of general-purpose breeds of which White Wyandottes accounted for 13; of the purely egg breeds four lots were White Leghorns and one Buff Leghorn. With the yearly trials Leghorns took a much higher position, and are now equal with the White Wyandottes. These two breeds stand first in numbers and results attained.

"The Second Annual Competition arranged by the Utility Poultry Club was held in association with the Harper Adams Agricultural College at Newport, in the County of Shropshire, so continuing for five years, when that club, reconstituted as the National Utility Poultry Society, transferred its interest to the County of Suffolk. At Dodnash Priory, on land belonging to the Great Eastern Company, the last-named corporation erected a fine plant where tests or trials have since been conducted as a joint enterprise. Last year the scope was extended by inclusion of a competition for small, or back-yard poultry keepers, promoted by the Daily Mail, one of our London newspapers, and which awakened a large amount of interest owing to the high prizes offered by that journal.

"The Harper Adams College, however, continued the series of trials, and has not regarded these merely as competitions, of which there is all too much danger, but as a means of extending knowledge. Its monthly and yearly reports are especially valuable for a large amount of statistical information, in which respect there is much room for improvement at others.

"At Dodnash Priory all the pullets are in single pens. At Harper Adams and Burnley provision is made for both these and also for larger flocks. At Harper Adams the larger flocks consist of 56 birds each, so as to be on commercial lines, which the single pens are not. The results, so far as production is concerned, are in favor of the small flocks. For example, in 1918-19 at Harper Adams the average of all breeds was: small flocks, 186.92 eggs; large flocks, 120.66 eggs. How far management accounts for this big difference it is impossible to state."

Australia's Laying Contests

Contests have been organized in New South Wales, Victoria, South Australia, Queensland, Tasmania and New Zealand. The Australian contests, beginning with the one at Hawkesbury Agricultural College, N. S. W., in 1902 (the first public 12-month contest on record), have had a marked effect upon the industry in that continent. Some extraordinary records have been secured and the contests have always been watched with keen interest in this country.

At Hawkesbury the college furnished the ground and the buildings, while a local daily paper ran the contest. This arrangement we understand, is still followed. The first contests consisted of 38 pens. This number has been

FLOCK CONTEST HOUSE AT HARPER ADAMS AGRICULTURAL COLLEGE, ENGLAND
The first laying contest in England was held at Harper Adams College. Twelve-month competitions were inaugurated here in 1907 and have been conducted continuously ever since. Provision is made for both small and large pens. In the house shown above the flocks consist of 56 birds (several entries), the object being to have conditions more nearly approximate those that obtain in commercial poultry keeping.

gradually increased and since 1916, 70 pens or 420 pullets have been handled. In addition, 20 to 50 pens of "second-year" birds were carried for a number of years.

During the first of these contests the sentiment swung toward Leghorns, and rather small Leghorns at that, with the result that there was a gradual falling off in size of eggs, as was also the case at the English contests. This finally reached the stage where the authorities in control found it necessary to take cognizance of the fact and adopt corrective measures. One step in this direction was the organization of the contests in two sections—Section A, consisting of Leghorns and Section B, of the heavier breeds. In addition, a rule was adopted providing that "any bird whose eggs do not attain an average weight of 24 ounces per dozen during the first four months, or that does not lay during that period, shall be ineligible for a prize that year." This seems to have had the desired effect as three years later (contests of 1918-19) it is recorded that no birds in the light-breed section were disqualified, and only one entry (6 birds) in the heavy breeds.

TYPE OF CONTINUOUS HOUSE USED IN SOME CANADIAN STANDARDIZED LAYING CONTESTS

The house above is used in the contest at Agassiz, B. C. The comparatively mild climate in this section makes it practical to use extreme open-front construction.

Production Records at Hawkesbury Contest

No. of Competition	Winning Total	Average per Hen
1st	1,113	130
2nd	1,308	163
3rd	1,224	152
4th	1,411	166
5th	1,481	171
6th	1,474	173
7th	1,379	180
8th	1,394	181
9th	1,321	168
10th	1,389	184
11th	1,461	178
12th	1,360	177
13th	1,541	181
14th	1,449	192
15th (A	1,526	216
(B	1,479	192
16th (A	1,525	209
(B	1,613	202
17th (A	1,448	199
(B	1,517	189

A—Light breeds. B—Heavy breeds.

Second-Year Records

"The records made by second-year hens at Hawkesbury are especially interesting, this being probably the most extensive comparison of first and second-year production on record. The table shows a general increase in first-year averages, covering the 12-year period of these tests, and also an increase in second-year production.

Comparison of First and Second-Year Production at Hawkesbury

No. of Competition	Average per Hen First Year	Average per Hen Second Year
1st	180	124
2nd	179	127
3rd	190	140
4th	194	134
5th	184	140
6th	201	135
7th	194	145
8th	190	158
9th	201	161
10th	213	165
11th (D	241	164
(E	210	131
12th (D	231	162
(E	222	150

D—Light breeds. E—Heavy breeds.

Single-Pen Testing

"Australia has taken the lead in single-pen testing, meaning by this that each bird is given a separate compartment and kept confined to it throughout the entire period of the test. (See Chapter III, Part III, for details.) This method appears to have had a favorable effect upon the egg yield in most cases. At the Hawkesbury Contest, for example, the six-bird pens gave an average of 183.3 during three years, against an average of 202.1 in the single pens for the same period.

"At the 1919-20 Victorian Contest (where both single test and pen classes are provided) the highest individual record was made in a pen, but the average for all birds entered in pens of six was 206, against 212½ for the single-test birds. In the report of this contest, however, it is observed that "It is reasonable to assume that competitors endeavored to place their best birds in the single test, and subsequently selected the next best for the trap-nested pens."

Light and Heavy Breeds

The classification of contesting pens into light and heavy breeds at Hawkesbury has not indicated any marked difference in favor of Leghorns. The preceding table shows that in the last three years this breed was always a little in the lead, in average per hen, but in two of these three years, the winning pen was in the heavy section.

Victorian Contests

Contests in Victoria began in 1904-05, but after three years were discontinued, and again resumed in 1911-12, this time under direct government supervision. The 1914-15 Victoria Contest has the honor of making the highest pen record to date—an average of 283.1 eggs each for six White Leghorns. Here six-bird pen and single-bird records are kept, light and heavy breeds are treated separately

COMPETITION PENS AT VICTORIA (AUS.) CONTEST, BURNLEY

EGG-LAYING CONTESTS AND THEIR LESSONS

and a further comparison is carried out between wet and dry mash in the Leghorn section.

At the Tenth Contest (1920-21) the average number of eggs per bird was 212 2/3 in pen sections as compared with 222 in the case of single birds. In the wet and dry mash comparison the averages per bird were 227.7 (56 birds) and 223.7 (23 birds), respectively.

Two well-known semiofficial contests have also been conducted in this state—the Bendigo and the Geelong Contest. At the first Bendigo Contest four 300-egg records were made in one year: 315, 313, 307 and 303, while in the contest of 1917-18 a record of 332 was reached. The 1919-20 Geelong Contest reported a 339-egg Black Orpington, this being the highest individual record claimed by any Australian contest.

As this chapter goes to press the report reaches us that a pen of Black Orpingtons has just established a new world record in Australia, at the Bendigo (Victoria) Contest, laying a total of 1,750 eggs in 365 days ending with March 31, 1922, or an average of 291.66 eggs per bird. There were six birds in the pen, their individual scores being 307, 306, 301, 300, 270, 266. This is a truly remarkable performance and sets a new standard for U. S. contest managers to equal and, if possible, excel.

South Australian Contests

The first contest in South Australia (organized by the Royal Agricultural Society) was held in 1903-04. Beginning with 1908 the contests were conducted exclusively by the government poultry department under the supervision of D. F. Laurie. Later one or two other contests were started in this state. All of these were discontinued just before the war, with the intention of concentrating the poultry work at a common center. Since the war, laying contests have been resumed in this state but no official reports regarding them have reached us.

By way of summarizing results secured at Australian contests the following tables are quoted from the August 1920, issue of "Journal of the Department of Agriculture of Victoria."

Official Australian Contest Scores—Teams of Six

Locality	Year	Breed	No. of Eggs
Victoria, Burnley	1914-15	White Leghorns	1,699
" "	1913-14	" "	1,667
" "	1915-16	" "	1,661
Queensland, Gatton	1917-18	" "	1,652
Victoria, Burnley	1915-16	" "	1,638
" "	1915-16	" "	1,637
" "	1914-15	" "	1,633
" "	1915-16	" "	1,623
Queensland, Gatton	1919-20	" "	1,627
" "	1919-20	Black Orpingtons	1,619
N. S. W., Hawkesbury	1917-18	" "	1,613
South Australia, Parafield	1919-20	White Leghorns	1,603
Victoria, Burnley	1915-16	" "	1,601

Official Australian Individual Scores of 300-Eggs and Over

Locality	Year	Breed	No. of Eggs
Victoria, Burnley	1917-18	Black Orpingtons	335
Queensland, Gatton	1918-19	" "	335
" "	1917-18	" "	334
N. S. W., Hawkesbury	1918-19	" "	324
Queensland, Gatton	1918-19	" "	317
N. S. W., Hawkesbury	1916-17	" "	312
" "	1917-18	" "	312
Victoria, Burnley	1919-20	White Leghorns	312
N. S. W., Hawkesbury	1916-17	" "	308
" "	1917-18	Black Orpingtons	305
Queensland, Gatton	1918-19	White Leghorns	303
Victoria, Burnley	1919-20	" "	303
N. S. W., Hawkesbury	1919-20	Black Orpingtons	303
Queensland, Gatton	1918-19	White Leghorns	302
Victory, Burnley	1917-18	" "	301
N. S. W., Hawkesbury	1916-17	" "	300
South Australia, Parafield	1919-20	" "	300

Contests in New Zealand

The first contest in New Zealand was held in 1905-06 and some excellent records have been made in the three semiofficial contests held in different sections of this country. As far back as 1912-13 the winning pen at one of these reached an average of 272 eggs per bird in 51 weeks and the average for all the 360 birds entered that year was a trifle over 200 each. To the best of our knowledge the highest record yet reported from New Zealand is 317, made by a White Leghorn in 1917-18.

Ducks also are included in New Zealand contests and we are informed that at the contest at Papanuni, N. Z., promoted by New Zealand Utility Poultry Club, a record of 363 eggs in 365 days was made by a White Indian Runner.

International Laying Contest, B C., Canada

The first Canadian contest was held at Victoria, B. C., in 1911-12, with six birds to a pen and with separate classes for light and heavy breeds. We do not have the figures for maximum pen and individual birds but the accompanying table shows the average for each contest.

Contest	Average Production	Months Duration
1	109	12
2	145.7	10
3	164.2	11
4	165.4	12
5	151.6	11
6	159.8	12
7	154.7	11
8	158.2	11
9	148.7	12

In the last (ninth) contest conducted on the pen basis, the best record was made by a pen of White Leghorns with an average of 185.3. The best hen in this contest was a White Leghorn that made a record of 230 eggs, while a White Wyandotte followed with 228.

In the tenth contest (1920-21) a change was made to two-bird pens, each consisting of a lightweight (white-egg) and a heavyweight (brown-egg) bird. It is interesting to compare the results of this contest with the ninth, where the birds were kept in pens of six.

Year	No. of Birds in Pen	Average Production
1919-20 (12-month test)		
Lightweight breeds	6	150
Heavyweight breeds	6	147
1920-21 (11-month test)		
Lightweight breeds	2	183
Heavyweight breeds	2	151.4

Canadian Standardized Laying Contests

Beginning with 1919 the Canadian government put into effect an elaborate plan for conducting standardized laying contests to be held at Dominion Experimental Stations in the different provinces, the intention being to utilize these as a national medium for obtaining official records of performance, thus to be able to provide for official registration of high-producing birds. The plan is outlined in Chapter III, Part III.

The rules and regulations approved by the Executive of the Canadian National Poultry Association provide that ten birds are to be allowed to a pen; each contestant can substitute up to two birds. The contests last for fifty-two weeks, from November 1 to October 29, but in order to secure Record of Performance, which is the main object of the contest, each bird is to be allowed to complete her fifty-two weeks of laying—that is, her record does not finish until 52 weeks from the time her first egg was laid, providing she finishes her record by the 31st of December, which is the limit set for all birds.

Situation and Number of Pens in Canadian Contests

The situation of contests and number of pens in operation are as follows:

Canadian Egg-Laying Contest, Central Experimental Farm, Ottawa .... 50 pens
Ontario Egg-Laying Contest, Central Experimental Farm, Ottawa .... 24 "
Prince Edward Island Egg-Laying Contest, Charlottetown .... 25 "
New Brunswick Egg-Laying Contest, Fredericton .... 21 "
Nova Scotia Federal Egg-Laying Contest, Nappan .... 22 "
Quebec Egg-Laying Contest, Cap Rouge .... 24 "
Manitoba Egg-Laying Contest, Brandon .... 24 "
Saskatchewan Egg-Laying Contest, Indian Head .... 20 "
Alberta Egg-Laying Contest, Lethbridge .... 22 "
British Columbia Egg-Laying Contest, Agassiz .... 27 "

In the first year's contests the best pen average was 176.5, secured at the Manitoba contest, and the best individual hen record was 262, also at Manitoba.

In the second contest the high pen (White Leghorns) at Ottawa contest reached an average of 225.5 per bird and the highest individual (a member of this pen) made a record of 281.

Two hundred and four birds qualified for Record of Performance at the Ottawa contest by laying 150 eggs or over, and 15 for advanced record (225 eggs or over).

EGG-LAYING CONTESTS IN THE UNITED STATES

It was not until 1911 that the first public laying contest was held in the United States. That year the International (Storrs) and the National (Mo.) were organized. Since then the number of contests has rapidly increased and is still increasing, so that it is impossible at the present time to give a complete list. Details in regard to a few are here given to show how they are organized and conducted and to indicate the results secured.

International Contest at Storrs, Conn.

This contest, conducted under the combined auspices of the Philadelphia North American and the Connecticut Agricultural College, was started in 1911. After the first

TWO-PEN HOUSE USED AT CANADIAN LAYING CONTEST, OTTAWA

two years the entire control of the contest was assumed by the Connecticut Station and some changes made in details, since which time these have remained practically unchanged so that results of any given year are fairly comparable with others. Bulletin 100 of the Connecticut Station summarizes results secured during the sixth and seventh years of the contest, and gives some averages covering results of the first five years (1913 to 1918) under exclusive state management. During these seven years the best layer at this contest was a White Wyandotte, with a record of 308 eggs, during 1915-16. This is the only 300-egg hen that has appeared at the Storrs Contest to date.

This contest has proved rather exceptional in the fact that no one breed has had the lead. In the 1913-14 contest the winning pen consisted of White Leghorns and in the next year White Wyandottes. In 1915-16 the leading pen was among the White Wyandotte entries again, and the next year a pen of Barred Plymouth Rocks stepped to the front. Leadership went to a pen of Oregons the following year and to Barred Rocks again in 1918-19 and in 1919-20.

Beginning with the third contest all the birds were judged as to their exhibition quality and Bulletin 100 states that taken as a whole the record "Certainly constitutes an argument for the contention that showroom and utility qualities can be combined in the same strain of birds."

The following table gives the average feed consumption for the five-year period, 1913-1918; also the number of eggs produced and weight of eggs per dozen:

Breed	Eggs per pen	Wt. per doz. in ounces	Mash pounds	Grain pounds
Plymouth Rocks	1,593	23.9	623	411
Wyandottes	1,626	23.3	554	405
R. I. Reds	1,480	24.5	584	411
White Leghorns	1,624	24.1	494	402
Miscellaneous	1,588	25.0	529	408
All Breeds	1,589	24.0	544	406

This table affords a fairly good comparison as to the relative amount of feed required by fowls of different breeds, particularly Leghorns as contrasted with the larger breeds, the relative productiveness of these breeds, and the weight of eggs. It will be noted that while the Wyandottes led in average number of eggs produced they were also lowest in weight of eggs per dozen, for which reason the good showing made by them must be discounted considerably.

Bulletin 100 contains the following interesting information in regard to certain factors in production. "Three factors (number of eggs from 100 lbs. of feed, eggs required to pay for feed, value of eggs from each dollar spent for feed) have been computed with an idea of securing such a basis of comparing five years that the results of the comparison would not be materially influenced by a rise in egg prices or increase in cost of feed. One such factor is the number of eggs produced to each 100 pounds of feed consumed. On a basis of a five-year average this factor has been 158, 159, 182 and 195 for the Plymouth Rocks, Rhode Island Reds, Wyandottes and Leghorns, respectively.

"In a similar manner the number of eggs required to pay for the feed for a pen of ten birds has been, for the same breeds, 721, 695, 645 and 627, respectively. It will thus be apparent that when one considers the returns from 100 pounds of feed rather than the return per individual, Leghorns are exceptionally efficient in the matter of egg production. The last factor is the value of eggs returned for each dollar spent for feed. Considering the five-year average, Wyandottes and White Leghorns have been practically equal with $2.53 and $2.58, respectively. Likewise Plymouth Rocks and Rhode Island Reds have been very close together with $2.15 and $2.18, respectively."

At this contest it was found that the production of Leghorns was quite largely affected by temperature, the birds regularly making a noticeable drop with the appearance of severely cold winter weather, making up for this, however, during the summer months. At the end of the year they again showed a tendency to reach minimum production ahead of the other breeds.

The entry fee at this contest is $20 per pen and 100 pens are provided for, the contest covering an entire year. Four acres of land are included in the contest grounds and the birds are kept in two-pen houses. During the first eight years of this contest a total of 6,800 birds was entered, the average production of which was 154.8.

National Egg-Laying Contest at Mountain Grove, Mo.

The National Egg-Laying Contest was organized by the Missouri State Poultry Experiment Station at Mountain Grove, Missouri, in 1911, and during the first year there were five birds to a pen. During the second and third years there were ten birds to a pen, but in the fourth year the number was reduced to five again and has remained unchanged to date.

The following summary of results gives the first, second and third pen at each contest, the record of the highest bird and the average for the entire contest. This contest is supported by entry fees and a state appropriation. The entry fee is $10 per pen for residents of Missouri and $20 a pen for nonresidents. Five acres are in the contest grounds, the houses are 8 by 10 and provide for two pens as in the Storrs Contest. Double yards are provided for each pen.

Results of Nine Contests at Mountain Grove, Mo.

Year of Contest		Breed	No. of eggs laid
1911-12			
High pen		R. C. Rhode Island Reds	1,042
2nd		White Wyandottes	1,015
3rd		S. C. White Leghorns	991
High hen		White Plymouth Rock	281
Average of contest			185.6
1912-13			
High pen		S. C. White Leghorns	2,073
2nd		Buff Wyandottes	1,884
3rd		Silver Wyandottes	1,877
High hen		R. C. White Leghorn	260
Average of contest			142.6

1913-14		
High pen	S. C. White Leghorns	2,296
2nd	S. C. White Leghorns	2,104
3rd	S. C. White Leghorns	1,814
High hen	S. C. White Leghorn	286
Average of contest		157.8
1914-15		
High pen	Barred Plymouth Rocks	1,121
2nd	Oregons	1,085
3rd	S. C. White Leghorns	1,049
High hen	S. C. White Leghorn	255
Average of contest		152
1915-16		
High pen	Barred Plymouth Rocks	1,185
2nd	Oregons	1,159
3rd	S. C. White Leghorns	1,100
High hen	S. C. White Leghorn	275
Average of contest		164.7
1916-17		
High pen	White Wyandottes	1,226
2nd	White Plymouth Rocks	1,141
3rd	R. C. Rhode Island Whites	1,130
High hen	White Wyandotte	269
Average of contest		175.2
1917-18		
High pen	S. C. White Leghorns	1,171
2nd	Rhode Island Whites	1,125
3rd	White Wyandottes	1,120
High hen	White Wyandotte	286
Average of contest		171.8
1918-19		
High pen	Rhode Island Whites	1,217
2nd	White Wyandottes	1,165
3rd	R. C. White Leghorns	1,128
High hen	Rhode Island White	298
Average of contest		184
1919-20		
High pen	S. C. Rhode Island Reds	1,233
2nd	White Wyandottes	1,171
3rd	Barred Plymouth Rocks	1,168
High hen	S. C. Rhode Island Red	296
Average of contest		187.7

The North American Laying Contest (semiofficial)

First and second years of this contest were staged on the grounds of Storrs Agricultural Experiment Station, Storrs, Conn. Third year was staged at Downingtown, Penna., and the fourth, fifth and sixth years at Newark, Delaware, on the ground of and under the supervision of Delaware College Agricultural Experiment Station. During the four succeeding years the contest was operated by the State Board of Agriculture of Delaware on specially arranged grounds at Georgetown, Delaware.

Four records were established for the United States through these competitions. Lady Eglantine, a White Leghorn produced by Eglantine Farms, Greensboro, Maryland, laid 314 eggs (1914-15); a pen of five White Wyandottes, produced by Tom Barron, England, laid 1,305 eggs (1915-16). A White Wyandotte, produced by Pennsylvania Poultry Farms, Lancaster, Pa., laid 109 eggs in 109 days. In the 1920-21 Contest an English-bred Buff Orpington owned by Egg-A-Day Farm, Meriden, Conn., made a record of 343 eggs in 365 days, the greatest one-year record claimed to date.

One hundred 6 by 8-foot houses are used, these being equipped with the Stoneburn or Connecticut trap nest. Automatic feeders distribute dry grain, and dry-mash feeders are employed in all pens. Each house is provided with a 10 by 20-foot yard.

High Pens and Individual Hens at North American Contest

Year of Contest	Breed	Number of eggs laid
1911-12		
1st pen	White Leghorns	1,071
2nd pen	White Wyandottes	1,069
3rd pen	White Leghorns	1,042
1st bird	Rhode Island Red	254
1912-13		
1st pen	S. C. White Leghorns	1,190
2nd pen	S. C. White Leghorns	1,107
3rd pen	S. C. White Leghorns	1,029
1st bird	S. C. White Leghorn	282
1913-14		
1st pen	White Wyandottes	1,180
2nd pen	White Leghorns	1,139
3rd pen	White Leghorns	1,136
1st bird	Columbian Rock	286
1914-15		
1st pen	White Leghorns	1,211
2nd pen	White Leghorns	1,168
3rd pen	White Wyandottes	1,130
1st bird	White Leghorn	314
1915-16		
1st pen	White Wyandottes	1,305
2nd pen	White Leghorns	1,151
3rd pen	White Leghorns	1,147
1st bird	White Wyandotte	289
1916-17		
1st pen	White Leghorns	1,166
2nd pen	White Wyandottes	1,165
3rd pen	White Wyandottes	1,158
1st bird	White Wyandotte	294
1917-18		
1st pen	R. C. Rhode Island Reds	1,111
2nd pen	White Leghorns	1,026
3rd pen	White Leghorns	967
1st bird	R. C. Rhode Island Red	263
1918-19		
1st pen	White Wyandottes	1,171
2nd pen	Columbian Rocks	1,076
3rd pen	Columbian Rocks	1,049
1st bird	White Wyandotte	272
1919-20		
1st pen	R. C. Rhode Island Reds	1,126
2nd pen	Barred Plymouth Rocks	1,052
3rd pen	Columbian Rocks	1,035
1st bird	S. C. White Leghorn	287

Vineland and Bergen County Laying and Breeding Contests

The Vineland Contest, started in 1916, differs from the foregoing contests in that it is of three years' duration and is intended to serve as a breeding as well as a laying test. Single-pen houses, 8 by 10 feet, were provided and pens consisted of 10 pullets. These birds were trap nested during the first and second years. In the second year they were mated to male birds provided by the contestants and a sufficient number of eggs were hatched to secure 10 vigorous, healthy pullets from each pen, these being placed in the contest houses the third year. Eight acres were included in the contest pens and six acres for brooding and rearing the stock the second year. The following table gives the production each year during the first contest:

VIEW OF EGG-LAYING CONTEST GROUNDS AT STORRS, CONN., AND A PORTION OF THE COLLEGE POULTRY PLANT

Production at First Vineland Contest

Year	No. of birds	Average no. of eggs	No. of 200-egg hens
1916-17	1,000	161.8	184
1917-18	1,000	129.5	29
1918-19	1,000	179.	324

In the second contest, beginning November, 1919, the pens were increased to 20 birds, these to be culled to 12 for the second or breeding year and to be replaced by 20 of their daughters in the third year. During the first year of this second contest the average production was 159 per bird and in the second 138.7. This year there were 234 hens with records of 180 eggs and therefore eligible for registration in the American Record of Performance.

At the Bergen County Contest (also of three years' duration), the first year of which was concluded October 31, 1921, the production averaged 134.7 eggs per hen, this comparatively low average being attributed to an unusual amount of disease during the first part of the contest year. In spite of this, 114 pullets qualified for registration, each producing 200 eggs or over.

For full details in regard to American Record of Performance Council, see Chapter III, Part III.

The All-Northwest Contest

The All-Northwest Contest at Pullman, Washington, began in 1916 and continued until the fourth year, during which the buildings were completely destroyed by a tornado. Up to that time some excellent records had been secured as follows: In the first of these contests the high bird made a record of 237 eggs. This bird also was mother of the high bird in the third contest, whose record was 270 eggs. The high bird in the second contest was a Single Comb White Leghorn, No. 251, which made a record of 311 eggs in 365 days. She was bred and owned by D. Tancred, of Kent, Washington. This bird was one of 36 in a pen. The highest pen in the contest was a pen of White Leghorns, owned and bred by Paul B. Towne, of Tekoa, Washington. The five birds averaged 252.2 eggs per bird. The second high pen was a pen of Barred Rocks owned by the Oregon Agricultural College, which pen made a record of 1,258 eggs, or an average of 251.6 per bird. The foregoing were the highest official records in the United States up to that time. A pen of Wyandottes in the third contest held the highest pen record for Wyandottes in the United States.

Western Washington Egg-Laying Contest

The first year's contest at the Western Washington Experiment Station was started in 1919 with 60 American-class birds and 165 of the Mediterranean class. This contest differs from the other United States Contests previously mentioned in that the birds were kept in fairly good-sized flocks, the American-class fowls being kept in flocks of 36 and the Mediterranean class in flocks of 54. The same method of feeding and management was followed as advocated by the station in the handling of commercial flocks, including use of artificial light in winter months. The following is quoted from Monthly Bulletin, December, 1920, of that institution, giving final report of the contest:

"The two buildings used for this contest, each 110 feet long, are of slightly different construction; one has a feeding aisle with an entrance in the back of each pen. The other is an exact duplicate of the Puyallup laying house, with a continuous droppings board along the rear wall and a dust bath in front. The Mediterranean breeds were given about 5 1/5 square feet to each bird. The American breeds were given over 6½ square feet to each bird. This method of housing is identical with that practiced in handling our breeding pens here at the station, and is the most economical and efficient way we know of handling small units.

"The partitions have 24 inches of boards at the bottom to prevent fights between pens. The balance is 2-inch mesh wire. The doors in the partition walls swing both ways and are over a 12-inch baseboard, which is removable for convenience when the litter is changed. The yards alternate front and rear, and birds are sent outdoors when the litter is changed, so that the litter may be shoved along the floor, half of the house being cleaned from each end. The nests are under the droppings boards, but the space underneath is very well lighted from the rear windows, which are spaced 15 inches apart. There is no dark place or sheltered corner, which might tempt birds to lay outside of the trap nests.

"The water is fed in a painted wooden trough which is suspended outside the front of the house in such a way that the automatic faucets keep the troughs full of water without running over. The weight of the water, when the trough is full, closes the faucet. The shell and mash hoppers are carried on the partition walls. The outside feeding

BIRD'S-EYE VIEW OF HOUSES AND YARDS AT THE VINELAND, N. J., EGG-LAYING CONTEST

trough permits the sanitary feeding of lawn grass, mangels, kale leaves and wet mash. The birds reach through the 2 by 6 mesh openings in the front and help themselves. The milk is fed in earthen crocks, set in racks with narrow slats for footboards. The front openings of both houses, which face the south, are covered with the continuous muslin curtain, which rolls up from the bottom. The skylight ventilators in the comb of the roof complete the ventilating scheme. The inside arrangements and furnishings of the pens in both houses are identical except that the pens devoted to the large breeds have lower droppings boards and larger nests.

"The records at this contest give opportunity of comparing not only the number but also the size of the eggs laid by the different classes and breeds.

Production at Western Washington Laying Contest

	Total eggs	Standards	Pullets	Av. per bird
Mediterranean class—165 birds	35,559	28,992	6,567	215.5
American class—60 birds	11,841	10,067	1,774	197.3

"This shows that the Americans, including Rocks, Wyandottes, Rhode Island Reds, etc., produced 85 per cent standards, while the Mediterraneans, including Leghorns, Anconas, etc., produced 73 per cent standards.

"The large birds much prefer to eat their dry mash concentrates in the wet mash form and eat much less of the dry mash proportionately than do the Mediterranean breeds. The total bulks consumed of the three feeds—sprouted oats, scratch and the prepared mash—are about equal under a 60 per cent lay.

"The winning birds were uniformly medium-weight for their breed, the larger birds of each breed being inclined to take on more flesh and lay correspondingly fewer eggs."

GENERAL VIEW OF HOUSE AND YARDS AT NORTH AMERICAN LAYING CONTEST (DELAWARE)

At this contest the average per cent of production for the entire 12 months was 57.5, the average number of eggs per hen being 210.5. Of the 225 birds entered, two laid over 300, 17 over 275, 49 over 250 and 139 over 200. The high pen (S. C. W. Leghorns from D. Tancred) made a record of 1,306 eggs; the second pen (S. C. W. Leghorns from C. H. Burnett), 1,289; third pen (Oregons from Ontario Agricultural College), 1,285. Two Leghorns tied for first place as individual layers, with records of 312 eggs. Both the first and second pens finished with the five birds originally entered, the alternates not being required.

The second contest produced pen and individual records both exceeding any that have been secured at any other official contest in this country. The first pen, entered by D. Tancred, laid a total of 1,384 eggs as against the highest former American record of 1,319 eggs. The second pen laid 1,353 and the third pen, 1,337. In all three of these pens the eggs laid were produced by the original birds entered, no alternates being credited in the pen totals. These pen records represent averages per hen of 276 4/5, 270 3/5 and 267 2/5, respectively. Seven pens in this contest made records of over 1,300 eggs and 21 went over the 1,200 mark. In the individual records, the high bird made a record of 313 eggs; the second, 304 and the third, 302.

Nebraska National Egg-Laying Contest

The Nebraska Contest is conducted along lines designed to approximate conditions in commercial flocks. The entries consist of 10 birds each, and these are housed in flocks of 200, birds of similar type and variety being penned together so far as possible. All birds are trap nested. The average production in the second contest was 154.6 eggs. Seventy-six of the birds made records of 200 or better and will be registered in the American Record of Performance. The best pen at this contest was a pen of White Leghorns, whose production was 1,994 eggs; the second pen consisted of S. C. Rhode Island Reds, record, 1,848 eggs and the third pen, S. C. White Leghorns, record, 1,817 eggs. The high bird at this contest was a S. C. Rhode Island Red, which made a record of 272 eggs.

California Farm Bureau Egg-Laying Contest (Santa Cruz)

This contest started on November 1, 1919, consisting of 54 entries of 10 birds each. From the preliminary report of first contest we learn that the first pen averaged 244 eggs per bird, while the second pen made an average of 213.3 and the third pen, 229.9. The highest individual made a record of 298. Nineteen birds laid over 260 eggs each, and over 50 per cent of the pens entered averaged better than 190. All the leading pens and individuals were S. C. White Leghorns.

Farm Egg-Laying Contests

In the last few years there have been a number of farm egg-laying contests, some of them quite extensive in number of fowls entered, and while the birds are not trap nested the flock records are closely scrutinized by the State Agricultural College Extension Department and are believed to be reasonably accurate. These contests have brought out some extremely interesting facts in regard to the possibilities of egg production on farms and in commercial flocks.

The first laying contest ever held (to which we have already referred) was the one conducted by the "National Stockman and Farmer" in 1889. It continued eight months, participants reporting number and weight of eggs each week. A summary was published monthly. Three contestants continued for a year, at the end of which time a yearly summary was published which stated that "in the 12 months ending August, 1889, Mr. Baker's six Brown Leghorn pullets laid 1,335 eggs which weighed 162 pounds and were worth at the Pittsburg market prices $19.27. Miss Whitman's six White Wyandotte pullets laid 1,203 eggs weighing 151 pounds, 8 ounces, and worth $17.70."

Another contest was started February 1, 1894, entries being accepted up to March 1. The following breeds were represented: Asiatic varieties — White Langshans, Black Langshans, Partridge Cochins, Buff Cochins and Light Brahmas. English varieties—Red Caps and Black Minorcas. Continental varieties—Houdans, Silver Penciled Hamburgs and Silver Spangled Hamburgs. Mediterranean varieties—Black Leghorns, Buff Leghorns, Rose Comb White Leghorns, Single Comb White Leghorns, Rose Comb Brown Leghorns and Single Comb Brown Leghorns. American varieties—Black Javas, White Wyandottes, Silver Laced Wyandottes, White Plymouth Rocks and Barred Plymouth Rocks. Grades and mixed—grade Leghorns, grade Plymouth Rocks, crossbred and two divisions of mixed breeds. Entries numbered 224 and ranged in number of fowls from 3 to 200. Contestants were from Ohio, Pennsylvania, Maryland, West Virginia, Indiana, Illinois, Tennessee, Missouri, Washington, Michigan, Massachusetts and New York.

A summary of the yearly report gives the six highest producing pens, as follows:

1st.—8 W. P. Rock pullets, average, 286 eggs each; value, $5.02.
2nd.—8 crossbred pullets, average, 283 eggs each; value, $4.82.
3rd.—8 W. P. Rock pullets, average, 280 eggs each; value $4.90.
4th.—8 S. C. B. Leghorn pullets, average, 279 eggs each; value, $4.64.
5th.—24 S. C. B. Leghorns, average 277 eggs each; value $4.89.
6th.—12 B. P. Rocks, average 262 eggs each; value, $4.24.

The value of eggs was fixed by monthly retail prices in Pittsburg, which explains why the third surpassed the second, and the fifth the fourth, in value of eggs produced.

Twenty pens each produced over 200 eggs per hen during the year. These 20 pens included one of Black Minorcas, one of S. S. Hamburgs, one of Buff Leghorns, two of R. C. B. Leghorns, six of S. C. B. Leghorns, one of S. L. Wyandottes, two of W. P. Rocks, three of B. P. Rocks, two of grade Leghorns and one crossbred.

The Missouri Farm Laying Contest

This contest was started in 1918 and has been made an annual event. The following information in regard to this

VIEW OF ONE OF THE BUILDINGS USED IN THE WESTERN WASHINGTON EXPERIMENT STATION CONTEST

In this building the birds are kept in trap-nested flocks of 36 to 54 and it is here that some of the greatest official contest records to date have been secured. Note ventilator in roof of building and the windows along the north side under the droppings platform.

contest is condensed from a most interesting communication from Prof. T. S. Townsley of the Extension Service Department of the University of Missouri, who is in charge.

This contest was started in 1918 with 20 farm flocks competing, with an average of 114 hens in each flock. The records for that year show an average production of 100 eggs for each hen competing. Sixty-five flocks, averaging 134 hens each, competed in the 1919 contest and made an average record of 106 eggs each. In 1920 the number of competing flocks increased to 138, having an average of 125 hens per flock, and the average egg production for the year was 114. In 1921 168 cooperators carried on the contest work with an average-sized flock of 144 hens and the average egg production was 125. Several of the original contestants have competed during the four years of the contest, and their individual flock records show the same upward tendency as the average for all flocks. One cooperator, whose records have shown a consistent improvement, secured an average of 133 eggs in 1918 from each of her 174 hens; in 1919 her flock of 250 hens laid an average of 135 eggs each; in 1920 her flock increased to 367 birds and the average production was 148 while in 1921 from 362 hens she secured an average of 160 eggs. Her profits have also shown an increase from year to year as her egg production mounted. In 1918 she realized a net profit of $2.58 from each hen; in 1919 her profit per hen amounted to $3.08; in 1920 the average hen returned a net income of $3.80; and in 1921 the average profit for each bird amounted to $3.95.

The plan of the Missouri Farm Contest is to have one poultry keeper from each community in the various counties enter, these flocks being termed "Demonstration Farm Flocks." The contestants agree to care for their entire poultry flock as nearly as possible according to directions furnished by the Poultry Department of the University of Missouri and to furnish each month to the Agricultural Extension Service, through the farm bureau office, a complete statement of the expenses and income from the farm flock for the month.

BUILDINGS USED IN THE NEBRASKA EGG-LAYING CONTEST

Entries consist of 10 birds each and are housed in flocks of 200, thus securing conditions similar to those met by the average commercial poultry keeper.

The contest has shown a material increase in number of contestants for each succeeding year. From an initial start of 24 flocks containing approximately 2,000 hens, the contest has spread until in November, 1921, the first month of the fifth contest year, records were received from 293 flocks containing a total of 57,194 hens.

A summary of the 1921 contest follows: "By closing the year with a new high average for October production, a total record which exceeded that of any other year by more than 11 eggs per hen was established by the birds competing in the Missouri Demonstration Farm Flock Laying Contest. In this contest reports were received from an average of 168 farms each month showing the production from an average of 144 hens to each farm or a total of more than 24,000 hens. These flocks were located in 37 Missouri counties and each flock was handled by the owner under representative farm conditions. The average record of each hen for 1921 was 125.1 eggs. In the 1920 contest the average was 114 eggs and in 1919, 106.3 eggs, while in 1918 the production averaged only 100.6 eggs per bird. Thus the 1921 contest shows a 25 per cent increase in production over the first contest carried on four years ago. Many of the contestants who competed this year started in 1918 and have participated during each year of the contest, and the improvement in production can be traced directly to better methods of feeding, housing and breeding.

"The 1921 records show a somewhat smaller profit than for 1920 but larger than for any of the other years during which the contest has been in progress. The average farm reporting for 1921 sold $628.48 worth of eggs and fowls and fed $213.61 worth of feed, leaving an income over feed cost of $414.87 per farm or $2.88 per hen. This compares with a net return over feed of $3.41 in 1920, $2.39 in 1919 and $2.48 in 1918."

Comparison of Annual Summaries for Four Years—Missouri Farm Flock Contest

Year	Avg. no. of flocks	Avg. no. of hens per farm	Avg. no. of eggs per hen	Total income per farm	Feed cost per farm	Avg. profit per farm over feed	Profit per hen over feed cost
1918	20	114	100.6	$545.37	$262.19	$283.18	$2.48
1919	65	134	106.3	577.40	257.74	319.66	2.39
1920	138	125	114.0	695.21	268.57	426.64	3.41
1921	168	144	125.1	628.48	213.61	414.67	2.88

Indiana Farm Egg-Laying Contest

This contest, organized in a manner similar to the Missouri contest, started in 1920 and is one step in a definite plan of poultry extension work in that state. This contest started out with over 1,000 farms and total entries of 125,000 birds. Notwithstanding the fact that practically one-half dropped out before the year was over, the total number of birds at the end of the contest was still quite large and the records are believed fairly to represent the average of the better class of Indiana farms, as regards poultry keeping.

The following table gives the results secured in the first contest:

Indiana Egg-Laying Contest, 1920-1921; 28 Counties, 267 Farms and 18 Varieties Represented

Breed	Barred Rocks	White Rocks	Buff Rocks	White Leghorn	Brown Leghorn	Buff Leghorn	Black Leghorn	White Wyandotte	Buff Wyandotte	Columbian Wyandotte	Rhode Island Red	Buff Orpington	Light Brahmas	Black Langshan	Black Minorca	Ancona	Several Varieties	Mixed
Total no. farms	56	9	4	41	12	1	1	10	2	1	43	8	1	4	1	3	29	41
Average hens per farm	92.6	59	69.5	176.5	178.6	62	107	77.1	42	56	96.2	70.6	98	82	42	55.3	140.6	115
Average eggs per hen	104.8	113.8	86.2	138.4	134.5	95.7	101	108.9	126.3	120.5	108.4	121.7	85.2	121.4	95	122.6	101.6	109

CHAPTER II

Poultry Keeping on the Pacific Coast

The Development in Commercial Poultry Keeping in the Last Few Years, All Along the Pacific Coast from Southern California to British Columbia, Has Been Truly Remarkable—This Chapter Presents a Bird's-Eye View of the Industry in That Section and Includes a Number of Interesting Reports from Individual Producers

DURING the last decade there has been a remarkable growth in commercial egg production on the Pacific Coast, all the way from San Diego in southern California to and including British Columbia. While the various localities present somewhat different climatic conditions and requirements, no one section appears to have any marked advantage over another, and the methods practiced throughout this region have much in common. Nowhere else in North America has specialized egg farming reached so high a stage of development and popularity. It is stated on good authority that fifty per cent of the entire egg yield in California, for example, originates on commercial poultry ranches, as distinguished from general farms which in the East and the Central West provide the great bulk of all poultry products that reach our markets. With few exceptions the poultry plants in all this region are quite intensive and the White Leghorn is the general favorite. One thousand, 2,000 or even more birds to the acre are quite common in California. In this state "one-man" plants are the rule, yet there are many "ranches" carrying thousands of layers and, in the hands of experienced men, they are uniformly profitable.

The industry is characterized throughout this region by an unusual degree of concentration around local centers. This apparently is due not to special advantages in location but to the tendency of newcomers, interested in this line of work, to gather about some successful pioneer, and possibly still more to the fact that the prosperity of the Pacific Coast egg farmers rests quite largely with their cooperative associations which have here reached an effectiveness of organization that has rarely, if ever, been attained among poultry producers in other parts of this country. Through them the individual is able not only to market his products to the best advantage but also to reduce cost of production to a marked degree. These cooperative associations are particularly well developed in California, where it is stated that they handle fully 45 per cent of all the eggs produced on commercial plants. The influence of these cooperative associations in building up the industry on commercial lines can hardly be overstated, particularly in view of the fact that, at present, there appears to be decided overproduction, so far as local needs are concerned. The associations keep close tab on markets at home and elsewhere and by judicious handling of the surplus maintain prices at reasonable levels, where otherwise disastrous slumps would occur unavoidably at certain seasons. Enormous quantities of eggs are now shipped annually to eastern markets, chiefly to New York City. Owing to the careful grading and packing for which so-called "association eggs" are distinguished, and to shipment in fast trains of refrigerator cars, such eggs command prices practically equal to the best of those which are received for general shipments from near-by producers of fancy grades.

M. A. Schofield, one of the leaders in the development of the poultry industry in southern California and in the organization of the Cooperative Association of Southern California, is authority for the statement that shipment of eggs to eastern markets approximated 1,000 cars in 1920 and probably exceeded that number in 1921. The three cooperative associations of the state of California are capitalized, one at $750,000, one at $250,000 and one at $100,000. This liberal capitalization, which is necessitated by the magnitude of the business handled, has been provided by a modest tax of 1c per dozen eggs handled, which in a year's time added $200,000 to the working capital.

The Southern California Association is said to be particularly thoroughgoing in its activities. It provides its own trucks which go from ranch to ranch daily, collecting eggs which are hauled to Los Angeles headquarters and there are uniformly graded, packed and shipped to market. It also has its own feed mill where grains are handled by the carload, and feeds ground and mixed are delivered by truck to the various associated ranches. This not only reduces the feed bill and relieves the poultry keeper of every detail aside from the mere management of his flock and the gathering of the eggs produced, but it also enables him to dispense with the expense of maintaining a team or truck. Writing on this subject in Reliable Poultry Journal, July, 1920, Morris M. Rathburn of Los Angeles, says:

"The development of a New York and London market, with shipments of from five to eight carloads per week to these markets during the high-production months, is but one of the achievements of the poultry raisers cooperative associations in California. Before the formation of these organizations, shipments were not made by the individual producer except to points not too far distant to insure successful shipment by mail, express or freight. And not only have these markets been developed, but the price paid for California eggs shipped by the associations is from five to fifteen cents higher than that usually paid for local eggs.

"This demand has been created through a most intensive selection which the associations have inaugurated. The standards of the associations are inflexible. Show-

VIEW ON HIGHLAND PARK POULTRY RANCH, PORTAGE, WASHINGTON

One of the well-known, prosperous poultry plants of the Pacific Northwest, where commercial poultry keeping has reached a high stage of development. Long laying houses here illustrated are of the type developed at the Western Washington Experiment Station and widely adopted by poultry keepers throughout the state.

ing what effect this selection has had to increase the production of the best eggs we may state that in a year the eggs sent by producers to the associations were graded as follows: 11.31 per cent were extras, 21.06 were underweight and were graded as pullets, and the balance was made up of peewees, dirties, crax and leakers.

"These two organizations, which were formed in California at the suggestion of the state marketing director in 1917, have placed the industry in the Coast state on a firm foundation. Through this cooperation and specialization, the industry in California in a little over three years has been brought from a more or less unstable condition to a thoroughly established business where profits are certain and markets sure."

300-EGGERS ARE BRED IN SOUTHERN CALIFORNIA ALSO

This hen, bred and owned by England's Egg Ranch, Inglewood, Calif., has a trapnest record of 312 eggs in 363 days.

There is a pretty general impression among poultry keepers in other parts of the country that the extraordinary results secured by commercial poultry keepers on the Pacific Coast, notably in the Pacific Northwest, and particularly the extraordinary high egg records secured, is largely a matter of climate. This belief however, does not appear to be at all well founded as will be seen by the following article written very recently after a careful study of this phase of the problem.

NOT A MATTER OF CLIMATE

The Climate of the Pacific Northwest Where Such Notable Achievements in High Egg Production Have Been Secured Is at Best Only Moderately Favorable to Poultry—Some Serious Disadvantages To Be Overcome

By Grant M. Curtis

IT is our firm belief, based on nearly a year's residence in the Northwest, that it is breeding and not climate which has produced and is continuing to produce the wide differences in results in egg production there as compared with other sections of this country. As a matter of fact, northwest poultrymen "also have their troubles" the same as or similar to those met in other sections of the country—such as colds, roup, canker, rheumatism, chicken pox, intestinal worms, lice, mites, etc., and that also out there the price of success is constant vigilance to prevent something of this kind getting into the flocks. Questioning M. E. Atkinson, of Hollywood Farm, Washington, about the matter of climatic conditions, on the basis of his ten years' experience, also regarding various poultry ailments that are common to domestic fowl, he replied as follows:

Actual Weather Conditions in Pacific Northwest

"Regarding weather conditions against which we have to contend here in the Pacific Northwest, our worst trouble is with heavy fogs and the long, rainy season. We surely have an excess of fog and rain out here. Now and then we also have freezing spells and snow storms, for which we are ill-prepared. Twice here lately (during January, 1922) the poultrymen of this locality found themselves but poorly equipped for the unexpected cold spells, two in number, each of which settled on us for several days, approaching zero in temperature. During the four or five months of rainy season each year, we have our share of heavy winds and driving rains which bring on colds and threaten our fowls with roup, unless we are extremely vigilant. When we get these rains we have to change the litter more often than once a week; as was the case very recently. Also the heavy fogs settle down and drift into the open-front buildings, especially so in the case of poultry plants that are located in the valleys, where most of our really fertile land is to be found.

"Aside from the troubles that have to be avoided in this territory on account of bad weather, there has been considerable anxiety of late about chicken pox being brought up from the South and as a preventative at Hollywood we inoculate each fall every pullet on our plant, and have done this for two seasons. The serum is furnished by the state and comes in doses of one cubic centimeter to a bird, at a cost of 1½ cents per bird. We inoculate the birds just before putting them in the laying houses.

"The intestinal worm referred to is a round white worm, pointed at each end and a little larger around than a darning needle. There is also another variety about half that size. Sometimes they get into the intestines of the birds and must be watched for and fought to a finish, otherwise there will soon be a falling off in egg yield. No, we do not know the real source of them. They are not fatal, as a rule, but still are a nuisance and also costly, as affecting egg production. We have not heard of any canker in this section for a long time. Where birds are not well cared for it will appear with colds, therefore poultrymen need to take every sensible precaution against such ailments. Fact is, I do not know of any disease of a fatal nature to which poultry is subject in this climate, yet to be truly successful from the breeder's point of view, also financially, the poultrymen here as elsewhere must take precautions constantly."

Success at Hollywood Due to Breeding, Not Climate

At the close of an interview, which had to do mainly with the demonstrated value of actual breeding for prolific egg yield, as compared with climate, proper care, right feeding, etc., Mr. Atkinson said:

"You can state with all possible emphasis, so far as my personal view is concerned, that at Hollywood we are absolutely solid on this idea of systematic line breeding and I am convinced that our results from the start fully sustain this conclusion. In my best judgment, based on ten years of study and experience, also confirmed by the degree of success we have had, I am as positive as a man can be that any poultryman who makes the mistake of giving up his systematic breeding, based on trap nesting and pedigreeing, to adopt physical measurements or exterior appearances, or any other so-called fad or system, will meet with disaster. I do not think I can state this belief too

TYPE OF OIL-BURNING COLONY HOVER IN GENERAL USE IN PACIFIC COAST STATES

strongly. These so-called systems are the 'isms' of the poultry business just as they exist in all other affairs of our people in every generation.

"Let me cite the case of a woman customer of ours in this state. She used Hollywood males and got her flock up to where she had a pullet-flock average of better than 200 eggs. Had only about 500 birds. She entered a pen in the First Western Washington Egg-Laying Contest, 1919-1920, and was up among the winners. Then she was advised by someone to get some new or outside blood into her flock, which she did by introducing males of other than the Hollywood strain (do not know where she got them) and her flock average from the progeny went all to pieces, as shown by recent contest records made by her birds, these records actually dropping below that of the previous year to the extent of nearly one half. She came back to me for advice and asked what I did for new blood and I told her how I mate our birds, season after season, as I have fully explained to you. The last purchase of breeding stock made by us was from Mr. Padman in 1916 and we haven't introduced any new blood in our several lines or families since then—not a drop.

"I recall another instance that should be of interest and value to your readers, as showing that trap nesting and line breeding are necessary for real and lasting progress. This fact was borne out strongly by another customer of Hollywood Poultry Farm. This young man started out a few years ago with some Hollywood stock and won prizes in a contest, making a good showing. However, one of his birds, a high producer at the contest—up in the 300-egg class—happened to measure up to all the requirements of the so-called Hogan test, which made him think that this test really was 'the thing.' When I learned that this good friend of ours had become impressed with the Hogan measurements, I felt that if he adopted that method, rather than to rely on trap nesting, line breeding and careful pedigreeing, he would not hold his own in public competition. A recent contest seems to bear out this conclusion. In this case our fears in the matter appear to have been well grounded, as his public records have dropped to an astonishing extent."

Personal View of D. Tancred

February 5, 1922, writer asked Mr. D. Tancred of Kent, Washington, for his view about the climatic advantages of western Washington for poultry keeping, to which he replied without a moment's hesitation:

BROODER CHICKS IN YARD AT HOLLYWOOD POULTRY FARM, HOLLYWOOD, WASH.
Oil-burning hovers are used and chicks brooded in flocks of 1,000. Brooder houses are of the two-compartment type, each compartment being 14 by 20 feet in size.

"Advantages? You meant to say 'disadvantages,' didn't you? If we have a real advantage it is in our cooler summers, which no doubt prove helpful, but this gain to the poultryman is more than offset by the nearly intolerable four to six months of rain, fog and marrow-chilling cold we have in these parts every fall and winter. Also we are about as far north as one can be and still remain in the United States, which means that we have a longer period of short days, as regards sunlight, than is the case farther south, a situation that cannot specially benefit our fowls nor add to egg production.

"After more than sixteen years of experience, I regard our climate here as being among the worst known for poultry. For a dozen years I have wished that I had located in California, where much dryer conditions prevail during our long season each year of rain, fog and penetrating cold. Such weather is visibly depressing to the fowls and no doubt cuts down considerably their average egg yield, both as to individuals and also the flock averages. It was mainly on account of our longer winter nights that commercial poultrymen throughout this territory adopted electric lights, something I have never done in any form, as my farm is a breeding plant, first and foremost. Our home records, therefore, have been made in all cases without the help of artificial lighting."

Poultry Keeping in Washington

The poultry industry of the state of Washington centers chiefly about the Pudget Sound region. Here are located the famous Hollywood Poultry Farm and the almost equally well-known plants of D. Tancred, L. C. Beall, Jr., E. Morgan and others, whose accomplishments in breeding for high egg production have challenged the attention of the world. This state has two experiment stations, one at Pullman in the extreme eastern part of the state and one at Puyallup in the Puget Sound region, both of which have done much to foster the welfare of the industry. George R. Shoup, poultryman at the Puyallup station, has exercised a marked influence upon the poultry industry in the western part of the state, having brought about a degree of standardization as to housing and equipment and methods that, so far as we know, is equaled in no other section of the country.

We are informed that there are upwards of 200 commercial

LAYING HOUSES ON RANCH OF GEO. J. RICHARDSON, SAN GABRIEL, CALIF.
Flock of 1,500 laying hens in foreground. Throughout this entire section of the country highly intensive methods are generally practiced.

poultry plants, the owners of which derive their entire livelihood, or practically so, from poultry keeping, within a radius of less than ten miles of the Puyallup Experiment Station. The interested reader is referred to Chapter IV of Part II for a more extended description of the methods practiced at representative Washington poultry plants, notably at Hollywood Poultry Farm and the Tancred Farms.

Poultry Keeping in Oregon

Oregon also has a growing poultry industry and there are many successful plants, particularly in the Willamette Valley. Professor James Dryden of the State Experiment Station at Corvallis has done much to popularize poultry keeping in this section and to make it profitable. Breeding stock of his high-producing strains of White Leghorns, Barred Plymouth Rocks and "Oregons" has been widely distributed among commercial poultry keepers and has had a tremendous influence on the average of production obtained.

In writing to the authors in regard to the state poultry industry, Professor F. E. Fox of the poultry department at the State Agricultural College at Corvallis, says:

"I was very agreeably surprised on coming from the Middle West last year, to learn of the extent of poultry keeping here. Undoubtedly there is a big future for the industry in the Pacific Northwest. The climate, particularly here in the Willamette Valley, where the important producing centers are located, is mild in winter, and the birds can get out at practically all seasons of the year. There is an abundance of green feed and a noticeable lack of freezing weather, which is a handicap to winter production in other parts of the country. I have yet to see a frozen comb here in the valley.

"The number of laying fowls in the state is, in round numbers, 8,000,000. Our estimate is that 15 per cent of these are on commercial plants, of which there are about 500. The total value of eggs produced in the state last year is $12,000,000. The extensive method is practiced quite generally. Long houses, holding from 500 to several thousand birds are the general type. These houses have open fronts and vary in width from 16 to 24 feet.

"There are two cooperative associations in the state. These are: The Pacific Cooperative Association, located at Portland, and the Association of Southern Oregon, with headquarters at Medford, to which approximately all of the commercial plants belong. Probably one-half of the eggs shipped out of the state are handled by the associations, Swift and other concerns handling the balance. From the figures which I have available it would appear that 200 carloads of eggs were shipped out of the state last year."

The following account of the success of an Oregon poultry keeper with a one-man plant is of exceptional interest and shows not only what can be accomplished in this section by earnest effort, but also the comparative ease with which it is possible to secure what a few years ago would have been considered extraordinary production, by use of bred-to-lay foundation stock and subsequent careful culling based on external characters.

A COMMERCIAL FLOCK OF 200-EGGERS IN OREGON

Through the Use of Pedigreed Males, on Rented Farm, With $1,000 Capital, J. A. Hanson, Corvallis, Oregon, Builds Up a 200-Egg Flock in a Few Years, and in Four and a Half Years Saves $18,300 Out of Poultry Profits—In Third Year Reaches an Average of 219 Eggs per Bird with Flock of 1,000 Pullets

By Prof. James Dryden, Oregon Agri. College, Corvallis

BUILDING USED IN EGG-LAYING CONTEST, SANTA CRUZ, CALIF.

THAT a flock of 1,100 egg layers will lay for their owner seven hundred eggs lacking five, on Thanksgiving Day, with eggs at 75c a dozen, is not only a cause for thanksgiving but the performance comes pretty near demonstrating to those who understand something of the "egg-laying characteristic" the fact that the particular flock that performed the feat was a good-laying flock with very few slackers and many star performers.

The owner of the flock is a young man who at an early age was called before the authorities in his home city in Ohio, on complaint of some neighbors, for keeping chickens in the residence section. The authorities after questioning the boy evidently had a premonition that he was destined for a good poultryman for they let him go and told him that he could keep his pests even if the neighbors had to move to some quieter section. In the course of time the boy, now a young man, was graduated from the Missouri Agricultural College without losing his early love, and forthwith took a job on a Missouri poultry farm where he spent a year. Then he wrote a letter to the writer and he was told to come on, that we would put him to work at the poultry plant of the Oregon Agricultural College and help get him located. He came with but $25 in his pockets and a will to work and to study.

That was some six years ago. After a few months at the college poultry farm he found a place to work on a commercial poultry farm. At the end of a year he had saved $1,000 and he wanted to make a start for himself. In four and a half years from that time, as his assets show, he has laid by a nest egg of $18,300. This is made up as follows:

Personal stock, including an automobile, incubators, brooders, household goods, etc.	$ 3,150.00
Livestock, including poultry, 3 cows, 2 horses	4,230.00
Feed on hand	3,500.00
Cash in bank, Liberty Bonds, etc.	8,020.00
Life insurance paid	400.00
Total	$19,300.00
Began business with	1,000.00
Money accumulated	$18,300.00

The above represents actual accumulations. During the same period some of the household expenses were paid out of the returns from the farm. The rent, which was $500 a year, is included in the miscellaneous expenses of the farm, shown below, and the milk, butter, eggs and chickens were commandeered on the farm. The receipts

---

Reprinted in condensed form from The Country Gentleman, May 17, 1919, by permission.

which are given herewith do not include any part of the living that was secured by the family in this way.

The Cost and Income

The receipts and expenses cover the period from March, 1914, to November 1, 1918, or 4½ years. The following statement gives the receipts for the different years, the first covering a year and a half:

1914-15	$ 5,638.89
1915-16	8,225 04
1916-17	12,907.78
1917-18	13,493.75
Total	$40,265.46

BUILDING USED IN PETALUMA LAYING CONTEST
House here shown is of type quite generally adopted for laying houses in California. Muslin screens open outward to shade interior in hot weather.

The operating expenses which include feed, labor—not counting owner's—and miscellaneous items, were:

	1914-15	1915-16	1916-17	1917-18
Feed	$1,038.50	$2,141.91	$2,895.30	$5,747.90
Labor	181.65	549.00	1,203.83	1,505.90
Miscellaneous	735.00	823.35	1,207.98	1,310.00
Totals	$1,955.15	$3,514.26	$5,307.11	$8,563.80

If the total operating expenses of $19,340.32 are deducted from the total receipts there is a balance of $20,925.14, which represents the profits of the business for four years and a half and this is a record few have approached.

This is a good showing for a poultry farm of medium size. The number of layers in any one year did not exceed 1,400, and there were only two or three cows. The showing is so good that one might be tempted to enthuse and say that it is a splendid showing.

How was it done? The receipts came from sales of broilers and old hens, breeding stock, hatching eggs, day-old chicks, cream, fruit, market eggs. The amount of sales from these different sources totaled as follows for the four years and a half:

Broilers and hens	$ 2,139.94
Breeding hens, males and pullets	3,464.56
Hatching eggs	3,582.95
Day-old chicks	3,979.83
Cream	1,674.45
Fruit	468.30
Market eggs	24,955.43
Total	$40,265.46

A few cows were kept and cream sold to the creamery and the skim milk used for the chickens. A little fruit was also sold. These two items added some $2,100 to the total receipts for the four years. Some feed was raised on the place, and this reduced the amount of the feed charge, but no credit is taken for it in the receipts.

He started in the business in February, 1914, by buying eggs for hatching. Many others have started in this way and failed; but he went to a poultryman who had a fair-sized flock of White Leghorn hens and he contracted with him to buy his eggs if he would use males that Mr. Hanson would furnish. At that time he could not purchase enough eggs from pedigreed trap-nested flocks, but he understood the value of the male, and he figured that by using pedigreed high-producing males each year he would have good-producing pullets. Each year he has procured pedigreed males from the Oregon Experiment Station, and he attributes his remarkable egg yield to this fact. Of course he is a good feeder; otherwise he would not have secured the high egg yield. But he figured that he could not do good work with poor layers any more than a good carpenter can do good work with poor tools.

In 1914 there were 650 pullets that began laying in September. Beginning November of that year and continuing for twelve months, 121,462 eggs were laid. This is an average of 187 eggs a hen, counting the full number of hens at the beginning of the year. During the year, however, there was a loss of fifty. If the number is averaged at 625 the yield would be 194 a hen.

The 200-Egg Flock Has Arrived

In the following year with the same number of pullets the production was 124,818 eggs, an average of 200 eggs. In the third year with an average of 1,000 pullets the production was 219,009 eggs, an average of 219 eggs a hen. In the fourth year the production was 187,969 eggs, with 885 pullets as an average for the year, averaging 212 eggs a hen.

There is no doubt but that the 200-egg flock has arrived—the flock of 500 to 1,000 pullets—and it has come by the heredity route.

A PEN OF J. A. HANSON'S HIGH-PRODUCING WHITE LEGHORNS
First pen in California Egg-Laying Contest (1920-21) at Santa Cruz. Made a record of 2,440 eggs in twelve months—the highest record so far made by a 10-bird pen. Bred and owned by J. A. Hanson, Corvallis, Oregon. A brief account of his successful poultry experience will be found beginning on page 138.

SCENE ON WHAT IS CLAIMED TO BE THE LARGEST POULTRY RANCH IN THE WORLD
A partial view of the Corliss Poultry Plant at Petaluma, Calif., where 35,000-50,000 layers are kept.

As to Mr. Hanson's management—feeding and care—there is nothing strikingly different in it. Other poultrymen have no doubt given their flocks as good feeding and care, though getting fifty per cent lower egg yield than he.

The farm consists of 30 acres of rented land, the rental being $500 a year. It was an abandoned poultry farm when he took it. The buildings consisted of a good residence, a barn, colony laying houses, brooder and incubator houses. In the way of equipment there were incubators and brooders, wagons and tools found on the average farm. There was no outlay required for buildings and very little for equipment at the start. But here is where the management is somewhat different: He was not obsessed with the idea that he must have all modern contrivances; that his buildings and yards must be laid out according to pictures in books.

Making the Dollar Count

Mr. Hanson was after utility both in hens and in plant. He succeeded where many others fail in that rare accomplishment of keeping the overhead expense down to a reasonable limit. In other words, he made the dollar count. There are no expensive buildings or other equipment on the place. It is not a model farm by any means when it comes to buildings and equipment. The farm was taken as it was found, for better or worse, and Mr. Hanson's achievement was in making the best of whatever equipment, buildings and contrivances he found at hand there.

The method of feeding followed during the period covered by this article is given below, except during the period of the war, when wheat was under the ban. On that account very little wheat was fed during the last year though in other years wheat was the staple food. But little difference was noted in the egg yield due to the change from the wheat ration, showing that hens will stand a wide variation in the ration without resenting it. In other words, a good layer will stand a good deal of abuse in the way of feeding and care and still perform well. The importance of good feed and feeding cannot be overlooked, but the point should be continually emphasized that good feeding does not make good layers. Good feed will fill the egg basket only when combined with good breeding.

The layers are fed four quarts of gray oats to each 100 hens in the morning, in the straw, and wheat at night. The mash before them at all times is mill run, sixty per cent; coconut, oil meal, twenty per cent; meat scrap or fish scrap, twenty per cent. Skim milk, kale, grit and water are before them all the time. Green cut bone is fed twice a week—all that the hens will clean up in fifteen minutes. During part of the period a moist mash was fed. A very simple method and ration. It was fed with regularity and there was no stinting to save on the feed bill.

Petaluma—"The World's Egg Basket"

For many years Petaluma has had the distinction of being the greatest center of production in this country. The poultry population there naturally varies from year to year but Dr. L. E. Heasley, in an article published in Reliable Poultry Journal, date of July, 1920, is authority for the statement that Sonoma County, in which Petaluma is located, had at that time over 4,000,000 hens, an investment of $20,000,000 in poultry plants, from which was realized an annual income of over $20,000,000 for eggs alone.

Poultry keeping methods here have always been highly intensified, "plants" ranging from 3 to 10 acres in size, as a general thing, and carrying 1,000 to 1,500 layers, the latter figure being regarded as the maximum for one-man plants. A few years ago small houses were the rule, these often being assembled in batteries of half a dozen or more and to poultrymen from other parts of the country floor space appeared to be quite inadequate to the number of birds accommodated. The climate here is quite mild, however, and the soil on which practically all the poultry plants are located is sandy and well drained, so that houses are required chiefly to provide roosting and nesting quarters.

Another noticeable feature of poultry keeping in this section is its extreme specialization. For example, large numbers of commercial producers keep no breeding stock and make no effort to do their own hatching, but rely upon buying from hatcheries as many baby chicks as are required to furnish the desired number of pullets each fall. Others keep few layers but devote their attention almost solely to raising pullets which they sell to egg producers, while still others buy yearling and two-year-old hens with which to fill their laying houses. Among the hundreds of producers in this section it is scarcely worth while to attempt to present, in our limited space, representative individual records of achievement, though mention must be made of the fact that Petaluma has the distinction of possessing the biggest laying plant in the world, known as the Corliss Farm, where 35,000 to 50,000 laying hens are kept each year. No breeding is done on this plant nor are any chicks raised, the owner purchasing yearling and two-year-old hens chiefly, with which to fill his laying houses. By eliminating all incubating and brooding work and concentrating attention solely upon production it is possible to systematize the work so that this huge plant can be handled by six or seven men.

VIEW ON PLANT OF F. R. RICHARDSON, SAN GABRIEL, CALIF.
Mr. Richardson and his brother, Geo. J., are leaders among southern California poultrymen. Brief personal statements from both are given on pages 141 and 142.

To provide the great numbers of pullets annually required by Petaluma poultry keepers, many hatcheries are operated. Some of these are of immense capacity, a single one being able to turn out 2,000,000 chicks in a season. The total number of baby chicks hatched annually in this district is estimated at 15,000,000, sixty per cent of which are taken by local poultry keepers. The eggs used in these hatcheries alone exceed in number the total annual production of such states as New Hampshire, Vermont or Connecticut, as reported in the 1920 U. S. Census.

There are a number of other local producing centers around or in the vicinity of San Francisco, any one of which would be considered important if not overshadowed by the magnitude of Petaluma. All around the bay, poultry keepers are located in numbers and, as a rule, are prospering in proportion to their experience and ability. The fact that the local market is no longer able to handle the product is a matter of diminished importance, in view of the excellent facilities now available for making eastern shipments and the relatively low cost of disposing of them in this way.

The State University and Experiment Station at Berkeley, gives particular attention to the interests of poultry keepers and, among other activities, is conducting egg-laying contests at Petaluma, Santa Cruz (where another growing center is located) and Pomona, near Los Angeles. The Veterinary Division of the University also maintains a poultry pathological laboratory at Petaluma, where the work consists entirely of studying poultry diseases and disease control. There are two extension poultry specialists, while a number of county agents have organized local contests, demonstration farms, etc., all of which not only assist in increasing interest in the work but in adding materially to the financial prosperity of poultry keepers generally.

Poultry Keeping in Southern California

Here is now taking place one of the most rapid local developments in the poultry industry that has been brought about in the whole United States during the last few years. Latest information indicates a poultry population in this section of in the neighborhood of 3,000,000 hens (including the eight southern counties of Los Angeles, Riverside, San Bernardino, Orange, San Diego, Imperial, Santa Barbara and Ventura). R. B. Earson, Extension Specialist of the State University, is authority for the statement that in Los Angeles County alone there are over 1,300,000 hens. The Poultry Producers of Southern California, a cooperative association, has played a highly important part in this development, assisting materially not only in securing better prices for products but in reducing cost of production. In the laying season this association ships as high as eight carloads of eggs a week to eastern markets.

We are indebted to B. R. Holloway of Holly Poultry Ranch, Van Nuys, California, for the interesting information here given in regard to the development of the industry in this section, and regret the limitations of our space that make it necessary to omit much that he has so kindly supplied and that readers would have found intensely interesting. Mr. Holloway himself has had a very successful career as a poultryman, beginning in 1910 on a small scale and with but limited capital and practically no experience. By faithful and persistent work he developed a large and profitable plant, but gradually worked over into commercial hatching. He now owns one of the finest hatcheries in California with a total capacity of 130,000 chicks at one time. These chicks he distributes in his own specially constructed delivery trucks, carrying them anywhere within a radius of 100 miles—a method of delivery that insures their reaching the customer in the best condition practicable and with minimum loss of time. Mr. Holloway secures a large percentage of the eggs from which his chicks are hatched from the farms of F. R. and George J. Richardson, two of southern California's best known and most skillful breeders. The total number of chicks hatched by Mr. Holloway from eggs secured from these two farms alone approximated half a million in 1920.

Personal Statements from Successful Poultry Keepers in Southern California

F. R. Richardson.—"Fifteen years ago I left a good bank job in Denver to go into chickens. I didn't know one breed from another, but I obtained the best poultry journals available and began to study. I attended the poultry shows and at any and all times I 'button-holed' anyone who knew anything about poultry and 'pumped' him. My first venture was in a frontier sort of place in northern Idaho, where there were few opportunities of raising stock and less to market the eggs. As soon as possible I sold out there and together with my elder brother, spent some

ONE INCUBATOR ROOM AT HOLLY POULTRY RANCH HATCHERY
The remarkable growth in the poultry industry in California has brought about the development of mammoth hatcheries. Above is shown interior of Room No. 2, at hatchery of Holly Ranch, B. R. Holloway, proprietor. Capacity of this room is 70,000 eggs. Total capacity of hatchery is 130,000.

months studying conditions around Los Angeles—conditions of soil, climate and wind. I wanted to select a place reasonably close to a good market and one where I would be content to establish a permanent home. And I have never once regretted that the chosen spot was San Gabriel.

"I had $2500 to spend. Part of this went for the first payment on 3½ acres, the balance towards improvements. We built a barn-sort of building and moved in. This was our abode for four years. For use the following spring, I ordered the best chicks I could get and started to make improvements for them in the order needed. The first year I had 1,000 pullets and right then I started to work out one of my hobbies—RANGE. The soil was new and I gave them the whole place, and when those pullets began to lay they paid me back with interest, making a flock average during the period of high production of 80 per cent. That meant profit, which I put back into the business each succeeding year, doubling the stock and equipment yearly. Very soon I needed more room and I got it, paying a rental of $30 per year per acre for 12 acres. Two years ago I bought 17 acres adjoining my original place. This now is planted to walnut trees and equipped with permanent and substantial apparatus. At the present time I own 20½ acres and am occupying 24. The larger part is used as range for the growing stock, the yards being occupied by the fowls but a short time each year, and the balance of the time they are used to grow green feed. I am still a firm believer in range and plenty of green feed. During the hatching season we have from three to eight acres of barley and try to feed about a ton a day. We carry a minimum of 10,000 and a maximum of 16,000 layers and our plant and stock are covered by $30,000 fire insurance."

George J. Richardson.—"In 1910, having lived for 14 years at an elevation of two miles above the sea, in the famous Cripple Creek Gold Mining District, I decided to give up my position with the local light and power company, get to a lower level and go into farming. While debating upon this move we received word from my brothers, E. G. and F. R. Richardson, who had already arrived in California, suggesting that we 'come on out here and go into the poultry business, where you can live in the country and still have all of the comforts and pleasures of city life.' We decided to do this and enthusiastically entered into egg farming, which we have found quite profitable.

"Our location we chose with great care, getting a rich, sandy, well-drained soil, with lots of soft water to assure an abundance of greens for our poultry, as well as securing a moderate winter climate and one free from cold spring winds. Much of our success has been due to this fortunate selection. We were fortunate also, in getting in touch with a few men who were getting good results in this line, and it was by following their advice on important matters and not having to experiment, that carried us safely over some of the rough spots that every beginner sooner or later encounters.

"And so the dream we had has materialized—an ideal life in the country, electricity, gas, fine water under high pressure, mail, ice and grocery delivery, good schools and neighbors, a nice California home, thirty minutes by auto from the heart of Los Angeles."

F. M. Molby.—"Beginning nearly eight years ago, with five acres of land and baby chicks from a near-by hatchery, the Molby Poultry ranch now consists of 12 acres of rather heavy soil and is located eight miles southeast of Los Angeles. We have laying house capacity for 4,000 hens, brooding accommodations for more than 10,000 chicks, and during the height of the hatching season we hatch better than 5,000 chicks a week, all from our own eggs, totaling more than 80,000 chicks for the first six months of 1921. We ship baby chicks to all parts of California and the Southwest.

"Molby's record-strain S. C. White Leghorns are the result of carefully selecting the breeding stock year after year by those external characters that denote vigorous constitution and high-producing ability. We have not used trap nests in our work, believing that in flocks of large numbers it is hardly profitable to do so, providing a strain of vigorous high producers can be established in another way. High flock averages concern us more than a few exceptionally high individual records, for the same reason that they figure more in the income tax report. We have proved to our own satisfaction that we can select

PARTIAL VIEW OF POULTRY PLANT OF THE FONTANA FARMS CO., FONTANA, CALIFORNIA
Buildings shown above form a part of the central plant on which breeding stock is produced, and stock, day-old chicks, and eggs for hatching are supplied to members of the organization. The long laying house in the foreground is the type designed and recommended by the poultry department of the State Experiment Station.

"Starting in with baby chicks—the best we could buy from the best hatchery we could find—we raised 800 fine pullets the first year. They were a success from the first, making us a fine profit that fall and winter and through the summer. From this time on we had both feet on the ground.

"In 1913, having had such a demand for 'settings' from our neighbors, we began to think seriously of specializing in quality hatching eggs (something that would be dependable), so in 1914, after strenuous selection, we mated up a fine lot of hens. There was an instant demand for the entire output of hatching eggs. From this time on, our plans carried us out of commercial egg farming into a breeding plant for S. C. White Leghorns exclusively. Small yards were replaced with large ranges for our growing pullets and breeding cockerels. More facilities were provided for raising abundant green feed for stock of all ages, also increasing our brooding houses to get away from the crowding, which had sometimes seemed unavoidable, and which cut down our per cent of No. 1 stock. These changes and many others were needed to make vigor the big idea, from baby chick to breeding pen.

"In 1918 it became necessary again to enlarge, which we did by purchasing adjoining land, where we planted walnut trees for shade and which proved a valuable addition to our free range. On this new place—the 'annex,' we have named it—we also built a 6,000-capacity brooder house which gave us a total spring capacity of 6,000 pullets and a fall capacity of 4,000 pullets.

the best of the flock for breeding purposes and eliminate the poor producers from the flock, basing our selection entirely on observation of external characters. That we err in a few cases goes without saying, but neither is the trap nest infallible. In fact, we are doubtful if the trap nest alone tells the value of the hen as a breeder as truly as external characters, when read aright in the light of the latest knowledge on the subject.

"As far back as 1915-16 we had a flock average for our entire lot of pullets of 194 each per bird—and similar averages have been secured in succeeding years, thus proving that high-laying capacity is firmly fixed in our strain."

Numerous other breeders in this section besides the foregoing have developed high-producing strains, and numerous illustrations could be given of large flocks where trap nesting is practiced, such as that of George C. England, who traps around 3,000 birds annually and had a pullet average in 1921 of 220 eggs. Mr. England's plant is said to be the oldest and largest trap-nested ranch in California. He started in 1911 with only $207 in cash, began trap nesting in 1912, and now owns a $50,000 ten-acre plant, completely equipped. Others have developed high-producing strains largely by giving special attention to vigor and by selective flock breeding (meaning by this culling and mating on external characters). R. C. Gibson of Newhall, for example, has produced a strain which in 1920 gave him a flock average of 181 eggs per bird.

CHAPTER III

Trap Nesting, Pedigreeing, Certification, Etc.

In This Chapter Some Successful Types of Trap Nests Are Described and a Simplified System of Records for Use in Pedigree Breeding Presented—A Practical Method of Official Certification and Registration of Standard-Bred Fowls, Particularly High Producers, Is Greatly To Be Desired—Some Methods Now Being Tested Out in Everyday Use Are Outlined

THE breeder who wishes to practice pedigree breeding, or who desires to have exact information regarding the egg yield of individual layers, finds the use of trap nests unavoidable. Nests of this type are so arranged that only one hen can enter at a time, and, once in, cannot get out until released by the attendant, who, in doing this, marks up the hen on the daily egg record and, when pedigree breeding is practiced, at the same time marks the egg with the hen's leg-band number. By keeping each hen's eggs separate at hatching time, suitably banding the chicks and adopting an accurate method of recording numbers, the breeder is able at all times to determine the ancestry of each and every individual in his flock. Obviously, such information is indispensable in all systematic breeding.

There is much variety in the styles of trap nests in practical use and in a general way it may be said that it does not matter greatly which particular one is adopted, provided it is carefully made and kept in proper adjustment. Herewith are described three styles that are in more or less general use, any of which the reader will find reasonably reliable and convenient.

The "Gravity" Trap Nest

The trap nest here called the "Gravity" was one of the first of those now in use to be introduced to the public and by many it is still regarded as the simplest and least likely to get out of order. As will be seen by examining the two illustrations on page 144, there are no working parts aside from the revolving door which is held open by its own weight when pushed back against a nail in the side of the box. The rear of this door is low enough so that when a hen enters the nest she must raise the door a little—enough to throw it out of balance, when it drops down, effectually confining the hen and making it impossible for another to enter. The semicircular front is covered with hardware cloth, and the top of the nest may be covered in the same way or, for economy's sake, with ordinary one or two-inch poultry netting. However, where the nests are not protected by droppings boards or another tier of nests, the tops should be tight in order to keep the nests clean.

The construction is readily understood from the illustration. The entire nest box is approximately 25 inches long, 12 inches high and 12 inches wide for Leghorns, or 14 inches for fowls of the larger breeds. The sides of the door should be approximately one-third of a full circle, measuring 13½ inches on the curved edge and 7½ inches on the straight edges. The sides are held

A DOWN-TO-DATE POULTRY PLANT IN SOUTHERN CALIFORNIA
This well-built and well-kept plant is owned by R. C. Gibson, Newhall, Calif. Is a demonstration farm. The strain of White Leghorns bred by Mr. Gibson has made a flock average of 181 eggs per bird.

together by two strips of ½ by 1½-inch lightweight lumber. The sides and bottom, also the front strips and partial partitions, may be of ½-inch material. A metal washer should be placed between the sides of the box and the door, and the door must be carefully hung so that it will balance properly and revolve easily. Aside from the width, the dimensions of this nest do not require any modification for fowls of different sizes.

The New Jersey Trap Nest

The trap nest illustrated on page 52 is the one used at the Vineland and Bergen County (N. J.) Egg-Laying Contests and is of the same general type as those used at Storrs, the North American Contest, also at the United States Department of Agriculture Poultry Farm. It should be about 18 inches deep (or long), 14 to 16 inches high and 12 inches wide, all inside measurements. The general details of construction will readily be seen from the illustration. The shape of the trigger used varies somewhat at the different institutions mentioned. In general, all that is required is a ⅞ by 2-inch strip, 6 or 7 inches long, with a notch cut in the upper edge to hold the door when in "open" position. The trigger should be hung so that the back end will drop readily when the hen lifts the door out of the notch in entering, whereupon the door drops into "closed" position. The door may be hung on wires passing through pieces of scrap iron on either edge of the door as shown, or the wires may be tacked to the top or passed through screw eyes inserted in the upper edge.

The Improved Cornell Trap Nest

This trap nest, as illustrated in halftone photo-engraving on page 145, is exceptionally well ventilated and lighted. It is 20 inches in length, 11½ inches in width and 12 inches high, all inside measurements. The front board is 3½ inches in width and there is a 7½-inch

SPECIAL TRUCK FOR DELIVERING BABY CHICKS
This fine truck is one of two used by B. R. Holloway of Van Nuys, Calif., in delivering baby chicks from his mammoth hatchery direct to his customers, thus avoiding danger of delay and injury due to mishandling.

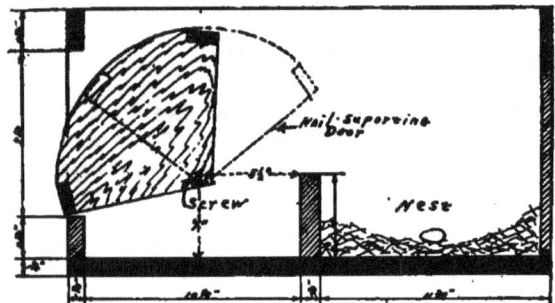

CONSTRUCTION OF "GRAVITY" TRAP NEST
This cross section through side of nest shows dimensions, method of hanging doors on screws, also location of nail which supports door when open. Hen in stepping over low board in front of nest lifts door and causes it to roll forward along dotted line. Courtesy of American School of Poultry Husbandry.

"GRAVITY" TRAP NEST WITH DOOR CLOSED
Photo from American School of Poultry Husbandry.

space in front, cut off from the nest by a strip 2¼ inches high. The door is 9½ inches in width, 8 inches in height and when open is held in place by a trip board suspended from the top of the nest and provided with a cleat at the lower edge so that the door is held securely open, but promptly released when the entering hen pushes the trip board back slightly with her back, as she must do in stepping over the 2¼-inch board which extends across the front of the nest. In the illustration the nest nearest the front is shown with the door held in place by the trip board, while the second nest shows a closed door with the trip board hanging down inside. There is a 6 by 6-inch footboard nailed on the outside of the nest, in front of each door, and a section of the back of the nest opens on hinges for removal of hen.

PEDIGREE BREEDING TO SECURE INCREASED EGG PRODUCTION

Plan Outlined That Will Bring Results with Minimum Use of Trap Nests—Beware of the Squawking Hen—How To Pick Breeding Males—Importance of "Depth of Body" in Both Sire and Dam—The Progeny Test Is the Only Sure Method of Knowing Results Obtained—Practical Pedigree System Described in Detail

By Dr. O. B. Kent, New York State College of Agriculture, Cornell University, Ithaca, N. Y.

POULTRYMEN generally desire to improve their egg production with as little extra labor and expense as possible. Probably the quickest and cheapest way is to buy cockerels from some breeder who has carefully pedigree-bred his birds for a number of years. When such birds are available they are worth a good price. But, unfortunately, it is comparatively difficult to buy stock that has been well bred in this respect, so the majority must breed up their own stock. The plan that is given here is suggested as one that will require the minimum of labor and yet get results.

Perhaps the clearest way of explaining the plan will be by giving a series of steps to be followed during a number of years. Of course, it does not make any difference what year the plan is started, but let us start with the fall of 1921, to make a definite program of breeding for the following five years. Five years breeding in the future may seem a long time, but five years past is a short time—and five years is a short interval in which to accomplish results in breeding.

Assuming that a man is keeping 1,000 hens or that he has a small flock and wants to get some first-class stock, in the fall of 1921 suppose he places ten yearling hens in each of five houses or places fifty hens in one pen together. The houses should be equipped with trap nests so that the birds can be trapped during the breeding season. It is intended that the birds should be trapped only during the breeding season, though it would not affect the plan in any way if one desired to trap them throughout the year.

If the birds are all put in one pen, provision must be made for stud coops for the males. These coops should be about 2½ by 2½, or 2 by 3 feet square, and 3 feet high. Preferably they should have the floor of the house for the bottom of the coop. Otherwise they should have a solid bottom.

The fifty hens must be selected with exceeding care. If the breeder were in New York state I would say, use fifty of the best Cornell certified birds. One of the factors to consider in selecting the birds is weight. A Leghorn should weigh 3.5 to 4.5 pounds at the time of selection, say November 1, 1921. If the bird is lighter than that it is liable to lay too small an egg and not have enough reserve fat and flesh to be a good breeder. It is very rare for a high-producing runt to be a good breeder. When you get beyond a 4.5 pound Leghorn you are beyond the efficiency point. It will not lay any more, in fact, probably not so many eggs as the smaller birds. Furthermore, a Leghorn that weighs over 4.5 pounds is liable to produce an egg that is so large that difficulty is encountered in getting it into a regular thirty-dozen case. There is no use trying to make a Minorca out of a Leghorn—certainly not until eggs are sold by the pound. The heavy breeds should weigh from 4.5 to 6.5 pounds.

Select birds that have laid at least into October. Birds that show by their wing molt that they stopped laying before the first of October should not be included. Select the individuals that are deepest in body in proportion to their length. All individuals should be in good flesh. An excessively thin pelvic bone is an objectionable point and not one to be desired. But do not take a bird whose abdomen is not soft and pliable. Then place a great deal of emphasis on the character of the hen. All the hens selected should be active, bright, intelligent individuals. A hen that squawks and yells when handled should be discarded at once. The hen that will come up and eat out of your hand and sing to you is the one to keep.

The hens that are still laying should be trapped for a day or two just to make sure that they are laying a good market egg. If a large number are not laying, a few extra hens should be obtained at this time to replace any that might be found laying a poor egg in the spring. Then try to get fifty birds that are as near alike as possible. It should always be the aim to get the best layers all the time.

Selection of the Males

It is equally important to select good cockerels to mate with the hens. Because there are a great many more cockerels than hens to select from, the chances for selection are much greater. Needless to say, the cockerels should be the best obtainable. If possible they should be pedigree bred with a good many years of high production back of them.

It is possible by using external characters to obtain a good idea of the quality of a male. In December, 1918, at Cornell University, eleven cockerels were measured that had been bred the previous spring. It was found by dividing them into two groups that the five cocks with the relatively deepest bodies had 14 daughters that averaged to lay 186 eggs the first year and the six shallowest bodied cocks had 48 daughters that averaged 155 eggs each, the first year. In this particular case the depth of body was a better indication of the production of the daughters than was the pedigree of the birds. Males have been measured each year since then, so in a few years more we will know just to what extent reliance may be placed on the shape of the male's body.

The first thing to consider in selecting a male, aside from condition or health, is depth of body. A good male is deep in breast and abdomen so that he looks rectangular. This is very noticeable while the cockerels are growing. The good cockerels have deep, full, prominent breasts, widespread legs, show good comb development early, have full, wide necks and wide-open, intelligent eyes. The poor cockerels are long-legged, long-necked, crow-headed, knock-kneed, spindle-shanked, roach-backed, shallow-bodied, triangular, nervous, and generally otherwise disagreeable fowls. While a squatty cockerel is not desired, a cockerel that gets its apparently long legs from a shallow body is exceedingly undesirable.

Temperament Is a Factor

Besides the physical qualifications of a good male one should consider the temperamental side. Select a male that you can talk to and that will talk to you. Of the various males that we have had I think that F-563, the crowing rooster (and by the way, he is still crowing), shows the best disposition of any male we have ever had. When he was in a pen with a number of other males he was boss of the flock. Not an ugly boss, but a boss that knew what he was doing. There was little fighting while he was in the house. A good male should be able to fight if necessary, but pugnaciousness is not a desirable quality to perpetuate.

This particular male seems to know just when he is being admired, for whenever you open his house to look at him he at once starts crowing. When taking pictures of him in 1918 it was a great deal easier to take a picture of him while he was crowing than at any other time. He is one of the friendliest males we have. Anyone can go to the house, hold out grain in his hand and get him to eat it from the hand. At the same time he is a very vigorous, active bird and produced good fertility when mated to ten pullets in his third year. His father, D-105, is a very similar male as regards disposition. D-105 in 1920 had nearly 50 daughters that had averaged over 175 eggs.

The extreme opposite of these two males are D-1452 and G-1874. Both are utter cowards. They both have a high-pitched, shrill voice that is extremely disagreeable.

THE LATEST CORNELL TRAP NEST
This trap nest may readily be made by anyone, following the instructions given in article on page 143. Photo taken at Cornell University.

I have never heard either one crow, though I suppose they do occasionally. They always act as if they were ashamed of themselves or as if they had a guilty conscience. Compare the body type of F-563 and D-105 with D-1452 and G-1874, and then compare their records.

Hens That Should Be Discarded

Assuming that we have selected the five pens, A, B, C, D, E, and have put the ten best hens and best males in A and the next best in B and so on, in 1922 during the breeding season they are all trap nested and their eggs pedigree hatched. Any hen that during the breeding season does not lay a first-class egg in size, shape, color and texture of shell should be replaced. Any hen that does not lay at least a three-egg cycle should be replaced. By three-egg cycle is meant laying three days before skipping a day. Any hen whose eggs candle out badly in the first and second tests should be replaced. It will be seen from this that it is quite necessary to have a number of substitutes to replace birds that may be thrown out for any one of a number of reasons.

After the chicks have been pedigree hatched and wing banded, no further observations are necessary until the cockerels are culled out as broilers. Keep at least one good cockerel from every hen in the breeding pens. Save the best cockerels from all the hens. These cockerels should furnish males for the general breeding flock the next spring. When the cockerels are culled a record of the quality of each should be made on the mother's record sheet.

Beginning of the Second Year

The first of November, 1922, the five breeding pens should be gone over carefully to sort out any bird that did not stand up well during the year. All birds that are in good physical condition at this time should be kept. At the same time at least fifty new yearling hens should be selected and marked so that they can be used to replace any in the breeding pens that are culled out later.

TYPE OF TRAP NEST USED ON HOLLYWOOD POULTRY FARM
The principle by which this nest works is the same as in "Gravity" nest shown on opposite page, but it is possibly a little easier to build.

On January 1, 1923, each pullet that came from the breeding pens should be carefully examined and their production as shown by the beak and shanks recorded on the mother's record. The method of selection by beak and shanks is shown in the book "Profitable Culling and Selective Flock Breeding," published by Reliable Poultry Journal Publishing Company. The approximate egg record, the weight, type of body, disposition and general appearance also should be entered on the mother's record. With these data in hand, study the breeding records of the hens carefully and save the twenty hens that have the best daughters, both in number and quality, and keep the two best cock birds as shown by the quality of their daughters.

Then put the twenty best hens

that have been saved in pens A and B, and put the best male with A and the other in B. The other three pens will be made up of thirty of the best yearling hens that were marked November 1, 1922. The males used will be the two best sons of the male put in pen A by the two best hens he was mated to, and the other male should be the best son of the male in B by the best hen he was mated to. The quality of a breeder is determined by the number and quality of progeny. In saying "mate the best son of the male in B by the best hen," it is meant that the hen that has the most daughters making a high record or having the most uniformly high daughters should be used.

The Third Year of Pedigree Breeding

We have now outlined the scheme that is to be followed for at least five years. The first of November, 1923, the yearling hens that are selected should be the progeny of the original fifty hens. At the time they are selected their quality should be recorded on the mother's record, to check up the first determination of the mothers' quality as recorded when their daughters were gone over on the first of January, 1923. In selecting these hens preference should always be given to large families of good individuals rather than an exceptional individual from a hen. The thirty birds that would go to the breeding pens on January 1, 1924, should come from not over ten hens out of the original fifty.

Each year the twenty best hens and the two best males would be kept, regardless of the age of either. If a hen will lay eggs that will hatch well and a male will get fertile eggs, it does not make any difference how old they are, except that the older they are the better.

The cockerels for these breeding pens should be used with the general flock so that in a few years the whole flock would be graded up. There is one very distinct advantage in not having a trap-nest record of the birds for the year. If you do not have a trap-nest record you will pay a great deal more attention to the quality of the individuals and the quality of the eggs they lay. A high egg record is very liable to warp one's judgment of the quality of a bird. If you do not have the record it will not worry you. Possibly one can progress faster with a trap-nest record, though I doubt if they will get quality as quickly. The other disadvantage in not having the birds trapped is the loss of possible advertising. But that can be made up by entering birds in one of the egg-laying contests. After the birds have been proven out as breeders, a pen of pullets from Pen A should give a very good account of itself anywhere.

ONE OF CORNELL'S BEST BREEDERS
Forty-seven daughters of this male (D-105) averaged to lay better than 175 eggs each. Was still a good breeder as a four-year-old. Sire of F-563.

THE "CROWING MALE"
The daughters of this male (F-563) averaged 197 eggs in their first laying year. He is splendid in type and disposition. Son of D-105.

When breeding for production it does not pay to know only the record of the hen you are breeding from. As shown by the Maine Agricultural College records, breeding the sons of high producers to fairly high hens does not necessarily increase production; apparently it may even decrease production. The only way is to know the sire and dam of every chick and then use as breeders only those individuals that have a record of being good breeders or that come from good breeders.

Some rather interesting results have been obtained in the breeding at this station in the last ten years. During the period from 1909-1914 no record was kept of the males except it was known that they came from high-producing hens. All the hens in the flock were bred and the best pullets selected. Beginning with 1915 all birds were stud mated so that the parentage of each bird was known. Gradually the birds that showed up as good breeders were used to a large extent. The records of the total were obtained by dividing the total egg production by the number of hens that started the year, no offset being allowed for mortality. If an allowance were made for mortality the only difference would be a slightly higher average production. The total is pretty strong evidence that using a bird because its mother made a good record is a poor method. Note the ten-year record herewith:

Year Hatched	First-Year Record
1909	148
1910	133
1911	128
1912	127
1913	111
1914	106
1915	120
1916	128
1917	127
1918	164

REJECT THIS TYPE
D-1452. The 34 daughters of this male have averaged to lay less than 100 eggs their first laying year. Disagreeable "voice" and in same class as G-1874.

POOREST BREEDER AT CORNELL
G-1874. The daughters of this male are making the poorest record of any male that we have ever had. Note the type and the actions. Poor disposition and "presence."

The second and third-year records are similar to the first year. They show a decrease up to the birds hatched in

1914 and then an increase from then on. If it has not been made clear so far it should be emphasized and reiterated here that the progeny test is the only sure method of breeding for egg production.

How To Pedigree Breed

There are two methods in handling hens in pedigree breeding. The first method, that of putting a small flock of hens in a pen with a male, is the most desirable. Where a male is running loose with the flock, slightly better hatches are generally obtained and there is less liability of the male getting out of condition than when he is penned up. This method, while giving slightly better results, requires a number of separate houses and yards, or a house with a number of small pens. It takes considerably more time and effort to feed several small pens than it does to feed one large one.

Where labor has to be kept at a minimum, the second method has been used with very satisfactory results. In this method the males are kept confined in individual coops and certain hens are mated with each male after they are taken from the trap nest. A very satisfactory method is to put an extra colored leg band on the hens to designate the mating. All the hens and the male of a mating are given the same color of leg band.

TAG SAFETY PINS USED IN CLOSING PEDIGREE EGG SACKS AT POULTRY DEPARTMENT, CORNELL UNIVERSITY

By using stud coops 2½ by 2½ feet, or 2 by 3 feet, on the floor, we have been able to keep the males in good condition through the breeding season and to obtain an average of over 90 per cent fertility for the season. It generally takes a week to two weeks for the males to become used to the coops so that it is advisable to have the coops in place and the males in them for two or three weeks before any eggs are saved for hatching. By having a battery of such coops it is possible to keep account of the matings in a pen holding a large number of hens. If it is not desired to mate all hens in the pen they can be quickly distinguished by not having a colored leg band.

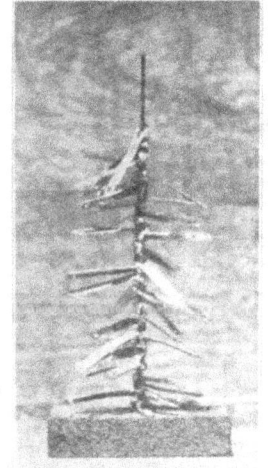

HAT-PIN SPINDLE
Showing leg bands arranged preparatory to pedigreeing chicks.

It generally is not safe to mate over ten females to a male in a stud coop without taking special precautions. It does not do to put three or four hens in with a male at one time. If it is desired to mate over ten hens to one male a record should be kept of when each hen is mated so that she will only be put in with the male after every second or third egg is laid. By only mating each hen once or twice a week it is possible to obtain good fertility and mate a large number of hens to one male. Unless it is a proven male that you know is a good breeder, it is desirable not to mate more than ten hens at the most to one male, for pedigree breeding consists in trying out the male just as much as the female. If too many hens are mated to one male few males will be tried out.

For pedigree breeding it is necessary to trap nest only during the breeding season and long enough prior to it so that the hens may be trained to lay in the nests and not on the floor. If the trap nests are in the pen and are fastened so that they will not retain the hen, little trouble should be experienced in getting the hens to lay in them when the traps are used. If a totally different kind of nest is put in, it will require a little patience to teach the hens that they are supposed to lay in the new nests and not on the floor. But most hens are obliging enough to lay in a nest regularly after they once find it or are put in there to lay, provided they can get in the nest all right.

Keeping Track of the Chicks

As the eggs are gathered from the trap nest the number of the hen and the number of the pen should be marked on the large end of the egg. The eggs from the different pens should be placed together and once each week or ten days the eggs should be set. It is easier to set the eggs once a week or every ten days rather than oftener because of the work of sorting and recording.

Preparatory to setting the eggs they should be sorted out according to hen number on a recording board. After the eggs are all sorted and the board is full, the eggs should be numbered consecutively on the small end. It is quite necessary to number the eggs on the small end because when the chick hatches it is likely to break up nearly all the shell except the small end. Most chickens break through the middle of the side and if an egg is marked there the number is liable to be destroyed. In the thousands of eggs that we have incubated I do not remember a case of an egg where the number on the small end was destroyed. There may have been some but it certainly is very rare, and in that case the hen number on the large end would be a sure means of identifying the egg.

After the eggs are numbered a record should be made of the egg number and the hen number using a sheet suitably ruled so that the eggs from the hens are arranged in order numerically according to pen number, hen number and date laid. The latter is not necessary for ordinary incubation. When the eggs are candled at the end of the first or second week a record should be made as to whether the eggs were infertile or dead. At the time the eggs are

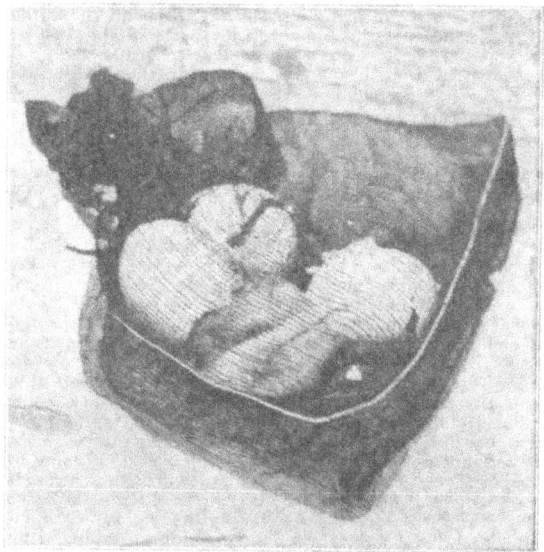

PEDIGREE SACK USED IN HATCHING
Photo from Cornell University.

to be turned the last time they should be sorted in 1, 2, 3 order according to the number on the small end of the egg. The eggs can then be quickly sacked, all the eggs from one hen being put in a sack. The sacks can be fastened with a safety pin or a piece of string. The safety pin is the quickest way. The sacks can then be arranged in the machine in regular order for hatching.

To be sure that the sacks are put in order a numbered slip can be fastened on the safety pin. A very efficient tag can be made with a little square of thin cardboard, such as a filing card, an O. K. fastener and a safety pin. By numbering the tags and sticking the pins into a pasteboard box the pins can be taken off in regular order. The pins can be sorted out, stuck back on the box and used over again.

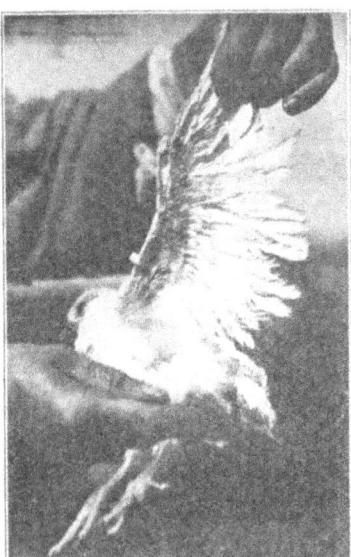

PEDIGREE BAND IN WING
Four-weeks-old chick showing wing band in place, where it serves as a permanent record. Photo from Cornell University.

It may seem like a good deal of extra labor to number the small end of the eggs and to use numbered safety pins, but it saves a deal of time in looking up the records of the birds later. By taking these two extra steps it is possible at hatching time to have the numbers of the chicks incubated run in consecutive order down the page. That is, Hen No. 1 has Chick No. 1, 2, 3; Hen No. 2 has Chick No. 4, 5, 6, 7; Hen No. 3 has Chick No. 8, 9, etc.

Convenient Marking of Chicks

When the chicks are dried off they should be taken from the sacks, their number recorded on the pedigree sheet and a band wrapped around the leg. We have found a No. 2 double-clinch pigeon band very satisfactory. When the chick is three to five weeks of age the band should be taken off the foot and clinched into place in the web of the wing. That serves as a permanent record of the bird. Be sure not to let the bands stay on the shank after the chick is five or six weeks of age for the band will be liable to cut off the leg of the chick or else drop off and be lost. If the pullets are to be trapped the entire year a leg band bearing the same number as the wing band should be used. With a small set of dies the bands can be quickly and plainly numbered.

To keep the years separate a different letter should be given to each year. For instance, let the chicks hatched in 1922 have A bands, 1923, B bands and so on, then when a bird is picked up it is possible to tell at a glance just how old it is.

The sacks used in hatching should be made of strong, heavy-grade mosquito netting. Sacks to hold a week's eggs should be about 6 by 8 inches. It saves space to have a few small sacks 4 by 5 inches to hold individual eggs in case the others have been candled out.

A convenient way of keeping the bands in order is to string them on a wire. The spindle shown herewith was made by driving a hat pin into a small block of wood and then cutting off the head of the hat pin. The pin very conveniently holds fifty bands. By putting the number 50 at the bottom they can be placed so that they will be taken off in 1, 2, 3 order.

In order to keep account of the matings a sheet should be made out for each male, the male's number to be at the top and the hens' numbers along the side. There should be room for notes in regard to each chick as well as the band number. If care has been taken to see that the band numbers come in numerical order such a sheet will be found to be an extremely convenient method of keeping pedigrees. See sample sheet herewith.

Male No. 1

Hen No.	Date hatched 1st hatch Chick No.	Date hatched 2nd hatch Chick No.	Date hatched 3rd hatch Chick No.	Date hatched 4th hatch Chick No.
A1	B1	14	......	......
	2	15	......	......
	3	16	......	......
	..	17	......	......
A2	4	18	......	......
	5	19	......	......
	6	20	......	......
	7	21	......	......
A3	8	22	......	......
	9	23	......	......
	..	24	......	......
	..	25	......	......
A4	10	....	....	......
	11	....	....	......
A5	12	...	....	......
	13	....	....	......

THE REGISTRATION OR CERTIFICATION OF STANDARD-BRED POULTRY

Outline of Plan Adopted by American Association of Instructors and Investigators in Poultry Husbandry for Issuing "Certificates of Egg Production" Through an "American Record of Performance Council"

THE official registration of poultry has been under consideration many years. It was undertaken by private interests thirty-five years ago or more, also in the recent past, but thus far without satisfactory results. It would seem that the American Poultry Association is the logical medium through which to introduce a general scheme of registration of standard-bred poultry, and steps have already been taken in this direction. At present the only "official" registration plan in use is the "American Record of Performance Council" established by the American Association of Poultry Instructors and Investigators. This has for its chief objects the keeping of records relating to prolific egg production at recognized egg-laying contests and the issuing of certificates of egg production in cases where the birds are found to be entitled to same, under the rules to be adopted from time to time by said council.

Professor Harry R. Lewis, secretary of the American Record of Performance Council of which Professor W. F. Kirkpatrick of Connecticut Agricultural College, Storrs, is president, has supplied a photograph of the "Certificate of Egg Production," as adopted by said council, a half-tone reproduction of which will be found herewith. Certificates are issued in the usual form, bound in books, and a stub is left in the book on which is kept a duplicate record of what each certificate represents, to whom it was issued, etc., for the information of the council and its officers. Writing under date of April 12, 1920, Professor Lewis said:

"I am sending you, under separate cover, a photographic reproduction of the certificate issued by the American Record of Performance Council. For your information will say that the Council is composed of one representative (usually the superintendent or supervisor) from each of the officially supervised egg-laying contests and breed-testing stations in the United States and Canada.

"We are recognizing and issuing certificates for all pullets which laid two hundred eggs or over at official contests, and for all yearling hens which laid one hundred and eighty eggs or over, provided the birds making said production are standard bred, scoring at least seventy-five and free from standard disqualifications."

Believers in and advocates of standard-bred poultry—meaning domestic fowl bred in conformity with the requirements of the "American Standard of Perfection"—will note with special interest and appreciation that pullets by interested persons on application to Dr. W. C. Thompson, Poultry Dept., State University, New Brunswick, N. J.

Our readers readily will understand the main objects of the American Record of Performance Council in issuing these certificates of egg production. In addition to encouraging our egg-laying contests, which are growing

CERTIFICATE OF EGG PRODUCTION
Photographic reproduction (much reduced) of attractive, large-sized certificate that is issued by "American Record of Performance Council" to owners of pullets that lay 200 eggs or better and of hens that lay 180 eggs or better in 365 consecutive days at recognized egg-laying contests. Birds to be admitted to these contests and entitled to a certificate must be standard bred, free from standard disqualifications and score not less than seventy-five points, regardless of condition.

and hens to be entitled to certificates of egg production, as issued by the American Record of Performance Council, must be standard bred, free from disqualifications and score at least seventy-five points. In other words, this is an official recognition by the American Record of Performance Council of standard-bred fowl, as promoted and governed largely by the American Poultry Association.

Professor Lewis also supplied a printed copy of the "Application for Admission to the American Record of Performance Council" and a copy of the "Official Report Records of Performance," copies of which can be secured steadily in popularity, they wish to place something tangible in the hands of the owners of birds that make creditable records—something these poultrymen can use for advertising purposes, thus to add value to the ownership of their birds. It is a move in the right direction and undoubtedly will meet with success, especially so as the whole proposition rests in the hands of the men who have control of the contests. Both control and responsibility are in their keeping; therefore they are in a position to govern matters satisfactorily, both for the welfare of the contests and in the interests of those who take part.

PRACTICAL STATE-WIDE POULTRY PROJECT FOR NEW YORK

Is a Logical and Consistent Development of New and Old Methods of Poultry Management and Breeding for the Much Desired "Best Results"—The New York State Method of Certifying Breeders—and Now They Are Establishing a "Proving Station for Breeders," Which Will Complete the Circle of Good Service

By Grant M. Curtis

GROWTH or development in private enterprise, in industrial progress and in world affairs should be logical and consistent—"a step at a time," if real and permanent benefits are to be obtained. And as a rule this is precisely how worth-while development is achieved. Take the flying machine, as an example. First, was the discovery of a new type of engine—that which utilized internal combustion. This made possible a small, lightweight engine of astonishing power. Then we had, in logical order, first the automobile, next the motorcycle and then wings were attached to the vest-pocket, motorcycle engine, which gave us the biplane, the monoplane, etc.

This same course of logical growth and consistent development is being followed to excellent advantage by the ten earnest workers who represent the man-power of the Poultry Department of the New York State College of Agriculture, at Cornell University, Ithaca, N. Y. Frankly, it reads almost like a romance and represents what R. P. J. considers to be one of the best poultry projects—if not the best—that to date has been worked out and "put across" in the interests of poultry culture as a science and for the welfare of the poultry industry in this great agricultural country of ours. All due credit to Professor James E. Rice and his capable, hard-working, loyal assistants. Following is the story—and it is one of genuine importance to many R. P. J. readers. Fact is, every state in the Union, meaning in particular the states in which agriculture (including live stock) is the foundation of general business and the bulwark of prosperity, should adopt this same method or a quite similar one, without unnecessary delay, the work to be done by the state agricultural colleges, respectively.

First, we have this lately discovered and expertly elaborated method of practical culling, with the object of eliminating the nonproducers, poor producers and early quitters, thus to protect the feed bin and the bank account of the owner of the fowls, few or many.

But that was not enough—was only one step in the right direction. The next step, in logical order, is "certified breeders." Of what real permanent use would it be if the field-extension men of the Poultry Department at Cornell were to continue to visit the farm flocks of New York State every late summer and fall to "cull out" the poor producers, the early quitters, etc., if no real progress was to be made by the owners of these flocks along the line of better breeding and better stock, both as regards standard qualities and high egg production?

And after the plan of selecting out good breeders from every culled flock and styling them "certified breeders" had been followed to a practical stopping place, what then? Was this ALL that needed to be done—all that could be done, in the best interests of earnest poultrymen and poultrywomen in New York State who look to their flocks for profitable returns?

No, there was another logical step to be taken—and now it is to be taken. Briefly it is this: the powers that be have made a special appropriation of funds for the Poultry Department of the New York State College of Agriculture, by the use of which a "Breed Proving Station" is now being established. New buildings that will contain twenty to twenty-five breeding pens of moderate size are to be erected as the nucleus—as the starting point of a proving station for breed or strain testing that before long is expected to be a big establishment, a large and vitally important part of the well-equipped and expertly managed Poultry Department at Cornell University.

After this Breed Proving Station has been put into operation, the course of procedure in the Empire State, as regards this line of work or "project" of the Poultry Department of the State Agricultural College will be substantially as follows: First, expert culling of good-sized flocks throughout the state to be done in July, August and September, by field-extension men of the Department, these men to instruct the owners of the flocks respectively how to do the work themselves; second, these same experts in the months of October, November and early December, will select from the best birds, both male and female, of each owner's flock those that should be used for breeding purposes, these specimens to be "certified," to be given leg-band numbers, to be kept track of, etc.; third, and later, after the owner of "certified" specimens has shown fully his or her active and intelligent interest in better poultry and more of it, two or more such flock owners in each county of the state will be invited to send a pen of these "certified breeders" to Cornell University, where they will be placed in the Breed Proving Station, kept under trap nest and be line bred, pedigreed, etc., for a considerable length of time, at state expense.

The foregoing is the order of procedure and we submit to the interested reader that it is a remarkably fine project—and for sound reasons. Its chief virtue, perhaps, exists in the fact that all these flocks belong to the people who are to be relied on to take care of them, help improve them and be responsible for feeding, housing, management, etc. The flocks also will be widely separated and until a limited number from each flock is sent to the Breed Proving Station, all responsibility or practically so, is to rest with the owners of the birds.

A second valuable element in the plan is that numerous earnest, intelligent and progressive poultrymen and poultrywomen of New York State will be getting the full benefit, if they so desire, of all knowledge, experience and down-to-date advice that the Poultry Department at Cornell University, with its ten capable workers, headed by Professor Rice, can give them, and this extraordinary help will be brought right to their farm or poultry plant. Such visits should be worth a great deal to every poultry keeper in that state who carries several dozen, several hundred, or several thousand head of fowl for productive purposes. Also it will help these field extension men who are working to improve not only the breeding stock and layers, but also the methods in management in caring for them, to obtain really profitable results.

Finally, this plan will allow a limited number of poultry keepers—taking New York State as a whole—to adopt and follow up intelligent, down-to-the-minute breeding methods, so that they can get the benefits of line breeding, of pedigree work, etc., and they will be getting this invaluable help at public or state expense. It is help that perhaps they could not get otherwise. Line breeding, trap nesting and pedigreeing represent expensive work. Only poultrymen regularly in the business can afford to do this and do it right, as a general rule. But the Poultry Department at Cornell can do such work and do it with thoroughness.

On the other hand, this work is destined to be of immense benefit to poultry culture, on practical lines, in New York State. As time goes on every one of these strains of line-bred stock, based first on selective flock mating, second on the use of "certified breeders," third on the benefits of "proved strains," as tested and demonstrated at the Breed Proving Station at the State College of Agriculture, will be just that many sources or near-by "fountains" where the ninety and nine interested poultry keepers of New York State can obtain surplus breeding males, small pens of breeders, eggs for hatching and day-old chicks, this stock and the products therefrom to be of decidedly superior value to the average run of present-day farm poultry, even in New York State.

CANADIAN RECORD OF PERFORMANCE

The Poultry Division of the Canadian Department of Agriculture Is Conducting a Record of Performance in Two Sections—"A" for Trap-Nested Flocks on Individual Plants and "AA" for Fowls Entered in the Various Standardized Laying Contests Conducted Under Government Supervision

THE Dominion Department of Agriculture, through the Poultry Division of the Live-Stock Branch, in 1919 put into effective operation an official "Record of Performance," this being a definite policy "planned to stimulate and facilitate the breeding of standard breeds of poultry along lines of greatly increased individual and flock production. * * * * Authentic trap-nest records, used as a foundation for intelligent pedigree breeding, form the basis of such work. Record of Performance has for its object the testing of purebred birds and flocks for the purpose of securing for poultry breeders reliable information as to sources of high-producing stock.

"Record of Performance 'A' consists of the inspection of trap-nested flocks on individual poultry plants, and is similar in form to the Record of Performance for dairy cattle. It is open to any owner of purebred poultry in Canada who wishes to enter, and is under the supervision and inspection of officers of the Poultry Division of the Live-Stock Branch.

"The regulations provide for the issuing of certificates for birds producing one hundred and fifty eggs or more in fifty-two consecutive weeks as required by the Record of Performance."

The principal requirements for entry are as follows:

Entries

"Applications for entry must be received at Ottawa, at least one month in advance of the date it is intended the records shall commence. No entries will be accepted after November 30.

"Only purebred stock of standard varieties and free from standard disqualifications may be entered. Entries will only be accepted tentatively pending the first report of inspection.

"If, upon the first inspection, the birds are found to be diseased, the houses or equipment not in good sanitary condition, or the trap nests not of satisfactory design and in serviceable working order, the entry may be refused.

"The minimum entry shall not be less than ten birds from any one flock, and all birds entered shall be identified by sealed and numbered bands, which will be provided and put on by officers of the Live-Stock Branch.

"In the event of any entry being cancelled because of attempted fraud on the part of the entrant or failure to observe the rules, or in the event of certificates not being issued for the above reasons, the Live-Stock Commissioner may refuse to accept entries from the same breeder for a period of two (2) years thereafter.

"There shall be a minimum entry fee of five dollars ($5) for the first twenty-five birds entered or part thereof, and an additional entry fee of two dollars and fifty cents ($2.50) for each additional

Condensed from Official Rules and Regulations.

twenty-five birds or part thereof. Entry fees shall accompany the application for entry. Fees should be forwarded in the form of a post office money order or express order or certified cheque and should be made payable to the Accountant of the Department of Agriculture.

Recording Production

"All stock entered shall be trap nested during the period of the official test, which in no case will exceed fifty-two consecutive weeks.

"Records may commence from the date a bird lays her first egg in the trap nest on or after August 1, and on or before December 31, providing the requirements of entry have been complied

SINGLE-TEST PENS AND YARDS IN USE AT SOUTH AUSTRALIA EGG-LAYING CONTESTS

with. In the event of a bird commencing to lay after January 1, the record year of fifty-two consecutive weeks shall date from January 1.

"Only eggs actually found in trap nests shall be counted and the entrant shall record or cause to be recorded each egg as laid and shall keep posted for the information of inspectors and others a record of same.

Inspection

"All flocks will be placed under unannounced inspection. Official visits may be at irregular intervals.

"All equipment on inspected plants shall be kept in a clean and sanitary condition.

"In the event of any outbreak of disease upon a poultry plant, inspection may be discontinued.

"The inspector shall have sole charge of the trap nests and eggs during all inspections.

"At the end of the record period a statement of the complete record will be returned to the owner and the owner required to take an affidavit that the weekly statements sent in and thereby recorded in annual form are a true and correct statement of the actual number of eggs laid by the individual bird or birds referred to in the statement. Failure to send in this affidavit before March 1 of the succeeding year shall render the breeder liable to forfeiture of the right to receive certificates.

Certificates

"Record of Performance 'A' Certificates may be issued for all birds, not otherwise disqualified, that in fifty-two (52) consecutive weeks lay one hundred and fifty (150) eggs, and, Advanced Record of Performance 'A' Certificates, for those birds not otherwise disqualified, which lay two hundred and twenty-five (225) eggs in fifty-two (52) consecutive weeks, providing in both cases the quality of the eggs is not lower than that of the grade 'specials' in the 'Canadian Standard for eggs' and they average at least two ounces in weight."

At the end of the first record year, 1919-20, the department published Report No. 1, giving the names of breeders, official number of each bird and its record, etc., this constituting a permanent record of the work, and of the performance of individual fowls that succeeded in qualifying for certificates or advanced certificates.

According to this report 4,436 fowls were entered, consisting of 968 Plymouth Rocks, 490 Rhode Island Reds, 2,331 Leghorns and 45 of other varieties. Of these, 17.2 per cent qualified for Record of Performance, 1.8 per cent of these qualified for advanced Record of Performance. About 54 per cent were dropped during the year, for one reason or another.

In addition to this Record of Performance which is known as "A" there is also a Record of Performance "AA" for the birds that qualify for certificates at any of the department's ten official egg-laying contests conducted at the Dominion Experimental Farms, which are located in the various provinces. The contest birds must meet exactly the same requirements as to number of eggs laid, etc., as in the "A" section.

THE SINGLE-TESTING SYSTEM

Persons Who Are not in Position To Use Trap Nests May Find It Practical To Adopt This System—Has Special Advantages in the Case of Extra Valuable Birds

UNDER some conditions penning individual birds separately as a means of securing an exact record of egg production is practical and the method deserves serious consideration, particularly by those who, for one reason or another, find it undesirable or impossible systematically to trap nest their flocks. In a publication of the South Australian government, D. F. Laurie, government poultry expert and leading advocate of single-pen testing, thus describes its advantages.

"My reasons for advocating the single-pen system are:

1.—There is no mechanical device to frighten or injure the fowl.
2.—She is well-housed and has sufficient room for exercise.
3.—All possible errors in identification are eliminated.
4.—The general character of each fowl can be studied daily and without any trouble. This is, of course, a most important consideration.
5.—You are in a position to control her food supply, and, by comparison with others undergoing the test, you accumulate valuable data.
6.—By carefully studying the occupants of the various pens you will with greater certainty observe divergence from type, tendency to a general type, and other characteristics. This accumulated knowledge, especially if tabulated and recorded with pedigree charts and photographs of the individuals tested, becomes an invaluable record.

"Some people who have no practical experience of the system of single-pen testing have expressed the opinion that the health of the birds must suffer, and that their subsequent value as breeders must be lessened, if not destroyed. Accumulated experience teaches the opposite, but of course much depends on the construction of the house and yard, forming the pen, and also the method of feeding adopted. In South Australia the mild climate admits of very simple but none-the-less effective structures. The severe climates of some other countries necessitate modifications in construction. In all mild to warm climates the materials used and the method of construction should offer as little harbor for vermin as possible. For Australia and similar climatic conditions the framework of the houses should be of hardwood free from all cracks, and should be moderately smooth. The covering material may be of corrugated

CONVENIENT TYPE OF HOUSE FOR SINGLE-TEST PENS

Shows caretaker feeding green stuff. Immediately below the green-feed holder is a continuous water trough and below this small drawer-tight troughs (closed in photo) for mash feeding. This house is one of several on the contest grounds of the Auckland, New Zealand, Poultry Keepers' Association.

galvanized iron (narrow fluted), or compressed asbestos (fibro-cement) sheets. Weather-boarding and similar material offer harbor for vermin and are liable to crack, twist and warp. The single pens may be fixed or movable. The dimensions of the pens need be not more than 3 ft. by 20 ft., and the roosting and laying house 3 ft. square. To have the yards less than 3 ft. in width is inconvenient for the average person; any additional width adds to the expense of construction. Portable pens allow the ground to be changed daily, and where grass, clover, etc., are abundant, this method is much appreciated by the birds. Fixed pens are more convenient where large numbers of birds are being simultaneously tested. In all cases the pen, portable or fixed, should be numbered, and the hen therein should have a leg band with a corresponding number.

CHAPTER IV

Special Articles on Breeding for High Egg Production

The Articles Presented in This Chapter, Written by Recognized Authorities, Afford Much Helpful Information on Special Problems Connected with Breeding for Production—First Article Is by Dr. O. B. Kent, of Cornell University and Treats on Selection of Stock, Precocity, Age of Breeders, Quality of Eggs, Inbreeding, Etc.

BREEDING for egg production is an art. There are so many factors involved that are but poorly understood that it is at best a difficult art. When breeding for shape or color one definitely knows when the bird is grown just what is the shape or the color. But it takes a year after a hen is full grown before you know its quality as an egg producer, and two years after a male is full grown before one knows his quality, and even then there is always the chance that some change in feed or weather may interfere with the record of the bird.

Some people think that it is a very simple thing to breed for egg production because they have never tried it. Various people, in severe competition, have been able to win year after year in poultry shows, but no one in this country has been able to win year after year in an egg-laying contest. Tom Barron came the nearest to it, but that was at the beginning of our contests.

A "Cut and Try" Process

Breeding for egg production is a "cut and try" process. If one kind of mating does not work, try another; but if certain birds do make a favorable combination, keep them mated and do not try to improve on it. Try to get a better new mating, but do not try to better a good old one.

In selecting a hen for breeding purposes there are a number of factors to be considered. Two hens may lay the same number of eggs and be radically different in their quality as breeders. A hen's record is made up of intensity of production, regularity of production and length of laying period. One hen may lay two hundred eggs in twelve months and another do it in eight months. Which should you use as a breeder? In order to settle that question we must consider the relative importance of length of laying period and intensity of production.

The length of the laying period depends on two factors, the time of beginning and the time of stopping. The time of beginning, or precocity, depends on the condition of the parents at the time they are bred, the method of hatching and the method of rearing. If the parent stock is not in good condition for breeding the chick will not get a good start and consequently will not develop as rapidly nor begin laying as early. Right here we have the weakest link in the breeding chain.

It has been shown by Pearl from the Maine records and by Rogers from the Cornell records that a high-producing individual does not produce as good hatching eggs as a poorer producing hen, especially in the early season before the hens are running on grass. If the eggs do not hatch well the chicks that do hatch are obviously weaker. Being weaker they do not grow as fast and consequently do not begin to lay so early and hence do not make as high winter records. This may be the explanation of Pearl's statement that high winter production was transmitted by the male and not the female.

A male, of course, would not be handicapped by egg production and consequently would not be impaired as a breeder. A bird's winter production or its age when it begins to lay should be studied to determine whether it is due to failure to make rapid growth because of poor digestive capacity or because of a poor start. The results at Cornell would tend to confirm Pearl's result that the male has a greater influence on winter production than the female, but the female has a greater influence on the spring and summer production and the following year's production. The male has more influence on the length of the first laying period and the female the greater influence on intensity of production. A female that is a good producer may transmit high production if she is in good condition in the breeding season.

Reprinted from Reliable Poultry Journal, September, 1921.

Precocity an Important Factor

Precocity is an important factor to consider, especially in the heavy breeds, for if pullets can be grown to lay full-sized eggs in the fall it decidedly increases the returns because of the high prices at that season. But precocity is as much or more influenced by the method of rearing than by breeding. A flock of birds may be brought into laying at five months or eight months, depending on how they are handled. Consequently precocity is more a measure of rearing than it is of breeding.

The time of ceasing to lay in the fall is controlled by a number of factors, the most important of which is feeding. A healthy, vigorous, well-fed flock will lay well in the summer, while a thin, poorly-fed flock will stop laying early, but the best individuals in each case will tend to lay the longest because they have greater ability to digest and assimilate food and hence keep in laying condition. Of the two factors, precocity and intensity, the latter is a better measure of (1) "future production," (2) of the mother's production and (3) of the daughter's production than the former. In other words, intensity is a more reliable index of inheritance than is precocity.

The intensity of production is a factor that is frequently overlooked and yet it is one of the most important factors of all. Intensity is undoubtedly less influenced by feed or environment than most any other factor and hence is a better index of the quality of the hen.

If a hen can lay seven eggs in a week or twenty-five or more eggs in a month, she has the digestive capacity to produce a large number of eggs, but a hen that can lay only three or four eggs a week or ten to fifteen eggs a month, can never be a high producer. While usually an intensive layer is also a long layer, occasionally a slow layer is also a long layer. A hen of low intensity that lays over a long period and hence makes a moderately high record is rarely a good breeder. It is much better to breed from a hen that laid a good many less eggs but that laid more per week or month while laying.

Makers of High Scores as Breeders

A CORNELL HIGH-RECORD BIRD
F-253—laid 253 eggs in her first year and 220 in her second year. She was quite exceptional in that her daughters are making very creditable records.

It frequently happens that of two sisters having a similar intensity of production the one that lays fewer eggs, say from one hundred and fifty to two hundred, as compared with the other laying over two hundred, is the better breeder. It is comparatively rare for a high-producing hen to be a good breeder. Most of the high-record hens have been daughters of hens that laid around two hundred eggs rather than around two hundred and fifty eggs. It is a disappointing fact that there are comparatively few pedigree records where the hens have uniformly laid around two hundred and fifty eggs.

It would seem to be a self-evident fact that the better the hen bred from, the better her chicks will be, but unfortunately that does not seem to be true. The best three-year-record bird that we have had at this station, "Cornell Supreme," laying 222 eggs the first year, 223 the second year and 220 the third year, was a poor breeder both in number and quality of daughters. The best one laid only 158 eggs in her first year. Yet a granddaughter of "Cornell Supreme" through one of her sons, No. C6789, with a record as follows, 80 first year, 159 second year and 176 third year, has been a good breeder. The average of eight daughters by two different males has been over two hundred eggs. Other similar cases might be cited. Note the following table:

	Egg production	Per cent egg hatch	Record of daughters
C3018			
1913-1914	198		
1914-1915	150	64½	(102 incomplete)
1915-1916	160	61	137
1916-1917	163	66	194, 110, 209, 150, 143, 185, 145, 191, 187
1917-1918	158	81	171, 181, 131, 189, 177
1918-1919	146	71	
C7122			
1914-1915	203		
1915-1916	151	25	(105 incomplete)
1916-1917	146	40	198, 117, 66
1917-1918	155	14	
1918-1919	150	10	
D345			
1915-1916	179		
1916-1917	172	60	209, 154, 196 (72 incomplete)
1917-1918	180	60	204, 209, 207
1918-1919	212	72	
D1402			
1915-1916	210		
1916-1917	204	40	123
1917-1918	160	45	
1918-1919	137	34	

Nearly all the best breeders have been individuals that have laid between 175 and 210 eggs and have laid over 450 eggs in three years. This merely shows that under our present method of feeding and handling a bird that lays or is allowed to lay over 210 eggs is handicapped as a breeder, especially as a breeder whose daughters will make high records. Perhaps a hen that is known to be a high producer could be forced to take a rest and so be put in good condition for breeding. This can probably be done better the second year rather than the first.

Yearlings Better than Pullets

Besides the egg records there are a number of other things to be considered in the selection of a breeder. It is better to breed from a yearling than from a pullet because a yearling has proved her ability to live and lay. A bird lays a larger egg as a hen than as a pullet and consequently the chick gets a bigger start in the world. As a bird gets older the hatching quality of her eggs tends to decrease somewhat. While the eggs of some hens do hatch just as well when the birds are six or seven years of age as when they are one or two years of age, the average hen gets poorer in fertility as she grows older. But the hens that are good when they are young are also good when they are old. A hen that does not produce a good percentage of hatchable eggs when young is not liable to be any better later.

The hatching quality of a hen is a good index of what her chicks will be. A hen that does not produce eggs that will hatch is very unlikely to produce pullets that will lay well. A well-bred hen whose eggs hatch well generally will have daughters that lay well. Take for instance the hatching records of C7122 and C3018, or of D345 and D1402 with their records and the record of their daughters. It would be possible to find a large number of such cases. If a hen's eggs do not hatch better than the average of the flock, do not use her as a breeder the second time and do not keep any of her offspring. It is sometimes a bitter dose, but it is better to throw out the offspring early rather than late. If the eggs do not come on well in the early hatches do not put any in the late hatches. Occasionally a hen will be infertile—why, we do not know. We only know that it is more common with extremely high-laying individuals than with moderately high individuals. If such a case happens, use a stronger hen. It has sometimes happened that a hen would be infertile until she got out on the grass and then have good fertility but rarely good hatching quality.

At broiler time sort out the chickens according to pedigree and discard any families that contain only a few chicks. Usually the larger the family the more thrifty the chicks. There may occasionally be an excuse for keeping a cockerel from a high hen in a small family, but none for keeping a pullet.

The Quality of Eggs

Besides the record of a bird, the quality of the egg it produces must be considered. Size, shape and color of eggs are all inherited so that it is necessary that any hen that does not lay a first-class market egg should not be used as a breeder. It is surprising how easy it is to breed down the size of eggs in a flock by using as breeders hens that lay small eggs. It often happens that of two hens laying about the same total weight of eggs, one hen will lay a good many more. Do not use such a bird. The easiest way to avoid breeding from such hens is to throw them out as pullets just as soon as it is found that they lay a small, off-colored

CORNELL S. C. WHITE LEGHORN
G-351
First year record, 251 eggs. Good type and disposition.

or otherwise undesirable egg, due allowance being made for the fact that a precocious pullet usually lays a very small egg. A poor quality of market egg is one of the curses that almost inevitably seems to follow pedigree breeding. The only way to get rid of it is to cull out such individuals before they become numerous, and keep them out. It often happens that the eggs from a farm that is pedigree breeding for egg production contain as high a percentage of culls as the eggs from a farm that is breeding for fancy points alone. Do not let an egg record warp your judgment of the necessary standard to maintain in market eggs.

The Breeding Male

It is just as important and generally harder to pick out a breeding male than it is a female. From the trap-nest record you should know considerable about a yearling hen in the breeding season, but little is known of a cockerel except his age, vigor, production of parents, winter production of sisters, size of family and egg quality of mother. Of these the record of the mother is usually greatly overemphasized in selecting a male. Frequently a male is used as a breeder just because his mother made a high record regardless of how ugly a runt he may be.

First importance in picking out a male should be given to the production of his sisters and half sisters. A male that comes from a family of good producers is liable to be a good producer, but an occasional exceptional bird is not an indication of the average quality of the males. In studying the winter records of a cockerel's sisters see that they have been intense layers as well as early layers. It is probably even more important that they should be intensive layers rather than early layers. The more generations of high-producing blood there is back of a male the better, for the more liable he is to be more nearly homozygous (characters stable, having been received in the dominant form from both parents, or in the recessive form from both parents; breeding true to type) for high production. But if he comes from a high hen with a low or varied general ancestry he is liable to be heterozygous (containing one or more recessive characters and therefore unstable; progeny not all true to type) of production. It is especially important that the mother and sisters of a cockerel should be intensive layers. It is also desirable that there should be a large number of sisters as an indication of the vigor of the family. Shy breeders are not at all desirable.

Occasionally a male will be infertile. Where a number of hens are infertile with a male the quicker the male is changed the better. We had an interesting case of infertility in the breeding flocks the spring of 1920. Four cockerels that were either full or half brothers were decidedly infertile. The cockerels were apparently vigorous and active, but did not get fertile eggs. As soon as the two worst ones were replaced the hens began producing fertile eggs at once, showing that it was not the fault of the hens. The two that were left improved during the season but never became first class. Do not breed from extremely large or small cockerels. The smaller bird generally gets better fertility, but it can quickly be overdone. Select the males for quality of the mother's eggs even more carefully than the hens themselves are selected.

About Inbreeding and Outcrossing

Besides the selecting of individuals as breeders there are several other questions that need to be considered—for instance, the question of inbreeding. There has been a great deal of argument but very little data presented on the subject of inbreeding. Some state that it is a positive necessity and others that it is exceedingly undesirable. Our results would tend to show that there is a decrease in fertility or hatching quality for the first three or four generations, but after that the evil effects are not as evident. A number of cases have been published where well-bred males have been mated to another strain or even another breed with highly beneficial results in that the daughters have made uniformly higher production. Such a mating seems to produce slightly more vigorous offspring that mature earlier and make better records, especially better winter records.

An outcross gets vigor and fertility in that generation, but there is little to be gained and much chance for loss in continually outcrossing. The chicks do not become more vigorous and since little can be known of the parentage, unfavorable matings are liable to be made. Males that come from an outcross mating are not apt to give as good results as inbred males. We have mated a number of outside males with our stock and have just one instance where a son was up to the average as a breeder, though his sisters quite uniformly made high winter records.

From the practical everyday standpoint a breeder has got to do a considerable amount of inbreeding if he wants to fix anything. Buying outside blood even of the same breed is frequently as bad as crossbreeding. It may change the size and type and injure for years the market quality of the eggs. If new blood is introduced be sure to discard it at once if the stock is not an improvement.

Brother-and-sister matings are the closest form of inbreeding and are generally looked on as the most undesirable form of inbreeding, yet it has the advantage of disclosing any weakness quickly. In a series of brother-and-sister matings through a number of years, we have been getting better and better results each year after the first two or three years. However, the first two or three years were nearly bad enough to put an end to the experiment.

If one is intelligently to breed up a flock of birds it is necessary to inbreed or to line breed, a terminology which is almost a distinction without a difference. In order to avoid the objectionable point of outcrossing a number of families or lines should be carried, each of which would be inbred, but any one of which would be available in case of decreased vigor in any family.

Annual and Winter Production

The relative influence of sire and dam has already been discussed to a certain extent, but it is necessary to emphasize especially that Pearl's work was concerned with winter production alone and except for a fair correlation between winter production and total production it cannot be construed as applying to annual production. Each poultryman must decide for himself whether he is after winter or total production, as they are probably inherited to a considerable extent independently. In winter production the male seems to have the greatest influence and in summer and second and third-year production the female seems to have the greater influence. A Mendelian interpretation can be given to the inheritance of egg production, but a Mendelian hypothesis at best amounts only to a "cut and try" process.

In buying males or breeding stock on pedigree there are two kinds of pedigrees commonly furnished. The first merely gives the record of the mother. Practically speaking, such a pedigree is nearly worthless. The mere fact that a hen has made a high record is but little evidence of the value of her progeny. The second type of pedigree consists of records of the sire and dam on both sides back through several generations, at least three or four. Such a pedigree is of some value though it may amount to little.

To summarize briefly the methods to be followed when the stock is trap-nested:

1—Breed from mature stock.
2—Use only the individuals laying a first-class market egg.
3—Use those individuals that are intense layers and that produce over a long period.
4—Discard any breeders that do not produce eggs of good hatching quality.
5—Extremely high layers are only occasionally good breeders.
6—Use precocious, vigorous males—sons of hens that lay a first-class market egg.
7—Select males on the intensity of their mother's winter production as well as amount of eggs laid.
8—Practice intelligent systematic inbreeding or line breeding.
9—Do not waste money on a short pedigree.

HIGH EGG PRODUCTION TRANSMITTED BY BOTH MALES AND FEMALES

Persons Who Doubt Whether Males of Bred-to-Lay Ancestry Can Transmit This Character, and Those Who Question Whether Such Inheritance Is Solely Through the Male Line or Through Both Male and Female, Will Find the Data Here Presented of Great Interest

By Dr. B. F. Kaupp, North Carolina Experiment Station and State Agricultural College

THE work here recorded was begun by the writer and his coworkers at the North Carolina Experiment Station and has been continuously in progress since 1914. The birds used were those found on the plant when the work was taken over and the males were purchased from other workers in experimental breeding work in egg production. The houses are modern, open-front, shed-roof buildings. The feed throughout has been the same, and from year to year as well as from generation to generation. Care is taken not to introduce any factor different with each generation except the factor of breeding. Everything else being the same and varying only the breeding factor, a true test of fecundity should be obtained.

Results with Common Hens Bred to Common Cocks

There were 90 hens composing flock No. 1 in their pullet year. Taking into consideration the number of eggs laid on the floor and those laid in the trap nests these hens averaged for their three years 89 eggs per hen per year. These fowls were bred to cocks in their first year and their pullets are known as flock No. 2. Both males and females were from unculled flocks and the hens of all flocks were never culled so that we might study individuals as well as study the proposition from a flock standpoint.

In flock No. 1 there were only 8 hens that laid more than 100 eggs in a single year throughout the entire three years. There was only one hen that laid 150 eggs, the next highest being 128 eggs. In other words, 9 per cent were capable of laying over 100 eggs each per year, and one per cent capable of laying 150 eggs per year.

Flock No. 2, produced from common hens and common cocks, laid an average of 88 eggs per hen per year in the second year and 92 eggs per hen per year as an average for the three years. The highest annual egg production from a single individual of this flock was 180 eggs and the next highest hen was 154. Only two hens then produced 150 eggs or more in any one of the three laying years.

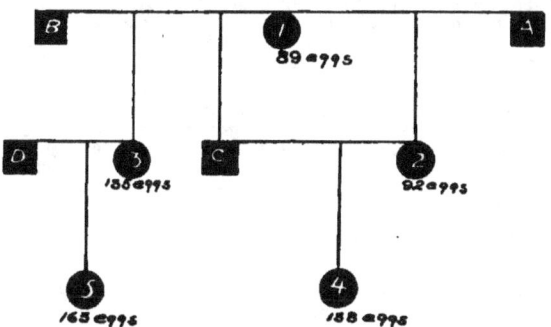

DIAGRAM SHOWING BREEDING OF EXPERIMENTAL PENS

"1" represents flock No. 1, that averaged 89 eggs per hen per year. This flock mated to common cocks (A) produced flock No. 2, which averaged 92 eggs per hen; No. 1 mated to pedigreed cockerels (B) produced flock No. 3, which laid 135 eggs per hen; No. 2 mated to cockerels (C), brothers of flock No. 3, and from pedigreed cockerels but common hens, produced flock No. 4 that laid 138 eggs per hen; No. 3 mated to a pedigreed cockerel—both hens and male pedigreed—produced flock No. 5, that laid 163 eggs per hen.

Results of Common Hens Bred to Pedigreed Cockerels

Flock No. 1, in its second year, was mated to 3 pedigreed cockerels, all full brothers, with the following pedigree:

Pedigree Chart No. 1

```
                    ┌ Sire—2154  ┌ Sire—unknown
         ┌ Sire 1661 ┤            └ Dam—unknown
         │          └ Dam—Lady Purdue ┌ Sire—unknown
Ckls.    ┤              229 eggs      └ Dam—unknown
4073     │          ┌ Sire 2276  ┌ Sire—unknown
         └ Dam 1740 ┤            └ Dam—unknown
           207 eggs └ Dam 166   ┌ Sire—unknown
                      175 eggs  └ Dam—unknown
```

There were 100 pullets in the first year of flock No. 3 from flock No. 1 bred to pedigreed sires, 96 completing the year. The average number of eggs per hen per year for three years was 135 eggs. Forty-six hens, or 48 per cent, laid 150 eggs or over. Twenty-three, or 24 per cent, laid 175 eggs or over in a single year, and 5 hens laid over 200 eggs in a single year. In this case the power to lay a large number of eggs was plainly transmitted from the sire to his pullets.

High Fecundity Is Transmitted from Sire to Son and from Son to Daughter

Flock No. 4 was produced by mating flock No. 2 with three full brothers from sires whose pedigree is given in pedigree chart No. 1 and from dam 64 that laid 128 eggs. Birds in flock No. 2, like their mothers, were low producers, laying only 92 eggs per hen as a three-year average. It follows that if flock No. 4 proves to be high producing, then high fecundity must have been transmitted from their ancestors who were the grandsons of Lady Purdue, and thus of high-fecundity lineage.

Flock No. 4 has just finished its second year with an average of 138 eggs per hen per year. The highest hen we have produced in this six-year series was from this pen this last year. This hen was A49, who laid 232 eggs. There were 23, or 56 per cent, that laid 150 eggs or over and 11, or 27 per cent, that laid 175 eggs per hen per year, and 4 that laid over 200 eggs in a single year.

Flock No. 5 was designed to show just what would happen when high-fecundity hens and cocks were bred together. It was produced by breeding the 12 highest producing hens from flock No. 3 with a cockerel with the following pedigree:

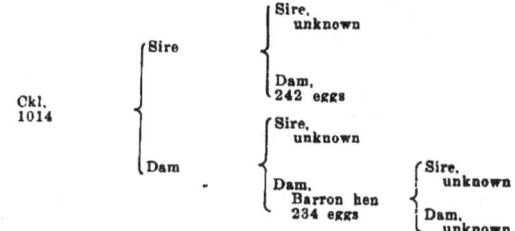

The 12 hens to which the cockerel was mated laid a total of 6,129 eggs in three years, an average per hen of 510, or 170 eggs per hen per year. The highest number of eggs laid by any one hen in a single year was by Lady Raleigh and was 223. Of the pullets produced by this mating, forming flock No. 5, 46 completed the first year, with an average of 163.2 eggs per hen per year. Twenty-eight, or 60 per cent, laid 150 eggs or over; 15, or 32 per cent, laid 175 eggs or over and 6 hens laid over 200 eggs. The highest hen in this flock laid 227 eggs in her first year.

Summary

The following table gives a summary of the breeding for egg production for the last six years as carried on at the North Carolina Experiment Station:

Kind of mating	No. eggs per hen per year of the progeny from this mating
Common hens	89
Common hens mated to common males	92
Common hens mated to pedigreed males	135
Common hens mated to sons of pedigreed males	138
Pedigreed hens mated to pedigreed males	163

By common hens is meant purebred S. C. White Leghorns, unculled, and bred in a "hit-and-miss manner" for generations.

The results of this breeding show that high egg production qualities are transmitted by the male to his daughters. It also shows that high egg production is transmitted from sire to son and from son to daughter as illustrated in the results obtained from flock No. 4. That we were dealing with unculled hens not bred to high egg production is shown by the fact that the daughters from flock No. 1, bred to cocks from a flock unculled or not bred to high egg production, produced another flock that as a three-year flock average, laid within 3 eggs per year per hen of what their mothers laid.

These preliminary tests also show that when pedigreed hens and cocks are bred together there is a still higher trend in egg production. In other words, it appears that high egg production is transmitted by both sire and dam and that when high-fecundity males are mated to high-fecundity females, the pullets will have greater power to lay

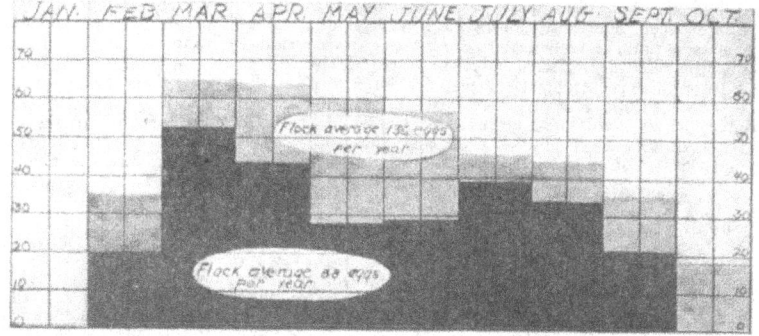

GRAPHIC PROOF OF VALUE OF USING MALES FROM HIGH-RECORD HENS

Chart showing the effect of mating common hens to males from high-producing females. The numbers along the side give the per cent of the flock that laid each month. Black shows production of common hens. Light-shaded areas indicate the production of pullets of the common hens bred to pedigreed males.

eggs than if only the sire or the dam was from a high-egg lineage. This statement is borne out in the results obtained from flock No. 5. However, we are perhaps limited by the factor of limitations, as it is not to be supposed that we can boost production each time we make a new mating till we get to the point where a flock average might go—we will say, over 300 eggs per hen per year. We do not know that anyone has definitely determined just at what point this factor of limitation comes into play, but it is perhaps near the 200-egg mark.

Experiments bring out an interesting point which is that to obtain a large number of eggs, hens must possess the inherited power of high egg production. If they do not it does not matter what, how much, or in what manner they are fed, housed and cared for, they will not produce large numbers of eggs. On the other hand, if their breeding for high egg production has been satisfactory it is, after that, a matter of health, care, feeding and management.

The first illustration is a graphic chart of the matings described and the results in egg production under the same housing, feed, care and attention, and with the same attendant throughout, so that only one factor was varied—namely, that of breeding.

The second illustration is a graphic chart in which the dark part indicates the number of eggs laid each month by the hens not bred to lay, and the black plus the light part shows the number of eggs laid each month by hens who had common mothers but pedigreed sires.

METHODS OF BREEDING FOR PRODUCTION AS PRACTICED IN AUSTRALIA

The Early Difficulties Met and Mistakes Made on the Other Side of the World in the Efforts of Poultrymen To Establish High-Production Strains of the Popular Varieties of Domestic Fowl by the Adoption of Systematic Methods of Breeding—There Was Danger at One Time of "Leghornizing" All Breeds, But Later That Peril Fortunately Was Avoided

By Vic Kappler, Adelaide, Aus., American Representative of the National Utility Poultry Breeders Ass'n. of Australia

IT might well be said that at the time that Australasian laying competitions were inaugurated practically nothing was known about breeding for egg production. The first contest gave a winning pen with an average of 185.5 eggs a bird, while the lowest yielded only 76.5 eggs each, and this ratio was general for the first few years in practically all contests and led to all sorts of theorizing and the springing into existence of schools supporting various views as to the whys and wherefores of such margins of birds fed and housed under identical conditions, and it is doubtful if we can be said to be really out of the pioneering stage, even after twenty years' experience, with the results of scores of contests before us.

We certainly are gaining information but as sometimes happens in man's efforts to control nature, the evidence is often apparently conflicting for the reason that the antecedents of the material used are not sufficiently known to enable us to account for different results from the repetition of a certain experiment. We really began groping in the dark and either our eyes are becoming accustomed to the gloom and we conclude we are seeing things better, or a few rays of light are breaking through.

Possibly the first lesson we learned was that some types were nonlayers or, to be more accurate, were poor layers. The massive, sluggish bird, with stately step, overhanging eyebrow, last off the perch in the morning, loafing around all day, the first to perch at night, was not the bird to represent the breeder in a laying contest, however well she might prove to rank in the exhibition class—nor was she of any use to the man who sought to make a living at egg production.

The leading pens always seemed to be the lighter, more active birds of a breed—with bright eye, clean face, neat head, close feathered, well balanced, of good breadth as well as length of body, muscular and free from surplus flesh condition, robust in health, and of a racy type.

Some breeds were noticed to have what might be described as a natural tendency to heavy egg production—notably the White Leghorn, Black Langshan and Silver Wyandotte. (This latter variety was first in several of the earlier competitions, but was soon displaced by the White Leghorn, which held undisputed sway until a few years ago when quite unexpectedly the Black Orpington flashed to the front and has not yet been overtaken.)

Early Stock from Exhibition Breeders

For the first few years the birds were almost all from the yards of exhibition breeders and while they may not have been classy enough to win in hot company, they were at least typical of their breed, and as no danger was apprehended of this standard being departed from it was not thought necessary to lay down a more definite regulation than one demanding that birds should be purebred, but so great was the interest created by these laying contests and so numerous were the people who had taken up poultry as the result—many with no experience in handling birds other than perhaps a nondescript flock that had possession of the farm or back lot—that the problem became not so much one to induce entries to fill competition pens, but to keep the number within the limit the facilities available made it possible to handle properly.

Many of these newest recruits to poultry keeping might fittingly be described as the rawest novices, their greatest virtue being their enthusiasm (their ranks including many potential breeders) but the majority never knew and never learned the most elementary lessons and, as fast as the fever left them, new victims came to light and before it was realized, especially by this new contingent of poultrymen, the obvious fact that the smaller and lighter birds were providing the competition winners had resulted in a policy with many of the competitors to take the active, undersized birds as the principal basis of their breeding operations; therefore size and type, let alone other standard qualifications, were rapidly being ignored, and if the quest for eggs had been allowed to continue along these lines it would have led to possibly one type of bird representing all breeds, the only distinction being that of color.

Indifferent as birds might be they could not be disqualified under the existing regulations, as it was only demanded that they should be purebred—and as they were descended from pure typical stock, and no infusion of other blood could even be suggested, it was realized that some other method than just demanding purity of breed was necessary and had to be adopted, really to work back to at least maintaining the size.

This tendency to produce lightweight, active birds really became a craze with us. Writers in the poultry press and in the week-end issues of the daily papers (which all give great prominence to aviculture) almost to a man were advocates of "the little hustler kind"—and as nothing really definite had up to that time been learned about breeding for egg production, this advice seemed as good as any other, particularly as the small, active Leghorn was at that time outstripping the medium and heavier sorts by winning every competition and providing almost always the best averages.

As the real object of the competitions was to give the farmer and others engaged in egg production as reliable and as prolific a layer as possible and to induce breeders to increase the productivity of their strains if possible, it seemed that the most promising results would be obtained with the small bird. The overwhelming proportion of poultry keepers had taken up the White Leghorn, other breeds being correspondingly neglected, and soon the markets were flooded with Leghorn cockerels and hens with an almost utter absence of the better-class table bird. This caused a cry that the table side of the industry was being neglected and it offered a very promising field to the breeder of the heavier kinds, if it were possible to produce a dual-purpose bird, and to what extent they have succeeded will be discussed later in this series as it is a matter bearing on the question of whether or not a laying type can be said to exist, or if standard-bred birds of all types can be turned into prolific layers without disregarding the requirements of the Standard.

Traveling "Head On" Toward Disaster

Without experience to guide them many breeders—unconsciously perhaps—were getting birds of several breeds down so fine that they defeated the very object they had in view, and instead of producing layers they were turning out a snaky-headed, narrow-bodied, knock-kneed weakling that could do no better than the heavy, massive type they wanted to get as far away from as possible. And with these undersized undesirables came another unwanted trait: the size of eggs fell off. It has been a rule from the inception of these contests that eggs to count had to be of marketable size and quality throughout the year and that at the end of the winter test, which varies from four to six months from the inception (usually the former period) the eggs had to weigh not less than 24 ounces to the dozen.

It was thus increasingly evident that a wrong track had been taken. Those conducting the competitions saw the danger and sought to minimize it with the least inconvenience to those concerned, and the first step in that direction was to divide the contest into two classes, one for light breeds such as Leghorns, Minorcas, Anconas, etc., and the other for heavy breeds. Then a weight limit was imposed and birds not reaching this—which was the recognized standard weight for pullets, or very close to it—were disqualified. More recently still, in one state at least, a demand has been made that the birds shall be reasonably good specimens of the breed they are supposed to typify, and birds have been sent back to the owners that did not make the grade, while in another state special sections for both light and heavy standard-bred birds have been provided. As there is no coordinated control of these competitions the regulations vary, but the downward tendency in size has been definitely abolished and a better size in birds is now apparent everywhere—and equally pleasing is the fact that the number of disqualifications for undersized eggs is also being reduced.

Not All Poultrymen Were that Foolish

Naturally all our breeders did not follow in the sheeplike procession that went in search of the small birds; some of them took the equally blind alley that "measurements of the body" has proved to be. For this fad we were solely indebted to our American friends. Wonderful discoveries were disclosed, under a bond of secrecy, and the gushing advertisements and glowing testimonials of satisfied users of these systems had their effect—it opened a new field of possibilities and to many it looked worthy of exploration, but the majority were soon convinced that this was barren ground and abandoned it. Some still hang on, however, but get nowhere.

One of our best known and outstanding utility breeders discussed this question with the writer some years ago and gave several illustrations as to how useless these systems are. One male bird belonging to this gentleman, according to the specifications laid down, was only fit for the pot and should long since have lost his head, but as a matter of fact he was siring world-record layers that were winning competitions in every direction and some of his progeny were testing up to 300 eggs—an unprecedented occurrence in those days—and incidently he was himself from stock with wonderful records on both sides.

Another case with outstanding features was the experience of a prominent South African breeder. He had been successful in laying contests in that country and sent to an Australian of repute for a high-grade male of unquestioned laying strain. On its arrival he subjected the bird to the measurements of a certain much advertised system and was so disgusted with the result that he felt he had been defrauded. The bird was a sad disappointment, according to the rule of thumb secret, but it carried the reputation of one of Australia's leading breeders and cost big money. These two facts saved it from early extinction and it sired the pen of pullets that won for him the African record in public competition.

Personally I have come across so many cases where the whole thing proved useless that I could give innumerable instances, but one more will suffice. It concerns a friend of mine, a very conscientious man but who, unfortunately, succeeded in humbugging himself that one of these systems that he has been tampering with for years is accurate and reliable, and so convinced was he of its merits that the efforts of his friends to put him on safer ground have been unavailing. Not only does he use it on his own flock, but considering he is conferring a boon on his fellow poultrymen he advertised his services as an expert and accepted engagements to put others on the right track by culling the poor layers and mating up the best for breeding stock.

Since I have been in America I have received the report of a certain laying competition in which my friend had a pen of birds and on going through the result I find that out of an entry of 90 pens my friend, with his infallible test, is not in the first 80—and some of those both above and below his representatives, had the misfortune to lose a bird during the contest.

No Safe Short Cut to Real Success

A few however—and it was very few—realized that there were no short cuts to the production of a reliable high-laying egg strain, if it were possible to produce such, and that it meant years of steady, patient work with possibly many disappointing experiences before even a solid base to work from could be secured. They reasoned that the methods of testing and selecting that had proved successful in producing beef and milking strains in cattle and meat-carrying types in sheep, etc., etc., could be used to produce high-quality layers on which it would be possible

THE PRODUCT OF SIXTEEN YEARS OF PERSISTENT LINE BREEDING FOR HIGH EGG PRODUCTION
This illustration made from three unretouched photographs shows a cockerel and two pullets bred and owned by D. Tancred, Kent, Washington. Back of these standard-quality birds are sixteen years of persistent and consistent trap nesting in one of the world's greatest high-production Leghorn strains.

to rely to reproduce the trait of high fecundity in their progeny, when once that characteristic became fixed in the strain. They were, however, in a position that no other breeder of stock who seeks to found a strain finds himself nowadays; they had really to evolve the ideal they were endeavoring to create—had to create it out of chaos.

Other studmasters when seeking to develop a particular trait can secure pedigreed stock behind which there are many generations—sometimes going back for hundreds of years—of foundation work already accomplished and in which the desired characteristic can be said to be more or less definitely fixed, and they carry the sequence forward and improve it if possible. Our poultrymen's situation was fundamentally different. There was no such basis to work from. True, our more remote ancestors must have, when domesticating some fowls and some ducks, made egg production a prominent feature of their operations. (Perhaps we do not realize that the European had but little to do in this work for we find the most prolific natural layers come from alien climates. The Langshan comes from China, the Runner Duck from India and the Leghorn, though it comes to us from the Mediterranean, possibly had its origin in Egypt for its prototype has been found on mosaics thought to be thousands of years B. C. and the Egyptians as well as the Chinese are known to have been aviculturists in the dim past, and practiced artificial incubation long before we resorted to it.) But there were no poultry available whose immediate ancestry contained an unbroken line of high fecundity. The sequence of laying had not only not been maintained, but it had been but of secondary consideration, if it entered at all into the reasons that led to the matings of their recent forebears—in fact they were actually handicapped to a degree that those who originally set out to increase the value of other kinds of stocks were not called on to face. They had to begin with stud stock that had been bred to conform to certain ideas of standards which may have resulted in a type of birds being created whose very physical structure was against the possibility of their being successful as high producers. (Perhaps the fact should be mentioned here that our fanciers of that time particularly were more prone to support the massive, square, sluggish types that were the ideals of the English fancier, rather than the medium, more active American standard.)

Our Lesson from Cattle and Sheep Breeding

The cattle breeders on the islands in the English Channel, for instance, when they set out to create the high-grade milkers had not a bulky beef standard drawn up for them. they produced their ideal first and standardized it afterwards. The same with the Merino sheep in Spain: the quantity and quality of fleece was assured—then came the fixing of the standard.

In another direction, but illustrating the same principle, is the wonderful sagacity in working sheep that the Scotch shepherd found and fostered in certain families of the Scotch collies, the most reliable of which have been line bred for many generations to retain that trait of what might be described as reasoning intelligence. Once having fixed this characteristic of working sheep the breed was standardized, but one wonders what amount of success would have been attained if the shepherds had only the bulldog to work on or the dingo.

Therefore, it readily can be seen that although possibly on the right track these poultrymen had an uphill task. The unknown ancestry was a big stumbling block. A person might start with almost any show breed at that time and unknown to him—and the danger unappreciated—the strain may have had, for example, a taint of White Malay blood in its composition. This was a favorite trick of the English breeder to get size into his stock, and the Malay being a very poor layer, the influence of its blood in the strain would possibly make his progress in developing good layers out of the stock he started with so slow that he would discard his birds, and either go out of poultry altogether or make a fresh start with stock from a breeder who had been more fortunate in his results.

So out of the few who did attempt scientific methods for breeding layers, fewer still got very far with it, and their number was reduced by the fact that some who had taken it up as a hobby had in the course of time to accept greater responsibilities in their chosen walks of life and therefore had not the time to devote to this exacting task. The Black Camel has knelt for some and others have defaulted for various reasons.

But the few have made good and new men are taking up the task. After all in a work of such magnitude one generation is not of any great moment. The Jersey, the Holstein, the Merino are the product of uninterrupted breeding of hundreds of years. We, today, see only the success, but the failures must have been innumerable with them, as they have been with poultrymen in their effort to fix high-record laying strains, but we can fairly claim that in Australasia we have succeeded to a greater extent, it would seem, than have other people. We have learned the right approach and discovered many pitfalls which we know enough now to avoid, also what avenues offer the greatest prospects of success.

Now Can Get Line-Bred, Pedigreed Fowls

Poultry breeders at present have their pedigree stock and can go on from now without that handicap of unknown ancestry telling against them. Where twenty years ago they were fortunate if they got better than 150 eggs per annum to work from, today they can easily get a base of 250 eggs laid by the females on both parent's sides for several generations back and the next decade will see an even more marked improvement, though it must be realized that the farther we progress the more difficult the advance is going to be, but the fixing of high-laying capacity with fewer deflections than there are at present, is assured.

It must be admitted that some of our most brilliant advances have been accidental, for more than once the mingling of bloods (or strains) has proved wonderfully successful—more often it proves disastrous—and as long as a breeder knew the matings of the parents responsible and could produce chicks from that mating his success was continued, but let it be not understood, or lost—then it was a case of the rocket over again—a wonderful spectacle, and oblivion!

One breeder, I recollect, in the earlier competitions was an instance of this. He had purchased a White Leghorn cockerel of a strain different to his own and mated him with some of his best tested hens. The pullets were ahead of the laying figures of either of the parent strains, and while he kept that pen intact and even when the daughters replaced the mothers he made good and had continued success. But after seven years in the breeding pen the old warrior died, and with him went the fame of that breeder. Then after vainly endeavoring to recover it by haphazard methods, out he went into another line of business.

Australasian laying records teem with instances of meteoric flashes. A new name comes to the front, stays a few years (maybe only one or two), then it ceases to exist, mainly for the reason that in the majority of these cases the underlying principles of breeding are not understood and they had not the knowledge necessary "to carry on," often falling victims to tempting offers for the very birds that were essential for their future operations.

All these short-lived successes, however, were not due to accidental matings. Many were the result of mated stock being purchased from some of the best of the methodical breeders, who, as often as not, received no credit and sometimes have even had to take second place in public competitions to birds produced from their own stock and sold ready-mated to produce layers. The newly acquired fame was quite safe with the purchaser as long as he kept that breeding pen intact, but as soon as he started on his own theory—goodnight!

It would be erroneous to deduct from these remarks that in Australasia we owe our success to good fortune and that there is a lack of stability in what we have accomplished. Even the results of the accidental matings which gave wonderful figures, though they failed in the hands of their original possessors, have not been lost; they have formed the basis of the operations of others, and while it cannot as yet be claimed that the tendency, even in the best strains, to revert back to lower levels of production has been overcome, it is unquestionable that both the individual productiveness and the general average have advanced to a degree at one time hardly thought possible; also there is no reason to suspect that the limit has been reached in either direction, and it is reasonable to maintain, with past results as a criterion, that our breeders will yet achieve still greater things.

BREEDING FOR QUALITY IN EGGS

This Article Gives Details in Regard to Experiments Conducted by Dr. Earl W. Benjamin, Cornell University, by Which He Successfully Demonstrated That Size, Shape and Color of Eggs Are Inherited Characters and Readily Controlled by Selection of Breeding Stock

By Miss Esther Cornwall, Cornell University, Ithaca, N. Y.

THE wide-awake poultryman of today finds many ways of improving his business methods. Ten years ago he had little knowledge of how a laying hen differs in appearance from a nonlayer. Today he can estimate within a dozen or so eggs just what a hen's production has been. Illumination has given him the power of controlling the length of the hen's working day and through it, the time of her production. Recent work at Cornell University has disclosed some of the results that may be obtained by the proper selection of breeding stock, and has further shown that by selection of eggs for hatching one may materially improve the eggs produced by one's flock. The results of this experimental work have been published by the Agricultural Experiment Station in Memoir 31, entitled "A Study of Selections for the Size, Shape and Color of Hens' Eggs." This is available upon application to the Department of Publications, College of Agriculture, Ithaca, N. Y.

Kind of Eggs Demanded by Markets

Dr. Earl W. Benjamin, author of this work, is one of this country's foremost authorities on eggs and egg marketing. In speaking of the importance of egg selection Dr. Benjamin says: "There is a certain type of egg which especially meets the needs of buyers in different markets. In order to get the highest price for his eggs the producer must meet this demand. It is usually not practicable to grade eggs closely, and so it becomes more economical to select and develop the flocks, thus reducing the proportion of eggs unsatisfactory to the customer to the minimum.

"The wholesale trade of New York City and its markets," continues Dr. Benjamin, "requires the size and shape of the eggs to be such that the eggs are not crowded, but fit snugly, in the fillers of the commercial thirty-dozen cases; this means an egg about 2⅜ inches long and 1¼ inches wide, and usually weighing from 2 to 2⅛ ounces when fresh. Shipping only eggs of proper size and shape insures less breakage, better appearance and a resulting higher sale value. The New York City market has a special demand for white-shell eggs and will sometimes pay from eighteen to twenty cents a dozen more for eggs having chalk-white shells than for those varying from cream-tinted to brown. It makes no difference if the producer be a farmer and seemingly receives the same price for big and little, brown and white eggs, still he receives a lower price for his good eggs in order to compensate the buyer for losses on inferior size and color. It is to the direct benefit of all concerned to keep the egg product of this country up to the highest possible standard."

Egg Characters Transmitted Through Either Sire or Dam

The fowls selected for the experimental work at Cornell were Single Comb White Leghorns, chosen because they are an important commercial breed, their eggs suit the New York City market, which consumes one-tenth of the eggs sold in this country, and they are further suited to the work since they are hardy, breed true to color and type and provide plenty of opportunity to select eggs of varying types. The records were started in the spring of 1911 and continued through 1919. The plan carried out was to select eggs of different types from this breed, incubate them and see how the type of egg affects the incubator record, the chick hatched and the eggs which that chick ultimately produces, if it be a pullet, with other studies of varying importance.

It was found that all egg characters studied were transmitted equally through either the male or the female parent. This means that in improving any of these characters in eggs, the breeder will obtain equal benefits through adding either better males or better females to his flock. The mating of two opposite characters, as a hen laying a white egg with a male known to transmit a brown egg-laying quality to his progeny, always caused the production of a medium character—as a creamy egg. In every instance the type of egg incubated and the average laid by the pullet hatched from it were similar for shape and size, but less stability can be placed upon a reproduction of color of eggs selected for hatching.

As might be expected, a selection of eggs through the entire year's production was found to give a more dependable idea of the shape, size and color egg laid by any hen than is selection only at incubation time, but the quickness and ease of the latter method make it worth while and by its use the commercial poultryman can quickly grade up the appearance of his eggs.

In looking over the experimental data it is easily seen that the size of the egg increased rapidly during the pullet year but there was little difference in size after this year. There was no difference between the shapes of eggs laid by pullets and those laid by them later as hens, but the eggs produced by hens are more likely to be darker colored than was their production during their pullet year. This darkening of shell color does not continue after the

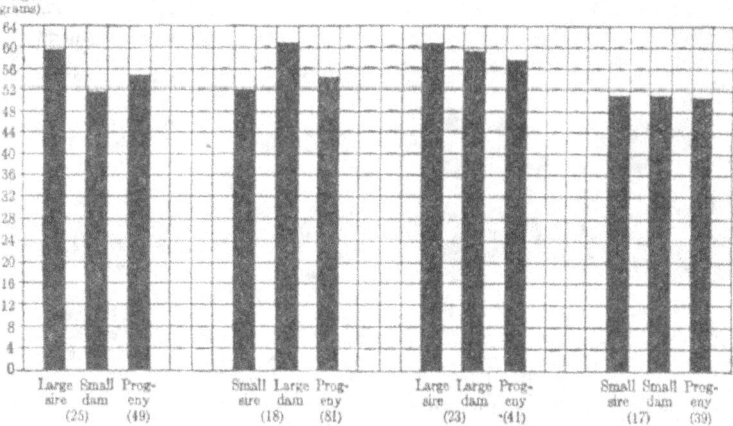

CHART SHOWING INFLUENCE OF SIRE AND DAM ON SIZE OF EGG LAID BY PROGENY

In experiments conducted by Dr. Benjamin, in determining the influence of sire and dam on size of eggs, it was clearly shown that females laying small eggs or males from small-egg females, invariably influence the size of the eggs laid by their progeny. The terms "large sire," "small dam," etc., do not refer to the size of the bird, but to the size of the egg laid by the bird or by its dam, or by the male's dam. Above chart indicates that the largest eggs were produced by birds that inherit large size from both sire and dam.

second year. There seemed to be a tendency with each year's production for the eggs to become whiter during the first five or six months of production, and then to become darker as the end of the laying season approached. But in spite of all such changes, there was still evident the fact that each hen tended to lay eggs of a certain characteristic shape, size and color. This was as evident in the production of the pullet year as in any later year and the fact that there was no more variability in this than in other years, does away, if this be true, with one of the old stock reasons for not using pullet eggs for hatching.

CHAPTER V

Special Articles on Selection and Management of High Producers, Etc.

How To Select Pullets at Beginning of Laying Period—Egg Shows and Scoring of Market Eggs—Possibilities in the Way of Increased Productiveness—How an Average of Ten Eggs per Hen per Week Was Secured—First Article Is by Dr. O. B. Kent of Cornell University, on Latest Discoveries in Science of Culling

THERE are two principal factors that determine the number of eggs that a hen will lay in a year. These factors are intensity and persistency of production. Persistency, because it is the most evident, has received the most attention—in fact, nearly all the attention until fairly recently.

Persistency or length of laying period is easy to determine because it is only necessary to be able to distinguish a hen that is laying from one that is not and to record the time that a hen begins laying in the fall and winter and the time she stops the following summer or fall. It is self-evident that the more months a hen lays the greater will be her chance of making a good record, or that the fewer months a hen lays the fewer eggs she will be likely to lay.

In order to obviate the necessity of a monthly examination of birds to tell those that are laying, various methods have been developed that help to determine how long a hen has been laying or resting. In the fall or winter it is possible to tell by the color of the shanks, beak and eye rings, how long a bird has been laying or, better yet, to tell very closely how many eggs have been laid. As shown in the book on "Profitable Culling and Selective Flock Breeding," published by the Reliable Poultry Journal Publishing Co., the color of the beak goes out from the base of the beak and gradually fades out at the tip, in yellow-skinned birds. It has recently been determined that not only does the depth of intensity of the yellow color affect the rate of fading, but the rate is also affected by the roughness of the scales and the lustre of the shanks. The beak of a bird that has a bright, lustrous, orange-colored, rough or bumpy shank will not fade as quickly as the beak of a bird that has dull, light-yellow, smooth, flat shanks.

A LOW PRODUCER OF SMALL EGGS
This hen's record is 76 eggs the first year, 78 the second year, 88 the third year. Two daughters averaged 52 eggs the first year that this bird was bred and a little over 100 eggs the next year that she was bred. Note the long, thin neck, small, dished-in face, shallow body, bowed wing and high tail.

The Molt a Reliable Indicator of Production

In the summer and fall there are a number of ways of telling how continuously a bird has laid or how late she has laid, such as the color of shanks and beak, the "Hogan System," the time of molting, vigor, condition and other factors. Of these, the condition of molt seems to be by far the most reliable indicator of how late the bird has laid or how continuously she has laid. While it is possible to give a figure that will be correct for the average bird in determining the amount of time lost from laying to molt each primary feather, it has been found that there is a good deal of individual difference and a considerable amount of difference due to the method of feeding. A high-producing hen kept under very good conditions may molt her feathers and renew them in one-third of the time that it would take a low-producing hen to do so under poor conditions.

In determining length of molt it has been found necessary to take into consideration the ability of the bird to grow feathers rapidly as well as the method of feeding. Knowing the method, by estimating the quality of the bird, it is possible, in September, October and November, to calculate within a few days the amount of vacation that a hen has had during the summer and fall. It has been found that certain good hens have the faculty of dropping two or three primary feathers as a unit instead of one at a time, consequently speeding up the rate of molt very considerably. This condition can usually be determined by the existence of two or more primary feathers of the same length growing in at the same time. In that case, the feathers of the same length should be counted as one feather in estimating the time lost for laying.

Value of Pigmentation Test

While pigmentation of the shanks does indicate summer production, it is not as reliable an indicator of production as it was at first considered. We occasionally find a hen that is capable of laying very heavily and at the same time retaining a certain amount of color in the shanks. We had a Barred Rock this year that laid over 243 eggs but retained considerable color in her shanks all the time. Another hen, a White Leghorn, that laid 250 eggs, regained a clear yellow color in her shanks in a short rest period, so that at the end of the year she was quite yellow in shank color.

Of course, this is not the usual condition found, but it raises a question as to whether it is not better for a hen to retain a certain amount of color—that is, will she not be healthier and be a better breeder? It has been my experience that the thin-skinned hen that lays late in the fall and winter, and becomes excessively bleached, is not liable to live long or be a good breeder the following spring. It is far better if the hen gets colored up before winter.

Intensity and How It Is Indicated

The second factor—intensity—which has really only received serious consideration lately, may be defined as the rate of a hen's production, or it can be measured by the number of eggs that a hen can lay in a month.

Of the two factors, intensity and persistency, the former is of far more fundamental importance and has a much broader general application than the latter. If there is a relation between body type and intensity of production, that relationship should affect the standard for judging any breed. Intensity of production depends to a less extent on management than does persistency. A good hen even under poor management will lay heavily for a short period, while it is only under the best management that a poor hen can be persuaded to lay for a long period. Since pigmentation, molt, etc., are merely factors determining condition, they should not be considered to any great extent in the formation of an exhibition standard. But since type probably can and does influence production, an exhibition standard should take as a model a bird that will give maximum production.

In studying the body characteristic that determines intensity, we should begin on the inside where the eggs are formed. Our measurements show that a good hen or male has a large heart, a large gizzard and a long intestine. Unless a bird has the proper digestive capacity it cannot produce an egg quickly. A hen that is capable of producing only one egg in two days cannot make a very high record. Perhaps the most striking example of the effect of size of organs on production that I have ever seen was the case of a hen that was in the Vineland contest, and was brought to the Cornell Judging School by Professor Lewis. The hen had laid about a dozen eggs up to the first of July, and was believed to be a nester. When cut open it was found that all the organs apparently were nor-

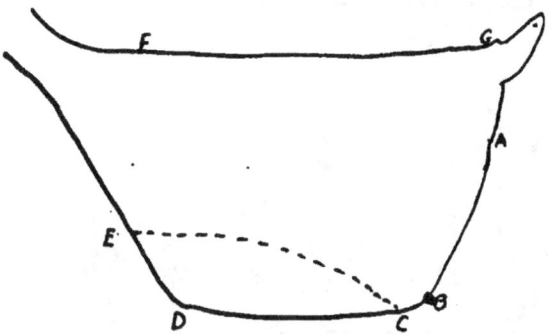

DIAGRAM SHOWING RELATION OF LENGTH AND POSITION OF KEEL TO SIZE OF ABDOMEN
In culling, a keel that drops from front to rear is regarded as desirable, but if the keel is short there may be less abdominal capacity than with a level keel, if it is of good length.

mal and there were no evidences of internal laying. On the contrary, the gizzard was so small (about as large as a small English walnut) that one wondered how she could lay at all. When this hen laid one egg a week she was probably producing her maximum.

A study of the relative length of bones shows that as far as our Leghorns are concerned the better birds are shorter and flatter boned. The poorer producing individuals tend to have long, round bones. This seems to be true of most of the bones of the body except the keel bone, and that is longer in a good bird than in a poor bird.

If a bird has a long keel bone it will tend to have a small abdomen, and yet a long keel bone goes with a good producer.

The size of the abdomen is affected by the shape as well as the length of the keel bone. A short keel bone that bends down will form a large abdomen without giving a large amount of room inside. Compare the difference in size of abdomen as shown in the diagram presented herewith. It is at once evident that the distance A—C is much greater than A—B, and yet the outline C—E—F—G does not give nearly as much room inside as B—D—F—G. Some strains that have been selected according to the "Hogan System" have been found to have keel bones shaped like E—C, and were only moderate layers. Incidently their pelvic bones were exceedingly thin.

Shape of Back Highly Important

If a large abdomen is not desired the question naturally arises as to what shape does indicate a good producer. In the first place a good producer will tend to be flat or slab-sided and deep-bodied. A good-laying hen is shaped like a good dairy cow and is not round and long, like a good meat animal. Above all, a good-producing bird should not be shaped like a good hog.

The keel should be full and deep and drop down slightly in going from front to rear with a slight upward curve at the rear end.

The shape of the back seems to be one of the most important indicators of the intensity of production. There are a great many different kinds of backs. Some are wide and flat, others are narrow and round, some are level and others are humped up. Some are cut in and others hold out well, some are bony and others are smooth. It has not been possible to determine all the desirable and undesirable forms of backs, but it has been found that the bony, humped-up back that cuts in is quite undesirable and that a good bird has a back that is level, fairly wide and that holds up well at the tail.

If you will spread your hand across the back and gradually run your hand from the middle of the back to the tail with your thumb and fingers pressing on the sides of the bird, you will at once get an impression of the shape of the bird. It is especially desirable that the whole pelvic girdle should hold out well or be well spread, clear to the rear of the fowl. A bird that cuts in decidedly is rarely a good producer. In making this test, it is well to bear in mind that the body and back will hold up better when the bird is in good flesh or is laying than when thin or not laying. We have found this to be the best single test of any that we have ever used in determining a male's breeding ability.

The fineness of head and skin are also qualities that indicate a bird's intensity. The clean-faced, prominent-eyed, thin, loose, velvety-skinned birds are better than the coarse-skinned, sunken-eyed birds.

No Uniformity of Shape in Heavy Layers

Just what influence the type and shape of feathering has on production has not been determined as yet, though there apparently exists a marked influence. The long, loose, fluffy-feathered birds whose wings cut into the sides so as to make a big roll of feathers above and below the wings are rarely good producers. Long feathers, narrow feathers and fluffy feathers seem to go with late maturity and moderate production.

A good bird will appear flat sided and deep. The breast will be full, prominent and angular. The abdomen will be slightly larger than the breast, but will balance well. The neck will be full and the back will be wide and hold up well at the tail.

A poor bird will tend to be long legged and long necked, shallow bodied and shallow breasted. The abdomen will frequently be much larger than the breast when the bird is laying or will look sagging and will not be balanced by the full breast. The back will be round and cut in at the tail and the wings and sides will be round and loose feathered.

Does this description mean that all good-producing birds will have the same shape? Not by any means. The shape of the birds may be just as varied as the shape of the eggs they lay and still be equal in egg production. The shape of a bird, as we see it, is influenced to a marked extent by the length and shape of the feathers. The length and shape of the feathers undoubtedly influence the rate of maturity, and production is influenced by the time the bird begins laying. From the commercial egg producing standpoint a poultryman may not be able to afford to keep a bird that costs a great deal to raise because of its slow ma-

IT IS DIFFICULT TO ESTIMATE A BIRD'S QUALITY FROM A PHOTO
This hen's first-year record was 253 eggs; second year, 220 eggs. The above photographic reproductions of the same bird show a flat-sided, flat-winged broad-backed bird of moderate depth. The first impression is that the bird is rather long and shallow, but the two photographs show a distinct difference in body type and bring out the great difficulty of attempting to estimate a bird's quality by a photograph alone.

turity, due to type of feathering, although such a bird may lay well after it is matured. Apparently, at the present time the loose-feathered, long-feathered strains of hens are not making especially creditable records in the egg-laying contests. Most all, if not all the birds making good records are fairly short in feather.

But aside from length of feather there is a great deal of difference in the shape of the bodies of high-producing hens, though they nearly all meet certain fundamental requirements. Some are large and some are small, some are relatively long and others are relatively short, and yet all may be good birds. The only explanation that has occurred to me is that the shape of the bird is indicative of the shape of the egg she lays. A hen laying a long egg may lay just as many eggs as the hen laying a round egg and their bodies be shaped relative to the shape of their eggs. It is interesting to note that this condition is true of breeds and strains to a large extent, though to what extent it is true of individuals has not been determined. Wyandottes certainly lay eggs shaped like Wyandottes. A Red lays an egg as much longer than a Wyandotte as a Red is longer than a Wyandotte.

HEAVY LAYERS SHOW GOOD CAPACITY EVEN THOUGH THEY MAY DIFFER TO A MARKED DEGREE IN TYPE OR GENERAL APPEARANCE

All the birds shown in above illustration were entered in the Storrs Laying Contest and their trap-nest records are as follows: Barred Rock in upper left-hand corner, 235 eggs, and the one in the middle, 234. Bird at the right laid 63 eggs in 63 days. The Single Comb Leghorns, reading from left to right, have records of 286, 260 and 255. The record of the White Wyandotte in the lower left-hand corner is 281, the Rhode Island Red in the middle, 258 and the Buff Wyandotte on the right, 246. Photos from Storrs Agricultural Experiment Station.

SELECTION OF PULLETS FOR HIGH EGG PRODUCTION

Early Laying Maturity as a Basis for Selection or Culling—Late Starters Found To Be Early Quitters, While Those in a Flock That Lay First Prove To Be the Best Layers—Early Laying Is a Matter of Breeding and Feeding—How To Get Rid of "Slackers" at the Beginning of the Laying Season Instead of Feeding Them at a Loss for Months

By James Dryden, Chief in Poultry Husbandry, Oregon Agricultural College, Corvallis

THE objective in poultry breeding has been an increased production of eggs as measured by the annual record of the individual hen or the average of the flock. In breeding from the best producers year after year—which, in the main, was the method followed in attaining the objective—some important results have been secured indicating that higher egg production is possible of attainment by methods of selection that do not call for the use of trap nests.

It should be understood, however, that it has been possible to make these discoveries only by the study of trap-nest records. Without the trap nest which gave us the records of individual hens they would have remained undiscovered or, at any rate, if they had been thought of at all, would have remained a contention and a controversy to bewilder and confuse present-day poultrymen until they had been followed by another generation suffering from a similar babel of tongues. Whether the trap nest is responsible for the discoveries or not, it would have been impossible without its use to prove or disprove the many theories that have been advanced for picking out the good hen.

Among the earlier results of a study of trap-nest records of good and poor layers at the Oregon Station, it was shown that the pullets which came to laying maturity early were good layers. A close relation was shown to exist between the annual record of the hen and the age at which she began to lay. That is to say, in a flock of pullets of the same age, hatched at the same time and having the same care and feeding throughout the summer, those that laid first were the best layers and those that laid last were the poorest. Pullets that laid at six months of age or less were good layers; those that did not lay till eight or nine months were poor layers, on the average.

The age at which pullets begin to lay, however, depends to some extent upon the feeding and care they get throughout the growing period. Laying maturity is hastened by good feeding and retarded by poor feeding. So that the age at which they begin to lay is not in itself a criterion as to laying ability. Early laying is a proof that the pullet is of good breeding and has had good feeding. Good feeding alone will not make the pullet lay early. A pullet that lays at six months of age or less has been well fed and well bred. A pullet that lays at seven months of age or more is not necessarily a poor layer, because her maturity may have been retarded by poor feeding. A well-bred pullet—that is, one of good-laying capacity—may not begin to lay till seven, eight or nine months unless she has had the feeding to make proper development.

Late Beginners Were Early Quitters

Some of the results of the Oregon experiments are shown in the accompanying charts. In Chart 1 the pullets are separated into three groups according to their first year's trap-nest records. These records are dated from the first egg laid by each pullet. In the flock there were 15 that laid less than 100 eggs in the year; their actual average was 59.4 eggs. In the first three months—November, December and January—they averaged only 4.5 eggs each, and in the balance of the year 54.9. The poorest layers in the first three months were the poorest for the rest of the year; in other words, the late beginners were the early quitters.

In the second group, which averaged 160.7 eggs, there were 162 pullets. They averaged 19.2 eggs in the first three months and 141.5 in the balance of the year.

In the third group there were 96 hens; they averaged 224.4 in the year, all laying more than 200 eggs. Their first three months' production was 37.8 eggs a hen and 186.6 the balance of the year.

The same lesson is brought out when the hens are grouped in another way. In Chart 2 they are grouped according to the age (in days) at which they began laying. Those that began laying at less than 200 days of age averaged 197 eggs in the year. They laid 33.08 eggs in the first three months of November, December and January, and 163.96 in the rest of the year. Those that laid between 201 and 250 days of age averaged in the first three months 28.35 eggs and 159.37 the rest of the year. Those that laid at 251 to 300 days of age, averaged 158.3 in the year and laid 19.12 in the first three months. Those that did not lay till they were passed 300 days of age averaged only 112.7 in the year and 9.3 in the first three months.

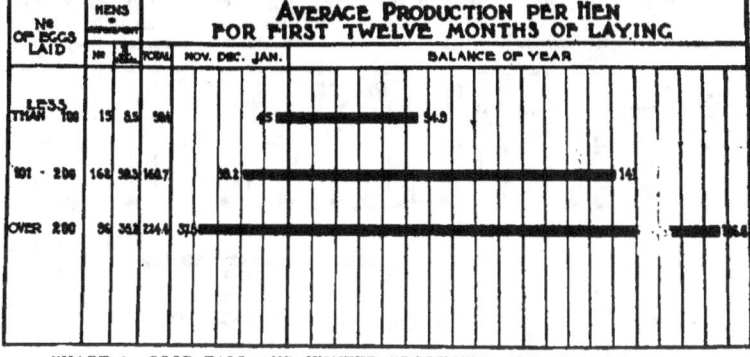

CHART 1—GOOD FALL AND WINTER PRODUCERS ARE THE BEST LAYERS

This chart records the results of some experiments conducted at the Oregon Experiment Station at Corvallis. The pullets were separated into three groups according to their first year's trap-nest records, which are dated from the first egg laid by each pullet. There were fifteen that laid less than a hundred in a year, their actual average being 59.4 eggs. Note that the birds that were the poorest producers in the first three months of their laying made the poorest record during the rest of the year. In other words, as Professor Dryden says, "the late producers were the early quitters."

By the simple process, therefore, of tabulating trap-nest records we make the discovery that the good layers begin to lay early in life. Rather it would be better to say the earliest layers of the flock because, as already explained, a hen may make a good record though she did not begin to lay early, due to methods of feeding that retarded development.

Discoveries of Great Importance

I believe these discoveries, if such they may be called, are of great importance. Our breeding work has shown that high egg production is inherited—that is, by breeding from good layers we get good layers on the average, and these charts show graphically one way in which high egg production is inherited, and indicate what is of practical importance to every breeder and commercial poultryman—two methods of selection or culling without the use of trap nests.

First, they show how culling may be done with the pullets at the beginning of the year or before they begin to lay and, second, how it may be done at the end of the laying period or year. At the first of the year the slackers may be picked out by evidences of their slow development. If they have not come into laying condition for two or three months after the other pullets in the flock have started to lay, they should be culled out. It is not necessary to use trap nests to tell which are laying and which are not. The development of the comb and wattles, especially in the Leghorn and other large-comb breeds, gives a pretty sure indication of laying or nonlaying. A large red comb and enlarged red wattles usually show that the pullet is laying or about to lay. In practice this usually will be all that is necessary to observe in picking out the best. To make sure, however, an examination of the abdomen and pelvic bones should be made. In the nonlayer the abdomen is hard and the points of the pelvic bone and keel bones are drawn together. In the layer the abdomen is soft and pliable and the pelvic bones are well spread and the distance between the pelvic bones and the point of the keel bone is much greater than in the nonlayer. An examination of a few hens, both laying and nonlaying, will readily show the difference.

To cull at the end of the year the same test for laying is made. In the nonlayer the comb and wattles have become shriveled and the abdomen becomes hard and the pelvic bones and keel bone draw together. In addition to this the early molter indicates the nonlayer and the early quitter. Then the hen, after she has discontinued laying for a few weeks, gets the yellow color back in her shanks and beak. These points help to pick out the early quitter.

Is Certainly the More Valuable

Between these two methods of culling, the first is certainly the more valuable, granting that both methods are equally efficient. If we can cull the slacker at the beginning of the year as well as at the end of the year, a great deal is saved in feed and in housing room for the slacker. Why feed the slacker all winter and half the summer if she can be spotted at the beginning of the year? The two methods of culling are practically the same and based on the theory that if the hen is not laying at a certain period

nest doesn't do all the culling on the farm. The pullets are culled as they are put into the laying houses in the fall, and as a result they average about 200 eggs in the year. The breeders are selected by trap nest, but there are always some poor producers among the best bred flocks, and these are culled at the beginning of the year as pullets.

If the conditions are right for pullet culling at the beginning of the year, I believe it can be done as accurately, if not more so, than culling at the end of the year. One of the factors that makes for inaccuracy in the summer and fall culling is that the conditions as to feeding and general care may have been such as to interfere with normal production. Irregularities in feeding, for example, will, we know, result in decreased production and at culling time many hens of good-laying ability have stopped laying and have yellow shanks and beak. Those will be culled out as slackers, though under proper conditions they would have made good egg records and would make good breeders. The fact that a hen may have yellow shanks and beak and shriveled comb and dry, puckered vent is not proof positive that she is a slacker; it may be that the poultryman himself was the slacker.

It is only when we know, if it can certainly be known, that the conditions have been right as to feeding and general care that culling can be done successfully in the summer and fall. To give a case in point, I know of a flock of pullets—pullets that were of a strain of heavy producers—that had laid heavily up till the middle of May. All at once the production fell off and in a few days it was down to forty per cent. It was then discovered that their water supply had been cut off and they had had no water for three days. This naturally stopped the laying of many of the pullets and if culling had been done in the next week or two a large percentage of the flock would have been discarded because they were not laying and had all the evidences of nonlaying. Even the yellow color of the shanks and beak was coming back. In spite of this interruption this flock averaged in the year more than 200 eggs. So that on account of irregularities in feeding and care the culling that is done in July and August, or at the end of the laying season, must necessarily show some inaccuracies. Some hens will be killed off that are really good layers.

Culling at Beginning of Year

To cull the pullets at the beginning of the year there are also certain conditions preceding the culling that must be observed.

First, is the matter of breed. The larger breeds naturally take longer to come to laying maturity, so that Leghorns and Plymouth Rocks, for example, could not be culled at the same time.

Second, the pullets to be culled should be of the same age, or practically so. If some of them were hatched in March and some in May,

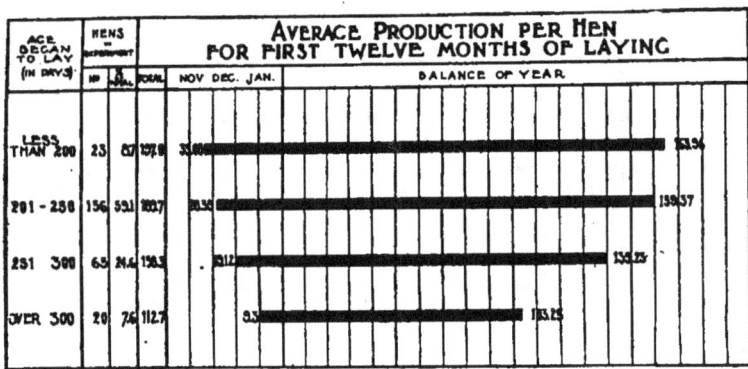

CHART 2—THE FIRST PULLETS TO MATURE ARE USUALLY THE BEST LAYERS

In this chart, prepared by Professor Dryden, the layers are grouped according to the age (in days) at which they began laying. Note that those that began laying first (under 200 days) laid the most eggs in the first three months and made the best records for the year. By tabulating trap-nest records Professor Dryden has found that the best layers in a flock are the ones that are youngest when they begin to lay.

she is a slacker, and successful culling is largely a matter of being able to pick out the hens that are not laying. If we can tell which are not laying at the beginning of the year as well as at the end of the season we should do it at the beginning of the year because we then get all the profit possible from the culling. The late starter is the early quitter, as shown by our experiments, and if we can get the late starter there should be no early quitters.

The summer and fall culling campaigns have been of immense value to the poultry industry, but I believe we should set ourselves to work with as much enthusiasm to perfect methods of pullet culling—killing off the late starters—as has been shown in the matter of culling the early quitters.

On Farm of 10,000 Layers

Culling the pullets has been successfully tried by one of the largest and best commercial poultry farms on the Pacific Coast, the Hollywood Farm at Seattle. This is a farm of 10,000 hens and one that has been highly successful. It has probably the largest trap-nested flock in the world, 2,700 being trap nested this year. But the trap

and their ages are not known at time of culling, the later hatched ones will be culled out though they may be good, because they will show the nonlaying conditions when the early-hatched pullets are laying.

Then there is a third point to be taken into account— that is, the feeding and care that are given the growing pullets. To make the culling accurate the pullets should have had the same feeding and care throughout the growing period. If some are forced ahead and some retarded and they are all culled at the same time without knowing how each was fed, the retarded pullets necessarily will go out as culls, though later in the year they will show the same percentage of good pullets as the forced pullets. It is possible of course to mark them so the factors of age and of feeding and care may be taken into account in the culling.

It will be a much simpler matter, however, if the pullets have all been hatched at about the same time, say within two weeks, and if they all have the same feed and care. The method of culling them will be to go over the flock when a number of pullets are showing laying condition and put them in the laying houses that have been provided for them. Then about a week later go over them again and pick out all that then show laying condition, and continue until all the promising pullets have been selected and those that are left have been sent to market.

Whether the culling be done once a week or once in two weeks or once a month or done all at one going over, is a matter that can be left to the judgment or convenience of the poultryman himself. Some of the pullets may start to lay in September, but unless he wants to grade them he may let the culling go till the first of November when it can all be done at one time. If he wishes to pick out the truly best he will do the picking in September, or when the early starters first begin to lay and they can be put in a separate house or marked.

The time of culling, or the age at which the first culling can be made, cannot be stated definitely because laying maturity varies in time as conditions of care and feed vary. With fairly good care but without forcing, the first "picking" will be at about the age of six months. If they have been forced the picking will begin two or three weeks earlier. If they have been retarded there may not be much to pick before seven months.

Which Are the Culls?

The pullets of the first selection will have the best laying capacity. Then the question is: at what age do we come to the culls, or how long will it be after the first culling before all the good layers are selected and only culls remain? There is room for further investigation here, but it would be safe to say that those that have not reached laying maturity in two and a half months after the first picking will be culls, and two months should be long enough. If one wishes to make a rigorous culling, and this should be done of course with the present high prices of feed and labor, the selections should stop earlier.

As to the number of culls to expect by this culling, that again cannot be stated definitely. Another factor enters here: that of the breeding behind the flock. If the pullets are from a strain that has several generations of heavy-laying ancestors, naturally there will be fewer culls. It is quite possible that when we have gotten our systems of breeding and feeding down to a fine point it will not pay to cull at all, for then the lowest producers among them may be profitable to keep. At any rate, good breeding lessens the necessity for culling.

In the flock shown in Chart 2 there were only 7.6 per cent that did not lay after they were 300 days old. There were 24.6 per cent that began laying between 251 and 300 days, but those averaged 158.3 eggs. That I would call a fair average for an average flock, but between 250 and 300 days there is too wide a margin. If these had been culled every week, or at shorter intervals than fifty days, some of them would have been culled out. In the first selection at 200 days there were 8.7 per cent laying. The largest percentage of the flock were laying between 200 and 251 days. More than half of the flock, it is seen from the chart, came into laying between those two ages, or 59.1 per cent. Roughly, about 75 per cent of the flock of good-laying pullets should come into laying before eight months of age under good but not forcing methods of growth.

There is need, as already indicated, of further investigation or further study of trap-nest records of good and poor layers along the line of selection of pullets or culling at the beginning of the year, but if the suggestions given above are followed carefully and conservatively the poultryman will be able to cull his flock at the time of beginning to lay as accurately as he can at the end of their laying period, with the certain advantage that he has disposed of slacker hens before they have eaten up a lot of good high-priced food.

Take another look at the charts and note how the black lines of the good layers are longer at both ends than the lines of the poor layers. As good layers inherit their laying qualities, it follows that the effect of breeding for eggs is shown both at the beginning as well as at the end of the laying period or year—the laying year being lengthened at both ends—and if we take proper note of the beginning of laying we can tell just as accurately what class of layer the pullet is at the beginning as at the end of the line.

WHAT IS POSSIBLE IN HIGH EGG PRODUCTION?

No Direct Correlation Between Capacity for Egg Production As Indicated by Visible Supply of Ovules and Actual Egg Production—Breeding, Feeding and Care of Test Birds—Attention Is Prime Factor in High Egg Production

By John H. Robinson, Associate Editor of Reliable Poultry Journal

OUR title is a question, and we discuss this subject with the object of arriving at a better understanding of its possibilities, and a better knowledge of the ways and means of pushing records ever higher, but with no thought of finding or fixing a definite limit beyond which high production cannot go.

In discussing the question here it is appropriate that I should first state briefly the viewpoint from which I see the subject. I am aware that to some of those who take a keen interest in the matter, my attitude toward some claims of high production has often appeared to indicate a considerable measure of skepticism in regard to the authenticity of the records. I am further aware that because I did not always accept claims at their face value, and said so in cases where there was occasion to do so, those who accept all claims have often concluded that I was denying the possibility of production reaching the figures in the case. In this and other matters where my attitude is misinterpreted, I have never thought it necessary to say just what that attitude was unless there was particular reason for doing so. In this case the particular reason is the fact that, while there have been many high records published that did not seem to me entitled to the serious consideration of any one who knew poultry culture and knew the circumstances of the making of those records, when it comes to the question of the utmost capacity of hens to produce eggs, I place the possible capacity of a hen far beyond any figures that I have ever seen suggested in print as attainable.

Relation of Ovules to Egg Production

In my mind the most important piece of work in poultry investigations that has been done to date was the counting of the ovules and minute ovules, or oocytes, in the ovaries of hens at the Maine Experiment Station. Before this was done it was the common belief among those well informed on poultry matters that the number of eggs a hen could produce in life was absolutely limited by an original number of ovules which ordinarily was not more than 600 to 700. It was supposed that in some hens the original number was larger, and that it could be and was gradually increased by selective breeding of the best layers. On this theory it was supposed to be good poultry husbandry in the management of hens for egg production only, to force the hens for rapid production, and "get the eggs out of them as quickly as possible."

This theory was developed from analysis of the egg production of hens kept through a long period of years, four to eight or more, which showed production, as a rule, gradually reducing year after year until the hen died or produced only a few eggs in the spring. Dr. Raymond Pearl and his associates at the Maine Station killed a number of hens of known age and laying records, and counted the ovules that could be distinguished by the naked eye and with the aid of a common reading glass. Fourteen hens were examined—mostly Barred Rocks, a few White Leghorns, and one Cornish, and one Cornish-Rock cross. The lowest number found was 914, the highest 3,605.

As far as the figures are given there does not appear to be any direct correlation between capacity for egg production as indicated by the visible supply of ovules and actual egg production, although it happens that the two hens with the highest records had the largest number of ovules. These were both White Leghorns, killed at about a year

Reprinted from Reliable Poultry Journal, July, 1921.

and a half old. One had then laid 198 eggs, the other 197. The latter had 3,605 visible ovules, the former 2,452. Another White Leghorn hen, killed at the same age, had laid during her lifetime only 2 eggs, yet had 2,143 visible ovules.

The two high producers mentioned, and the Cornish-Rock cross (which when killed and examined at a little under a year old had 124 eggs to her credit, and was found to have 2,000 visible ovules), were the only ones among the fourteen described in the list that could be called good layers—on their records of performance. The best record among the Plymouth Rocks examined was 34 eggs at 11 months and 1 week of age. This hen had 2,306 visible ovules. Hens that at ten months of age had laid only 10 to 15 eggs showed from 1,200 to 2,100 visible ovules.

The figures show that the possibilities of egg production latent in the undeveloped ovules of poor layers are far beyond the highest long-life records that have ever been published. They show that as far as producing the "seed" from which their eggs are developed is concerned hens have apparent capacity for egg production far beyond any recorded or claimed production. They completely demolish the theory that nature or heredity impose limits on production within any range that poultrymen have been accustomed to consider.

It should be noted further that the figures given are in every case incomplete. The ovarian tissue contains in addition to those ovules counted under moderate magnifying glasses, unknown numbers visible only with the aid of high-power microscopes. The ovary appears to be an organ for the production of seed indefinitely as well as for the development of the yolks of eggs from this seed, and the production of the germs in which the offspring originate. It may then be regarded as established that the ovary of the hen has capacity to produce ovules in much larger numbers than the hen, considered apart from the function of originating ovules, can grow and mature finished eggs. Hence the practical question in high egg production is to develop all the other functions of the hen with the object of securing the highest possible efficiency in egg making.

To Those Who Are Striving for High Egg Production

In considering what is possible in egg production we have to distinguish between actual possibility and what may be termed ordinary possibilities which are more or less governed by circumstances and restricted by the practical necessity for making a profit. It is a fact which those who are much more interested in egg production than in breeding to all the points mentioned in the Standard do not like to admit, that working for extreme high egg production is a fad. Here again we must distinguish between working for the utmost in egg production, and working for the best results to be obtained under any particular circumstances.

Our commercially successful breeders of good-laying stock have generally devoted themselves to the development of strains of more than average laying capacity under any ordinary good conditions and management. Most of them began with a small flock—perhaps only a few

HIGH EGG PRODUCTION CAN BE OBTAINED IN FOWLS OF PRACTICALLY ANY BREED
The high-producing birds here illustrated prove that no one breed has a monopoly on high-laying ability. Eight breeds and varieties are represented above, all members of contest pens at the National (Mo.) Laying Contest, and equally good records have been made by fowls of many other breeds and varieties. Reading from left to right, starting at the upper left-hand corner, the record of the White Leghorn is 286; White Wyandotte, 270; White Plymouth Rock, 302; Buff Leghorn, 247. In the lower row, the Silver Wyandotte has 230 eggs to her credit, the Barred Plymouth Rock, 249; the Ancona, 263; Rhode Island Red, 296.

birds—to which they gave very close attention. In general, the breeders who make reputations for strains of high producers get a larger proportion of high individual records in the early years of their work than when the stock has grown to proportions that make it impossible to give the attention to individual birds that they did in the beginning. Then it will usually be found that the highest records with a strain are made not by the originator who is now working with it on a large scale, but by some of his customers who have small flocks to which they give very close attention—people who can give the small flock or the individual hen any and every attention that will increase egg production, regardless of the cost of extra labor or extra feed.

With those who do this, working for extreme high production is a fad—a very creditable one, capable of being used greatly to the benefit of poultry interests, but still a fad. And the seeker after the highest possible in egg production should recognize this, and for the present, at least, should not concern himself with questions of profit and loss on high egg production. He is in the position of the inventor, or manufacturer, who makes his models regardless of ordinary economies, but when the model is completed to his satisfaction takes up matters relating to its manufacture and distribution on a commercial basis.

Of late years those interested in high egg production have worked mostly with the idea that degrees of fecundity are regularly inherited, and that breed, strain, family or line is the prime thing in securing high egg production. This is all right and good with qualifications. Type has its relations to egg production. Some breeds are easier to get high yields from than others. Some strains stand up under heavy egg production much better than others. Some families or lines are more reliable producers of heavy layers than others. Those who work for the utmost in egg production must take account of all these things. But to date the work has been mostly along the line of seeing what hens "bred for eggs" could and would do under ordinary conditions, and the result is that while we have many more high record hens than ever before, all our highest records still linger a little above the 300 mark, where the high mark has been from the earliest years of modern poultry culture—and I personally doubt not for centuries—if anyone took the trouble to keep records.

We Can Beat an Egg a Day

Now for any ordinary circumstance 300 eggs from a hen in a year is mighty good production. As far as I know the circumstances under which high records have been made, not even those most favorable to high production have brought into the case every possible factor in securing the highest egg yield of which a hen is capable. Leaving out of consideration now the relation of high egg production to breeding power and looking only to ways and means of keeping a hen laying as long as possible and as heavily as possible, let us inquire what the effect would be of applying to a carefully selected small flock of hens of a line known to be uncommonly good layers and long-lived layers, all the arts for increasing egg production known to poultry keepers. How far beyond figures reported to this time can we go?

The records are in the vicinity of 310-314 for well-authenticated recent performances, and up to the vicinity of 330 for less well-authenticated and more remote instances, some of which may be accepted as credible. My opinion is that it would be possible to take hens of some of the best laying stocks today and get up to 400 eggs or over in a year, and I can conceive of the eventual development of strains or lines that would produce some hens that with the necessary care and attention would go well up toward 500 eggs in a year. In saying this I am also frank to say that I think it would take years of building up the vitality and productivity of fowls to make these high producers of much value as breeders after they had made such records.

The present best authenticated high records are approximately at six eggs a week for the hen throughout the year, or an egg a day except while molting and an occasional egg during the molt. It is commonly regarded as remarkable for a hen to lay two eggs in a day, yet everyone who has closely observed heavy layers, or trap-nested many, knows that it is a frequent phenomenon. I have known heavy-laying hens well supplied with shell to lay many soft-shelled eggs apparently because they were producing eggs ready for the shell faster than they could cover them with shell. I recall a remarkable instance where a Black Langshan pullet I owned dropped one egg with a perfect shell, one with a very thin shell, one with a soft shell and two full-sized yolks within a few minutes. She had laid a perfect egg the preceding day. These five eggs were dropped early in the morning. Toward evening of the next day she again laid a perfect egg. Many hens that frequently lay double eggs appear to do so because the yolks are coming so fast that the oviduct takes them in pairs.

Breeding, Feeding and Care of Record Breakers

We do not know as much as we might about matters relating to the production of the egg in the hen, for the process is concealed and our observations are necessarily superficial. But it is reasonable to suppose that if the poultry keeper seeking to find the utmost in high egg production would make the same study of individuals, and take the same care of individuals, that the dairyman working for great milk and butter records does, much higher egg production might be reached than has ever yet been claimed. The practical difficulty is the low individual value of the hen and her product at market prices, and the fact that generally the eggs from her are valuable for other than food purposes for only a short season in the year, and that is the season at which even ordinary and poor layers lay well.

There is a further obstacle in the difficulty which most poultry keepers have in concentrating attention on a hen, or a few hens, or a flock, as closely as is necessary to insure that every possible help to high egg production is being utilized to full advantage. Those who work for utmost records in high egg production must make themselves "slaves of the hens" for the period, allowing nothing to interfere with their doing everything that will in any measure contribute to the end they seek. Most of the high records that have been made in the past are chance records in that they were not made by hens picked for the purpose, but by hens which it was discovered were doing remarkably well under ordinary circumstances, and which from that time were given closer attention than usual, yet rarely or never all the attention that would contribute to greater egg production.

To make records far surpassing those we now have there must be first, selection of a sufficient number of hens of the best promise to provide against accidents and losses during the test. These hens must not only be great layers, but they must have certain unusual combinations of characteristics. They must be hens of good size for their breed, but layers of only average-sized eggs. If of good size, hens of the Mediterranean class will generally have the advantage of others in making extreme high records, because they stand the hot weather better. It is quite the common thing for hens of this class to fail in egg production in ordinary houses in extreme cold weather, but it is possible to make houses warm enough at any low temperature, while there is no way of keeping heavy-feathered fowls, and those with a tendency to put on fat, in tune for high egg production in extreme hot weather. With circumstances favoring, hens of the medium to heavy breeds might make higher records than the light breeds, but all

OUTDOOR COOPS FOR BREAKING UP BROODY HENS
With fowls of the larger breeds broodiness may seriously affect average production and special attention should be given to preventing this unnecessary loss of time. Hens that are removed from the nest as soon as they show signs of broodiness, and are confined in comfortable coops like the above, with plenty of feed and water, will soon be ready to return to profitable production again.

possible contingencies can be met more certainly with the type best suited to work through extreme hot weather.

Conditions that Must Be Met

Space will not permit any extended or detailed statement of the things that must be done in the management of layers that are expected to eclipse all previous performances, but I will briefly outline the most important. They must have warm houses, with the ventilation as carefully regulated both winter and summer as in a well-ventilated living room. They must have artificial light not only in the short days of winter, but on dull days at all seasons. They must be "hand fed" and at the same time required to take enough exercise to keep them in good condition—and not more. Their rations must be carefully adjusted to all changes of temperature, and also to maintain their normal weight while producing heavily. They must have regularly much greater variety of feed than laying hens usually get— all the fresh meat and green cut bone they will eat when supplied daily, all the milk they want at all seasons, an abundance of succulent green feed (limited only in hot weather when the tendency is to eat more of it than is consistent with good egg production) and all the accessories of shell, water, etc. And it must be borne in mind that to get hens to do their utmost in egg production one cannot simply put these things where the hens can get them and leave them to themselves. Personal interest and attention count for as much with the hen as with the cow. Nearly all hens respond to individual attention from persons in whom they have confidence. From an economic point of view it is wise to reduce the work with hens to the minimum, but for the utmost in egg production it is necessary to fuss with the hens to any extent that they appear to like.

I do not know what financial rewards await the poultry keeper who some day will succeed in getting more eggs from a hen than there are days in a year. He will certainly be famous, and his accomplishment will undoubtedly have remarkable influence as a practical demonstration of the fact that attention is after all the prime factor in high egg production. All the other factors ultimately can be analyzed into that.

In conclusion, I would urge upon everyone who undertakes to make high egg records, the importance of absolute accuracy in the records, and also of making provision for a check against errors and for some sort of trustworthy corroboration of his figures. The simplest method is to mark all eggs with the date as laid, and then have a disinterested referee inspect the eggs and check the record at stated intervals. A hen's eggs usually have enough individuality to make it impossible to pass any substitutes off on a keen inspector. As between two nearly like eggs mistakes in identity are possible, but with several eggs known to be from the same hen it is not often possible to pass an egg of another hen as laid by her.

TEN EGGS PER WEEK PER HEN

Three Experiments Were Made in Three Consecutive Years, Extending from December 1 to April 1—Description of the House, the Care and the Feed—No Pullets Were Used—Daylight Was Artificially Prolonged—Natural Climatic Conditions Secured by Hot-Water Heating System

By E. C. Waldorf, M. D., Buffalo, N. Y.

ACTING on your request to give the readers of your Journal the method and results of my forced egg production, I herewith present the plan from the data still preserved, together with facts as I recall them. It might be well to state at the outset that the primary reason for attempting forced ovulation was not to obtain more eggs from a given number of hens in a specified time, but to secure eggs of the highest hatchable quality.

This suggestion developed from the fact that the eggs from hens laying five to seven eggs per week hatched much better than did the eggs from the same hens when laying fewer eggs per week. Close observations along these lines were made for two seasons. It was also observed in connection with these experiments that only the dense eggs hatched in high percentage and that the density of the eggs was not lessened while or during increased egg production. With these facts established, I began my first attempt in forced egg production in December, 1889.

These experiments were made on the premises of Patrick Kinney, 56 York St., Buffalo, N. Y. A henhouse for the purpose was erected, measuring 12 by 16 feet, inclining to the southwest and northeast. The height was 16 feet on the south side and 21 on the north. This gave a sloping roof to the south, which was glass, after the order of a greenhouse roof. In the center of the glass roof was placed a ventilator stack one foot square and three feet high. The top of this shaft was closed, having six one-inch holes on each side near the top. Aside from the door entrance there were no other means of ventilation.

The entrance door was placed in the middle of the south end, leading into a hallway three feet wide, extending to the north end and as high as the skylight. This hallway was made of matched lumber, provided with four matched doors opening into each individual pen.

The house rested upon the ground, which was a sandy loam covered with fine gravel three inches deep. A trench 18 inches wide and 3 feet deep was dug the full size of the house. This trench was covered with loose boards, sufficient to sustain banking but open enough to permit drainage, thereby ensuring against moisture in the ground floor.

It may here be stated parenthetically, that several previous years' experience in poultry raising had taught me that success in any particular branch of the business was in direct proportion to attention to detail; therefore, in an attempt to realize the unusual, no important feature to that end should be omitted. To proceed, an excavation three feet wide, four feet long and three feet deep was made, bisecting the house from the east side to the west side. The excavation was lined with concrete, into which was installed a hot-water heating system with natural gas as fuel; also illuminant for the house, to be explained later.

Housing Conditions Highly Important

The house was divided into three floors or sections from ground to roof, each subdivided as follows: Ground sections seven feet high, each with a matched board ceiling. Six feet above this a similar matched ceiling was placed, leaving the top divisions, or third floor rooms or pens, three feet high on the south side and eight on the north. It will be recalled that the roof was of glass and it was here that most of the sunning and dusting was done, the floor being covered with dry dust and fine sand. Three inches of cut straw covered the sand and dust on this, the top floor. The second floor was covered with chaff and fine straw, to the depth of one foot. The ground floor was covered with oat straw one foot in depth, which was renewed once a month.

This house was provided with windows on all sides for the first and second stories, but high enough to prevent the fowls looking out from the floor. On the ground floor each sash was 18 inches wide, extending the full length of the pen and placed three feet above the ground. I might say in passing that some fowls become uneasy when they can look out, but are unable to get out. Therefore all windows were placed too high for the hens to see the outside ground from the floor. I had learned through previous experiments made with the same hens that where the windows were close to the ground and there was no chance to get out, a loss of 30 to 50 per cent in egg production would result.

Wooden troughs one foot wide and six inches deep were constructed for each of the four lower pens and placed under the windows. The hot-water heating system consisted of one-inch pipes just above the windows on the ground floor and extending completely around the coop, returning to the heater along the floor of the wooden troughs. The troughs were filled with fine dusting earth

---

Reprinted from Reliable Poultry Journal, February, 1915.

and wood ashes. Four small V-shaped troughs, one for each pen, 6½ feet long, were placed in the partition and pivoted at the ends so as to appear in the aisle for filling and in the pen for feeding as desired. The troughs were suspended just high enough for feeding the daily mash. Two roosts 5½ feet long for each pen were placed 4 feet above the ground, above the feed troughs. The droppings boards were placed six inches below the roosts and nailed tight to the partition. A snuggly fitted board six inches wide in the aisle partition, on hinges, facilitated cleaning the boards from the aisle. Four automatic nests for each pen were placed 18 inches above the roosts and made accessible from the aisle by hinged boards. Narrow doors for entrances at the corners where the aisles met completed the woodwork, except the storehouse 6 by 10 at the entrance, in which the feed was stored. The second and third floors were reached from openings cut in the floors at the ends.

I wish particular emphasis to be laid upon the tight board partitions, cutting off all view from any pen to another, as this arrangement is an essential factor to contentment. Nine hens and one rooster occupied each pen. They were locked in on December 1, and released April 1 for three successive years. No pullets were used. The stock was purchased in the summer, indiscriminately, at the open Buffalo market for killing, only early molters being purchased. Both Mediterranean and Asiatic breeders were used, mixed or otherwise—range and health being the only requirements.

What They Were Fed

From this heterogeneous flock of 100 fowls the selection was made, using the same hens in many instances for two winters. Feed consisted of cracked corn, oats, wheat and barley, each one-fourth. This grain was strewn liberally in the litter on the ground and second floors, after fowls were on the roosts, and then only. Moist, hot mash was fed daily at 10 a. m. The base of this mash was boiled barley and wheat, equal parts, in water which contained cut clover, cabbage and fresh beef bones cut into small pieces. This mixture was prepared daily and kept on the stove constantly and was boiled in a large copper kettle and used after twenty-fours boiling and simmering.

Sufficient ground oats and middlings, half and half, were added to make just a moist mixture. Twice a week ground oyster shell and a tablespoonful of Cayenne pepper were added to the mixture. Beef lungs were kept hanging within reach of the fowls at all times and were replaced with fresh when stale.

Six weeks were allowed for preparation, which was until January 15.

In order to give the fowls natural climatic conditions the automatic thermostats were set as follows: Temperature of dusting boxes, 75 degrees; ground-floor pen, 58 degrees; second floor, 65 degrees; third floor, 70 to 95 degrees, depending on sunlight for highest mark.

Polonged Their Day Artificially

The proper length of daylight was provided for by the installation of four 100-candlepower Argand gas burners suspended from the first ceiling one foot from the outside edge and five feet from the ground. Each burner was provided with a large reflector, throwing the light directly downward. These lights were controlled by an automatic time adjuster and were turned on at 3:30 a. m. and off again at 7:30. They were turned on again at 5 p. m. and off at 8 p. m. for the night. The results of these tests were published in The Clyde Times, Clyde, N. Y., in February, 1889, the complete daily record having since been mislaid or lost.

It is sufficient to say that ten eggs per hen per week was the average for three months, and very nearly so for the entire period, gradually falling off for the next two months and then ceasing altogether.

Any poultryman who will adhere to the general principles herein specified in every small detail can realize like results, and with well-bred, vigorous, 1914-1915 laying stock no doubt could obtain twelve eggs per week per hen. The remarkable activity and vigor of the hens was most surprising. The instant the clock mechanism turned up the lights all would bound from the roost, and in two minutes

A QUARTETTE OF HEAVY LAYERS AT THE NEBRASKA LAYING CONTEST
The records of these birds are as follows: White Leghorn, upper left-hand corner, 249; Barred Rock, 231; Rhode Island Red in lower left-hand corner, 227; Barred Rock, 230.

they were digging for food in the litter like so many machines. Between four and five o'clock one-half the hens, as a rule, had laid, and the others in the next two hours. In the same afternoon between four and six the majority would lay again, and so on, week in and week out.

Whenever the shells of the eggs from any hen were not firm and smooth, lime water was substituted for the regular drinking water until all eggs had firm shells. Solid cabbage heads were kept within reach of the hens when loss in shell or loss of density was noticed. At this time the percentage of wheat to barley also was increased.

CHAPTER VI

Opinions of Scientists on Inheritance of Fecundity

In This Chapter Are Presented, in Brief Form, Opinions of Leading Investigators on the Laws of Inheritance in Egg Production, As Set Forth by Them in Official Bulletins and Scientific Papers—Differing More or Less in Regard to Details They Nevertheless Confirm, in General, the Methods of Successful, Practical Breeders, and Explain Results Secured by Them

THE breeding of fowls, specifically breeding for increased egg production, has for a number of years engaged the attention of agricultural experiment station workers, and an extensive library on the subject is gradually accumulating. In this chapter the attempt has been made to reproduce such extracts from recent statements by some of these investigators as will serve to give the reader a general idea of the work done by them and of the results secured, though in the space at our disposal it is only possible to do this in a highly condensed form. In addition to the material here presented the reader is referred to other chapters of the book for special articles relating to similar work done at the Experiment Stations of New York, North Carolina, Wisconsin, Kansas and Washington.

The first experiment station to undertake systematic breeding of trap-nested fowls was that of Maine, where Professor Gowell began work along this line in 1898. In regard to the negative results secured by Professor Gowell poultry keepers generally are well informed. With the introduction of what is now known as the "progeny testing" method in 1907, under the direction of Dr. Raymond Pearl, a different result was secured. The results of the work done from 1907 to 1914 have been summed up by Dr. Pearl in Bulletin 231 of the Maine Station, from which the following is quoted:

IMPROVEMENT OF EGG PRODUCTION BY METHODS PRACTICED AT MAINE STATION

Extracts from Maine Experiment Station Bulletin 231, Giving Dr. Raymond Pearl's Views in Regard to Laws of Inheritance of Egg Production and Methods of Utilizing Them in Practical Poultry Breeding

THE following are simple statements of the actual results obtained in trap nesting Barred Plymouth Rocks and Cornish Indian Games, and all possible sorts of crosses between these breeds, over a period, collectively, of nearly fifteen years. The total number of birds involved in these trap-nesting operations has been large, aggregating, all told, between five and six thousand individuals. Out of these records the following facts clearly appear:

1. The record of egg production of a hen, taken by itself alone, gives no definite, reliable indication from which the probable egg production of her daughters may be predicted. Furthermore, mass selection on the basis of egg-laying records of females alone, even though long continued and stringent in character, failed completely to produce any steady change in type in the direction of selection.

2. Differences in egg-producing ability are, in spite of the above results, certainly inherited. There are two lines of evidence showing that this is the case. The first is that derived from the general observation that there are widely distinct and permanent (under ordinary breeding) differences in respect to egg-laying ability between different races, strains and breeds of fowls. In the second place, a study of pedigree records of poultry at once discovers blood lines in each of which a definite particular degree of egg-producing ability constantly reappears generation after generation, the "line" thus "breeding true" in this particular. With all birds kept under the same general environmental conditions such a result can only mean that the character is in some manner inherited.

3. The number of visible oocytes on the ovary bears no definite or constant relation to the actually realized egg production.

4. This can only mean that observed differences (variations) in actual egg production depend upon differences in the complex physiological mechanism concerned with the development of oocytes, and the separation of them from the ovary and the body (laying).

For reasons which cannot be gone into fully here on account of lack of space, attention has been focused during the later phases of the study, on winter egg production.

5. It is found to be the case that birds fall into three rather well-defined classes in respect to winter egg production. These include (a) birds with high winter records, (b) birds with low winter records, and (c) birds which do not lay at all in the winter period. The division point between a and b for the Barred Plymouth Rock used in these experiments falls at a production of about thirty eggs.

The next step is to inquire for each of these classes separately how egg-producing ability is inherited within the class. We may first deal with high production.

6. High productiveness may be inherited by daughters from their sire, independent of the dam. This is proved by a mass of detailed evidence presented in the complete paper. This evidence consists of the results of mating after mating, in which the same proportions of daughters of high-laying ability are produced by the same sire, whether he is mated with dams which are poor layers or with dams which are high layers.

7. High-laying ability is not directly inherited by daughters from their dam. This is proved by a number of distinct and independent lines of evidence, of which the most important are: (a) that continued selection of

TWO GENERATIONS OF LAYERS WITHIN TWELVE MONTHS
White Orpington pullets illustrated above were hatched quite early in the season and as soon as they came into laying were mated and from this pen were raised pullets, two of which are shown on the next page, and which came into laying at 17 weeks of age, the entire time from the date when the first lot of pullets was hatched until the second lot came into laying being less than twelve months.

high-producing dams does not alone alter in any way the mean egg production of the daughters. If an alteration does appear in any case following such selection, further analysis shows that some additional element other than the dam's egg record came into account in making the selections of breeders; (b) the proportion of high-producing daughters is the same whether the dam is of high or of low fecundity, provided both are mated to the same male; (c) the daughters of a high-producing dam may be either high layers or poor layers, depending upon their sire; (d) the proportion of daughters which are medium or poor layers is the same whether the dam is a high or a poor producer, provided both are mated to the same male.

8. Mediocre or poor-laying ability may be inherited by the daughters from either sire or dam, or both.

A Plan for the Practical Breeder

It may be helpful to draw out from these results some general principles in breeding for egg production, which every poultryman can apply. What then are the basic elements in a well-directed effort towards the improvement of poultry in egg production by breeding? I should put them in this way.

1. Selection of all breeding birds first on the basis of constitutional vigor and vitality, making the judgment of this so far objective as possible. In particular, the scales should be called on to furnish evidence. (a) There ought to exist, for all standard breeds of fowls, normal growth curves, from which could be read off the standard weight which should be attained by a sound, vigorous bird, not specially fed for fattening, at each particular age, from hatching to the adult condition. These curves we shall sometime have. (b) Let all deaths in shell, and chick mortality, be charged against the dam, and only those females used as breeders a second time which show a high record of performance in respect to the vitality of their chicks, whether in the egg or out of it. This constitutes one of the most valuable measures of constitutional vigor and vitality which we have. If for no other reason than to measure this breeding performance, a portion of the breeding females each year should be pullets. In this way one can in time build up an elite stock with reference to hatching quality of eggs and viability of chicks. (c) Let no bird be used as a breeder which is known ever to have been ill, to however slight a degree. In order to know something about this, why not put an extra leg band on every bird, chick, or adult, when it shows the first sign of indisposition? This then becomes a permanent brand, which marks this individual as one which failed, to a greater or less degree, to stand up under its environmental measures of constitutional vigor. (d) Many of the bodily stigmata by which the poultryman, during the last few years, has been taught to recognize constitutional vigor, or its absence, have, in my experience, little if any real significance. Longevity is a real and valuable objective test of vigor and vitality, but it is of only limited practical usefulness, because of the increasing difficulty with advancing age of breeding successfully on any large scale from old birds of the American and other heavy types.

2. The use as breeders of such females only as have shown themselves by trap-nest records to be high producers, since it is only from such females that there can be any hope of getting males capable of transmitting high-laying qualities.

3. The use as breeders of such males only as are known to be the sons of high-producing dams, since only from such males can we expect to get high-producing daughters.

4. The use of a pedigree system, whereby it will be possible at least to tell what individual male bird was the sire of any particular female. This amounts, in ordinary parlance, to a pen pedigree system. Such a system is not difficult to operate. Indeed, many poultrymen, especially fanciers, now make use of pen pedigree records.

5. The making at first of as many different matings as possible. This means the use of as many different male birds as possible, which will further imply small matings with only comparatively few females to a single male.

6. Continued, though not too narrow, inbreeding (or line breeding) of those lines in which the trap-nest records show a preponderant number of daughters to be high producers. One should not discard all but the single best line, but should keep a half dozen at least of the

WHITE ORPINGTON PULLETS LAYING AT 17 WEEKS
These pullets were hatched from eggs laid by birds illustrated on preceding page. They came into laying at 17 weeks of age —less than 12 months from the time their dams were hatched.

lines which throw the highest proportions of high layers, breeding each line within itself.

Items 4, 5 and 6 imply the carrying over of a considerable number of cockerels until some judgment has been formed of the worth of their lines, through the performance at the trap nest of their sisters.

Item 6 assumes, as an absolutely necessary prerequisite, that item 1 will be faithfully and unfailingly observed.

The whole system of breeding here outlined is an application, in the simplest form possible, of two principles, one general and the other special to the present case. The first is the general principle of the progeny test in breeding for performance. This is the principle which has led the plant breeder to such notable triumphs during the last fifteen years. In my judgment no system of breeding for performance in animals not fundamentally based upon it will ever achieve any permanent success. The second principle is the recognition of the significance of the male in breeding for egg production.

INHERITANCE OF FECUNDITY IN FOWLS

Professor James Dryden of Oregon Experiment Station Has Been Exceptionally Successful in Breeding for Increased Egg Production—This Article Consists of Extracts from Oregon Bulletin 180, Which Contains the Latest Published Data in Regard to Professor Dryden's Work, and Outlines Their Bearing on Methods of Practical Poultry Breeding

(NOTE:—Professor James Dryden is one of the foremost investigators along the line of breeding for egg production and his success in developing high-producing strains of Barred Rocks and White Leghorns and in producing the new heavy-laying breed "Oregons" has been extraordinary. The experimental data accumulated by him, covering the period 1908-1918, have recently been published in Oregon Station Bulletin 180, entitled "The Egg-Laying Characteristics of the Hen," from which the following extracts are taken.)

Summary of Experiments with Barred Rocks

A SUMMARY is here given of the Barred Plymouth Rocks, showing egg records for first twelve months of laying of the pullets, their dams' dams and sires' dams, where known for the different years. This gives the record of 517 pullets in their first year, and the average for all of the years, for eight generations, including the foundation stock, together with the records of ancestry. Beginning with the foundation stock in 1908-09, and ending with the eighth generation, there is an increase in the production per hen of 128.49 eggs, or a percentage increase of 149.16. If we make the starting point 1909-10 in view of the fact that the pullets of the previous year were selected from different breeders and were not reared at the Station as all subsequent pullets were, we find the increase to be 93.95 eggs, or 77.78 per cent.

Summary of First Year's Production of Barred Plymouth Rocks (in average per hen)

Yard	Year	No. of birds	Pullets	Unidentified eggs added	Best two months	Dams	Dams' dams	Sires' dams
4 and 5	1908-09	92	82.67	86.14	31.29			
6	1909-10	28	117.64	120.68	33.71			
15	1910-11	42	161.78	164.28	38.48			
6	1912-13	38	174.13	178.16	44.74	210.94		194.79
7	1912-13	37	177.43	180.97	45.22	181.37	202.44	185.96
8	1912-13	33	171.69	174.18	44.30	196.76		183.85
R	1913-14	52	178.75	185.00	46.88	193.86	215.05	220.52
C	1913-14	51	169.41	175.59	46.57	183.32	205.75	224.57
L	1914-15	83	176.36	185.78	45.27	181.33	208.71	216.19
P	1914-15	38	184.06	189.01	46.69	218.84	214.81	231.68
18	1916-17	11	194.45	201.90	51.18	247.36	197.44	223.13
E	1917-18	62	202.40	214.63	47.19	200.23	229.47	244.58
Average		517	158.57	164.09	42.25	198.11	212.45	217.29

Summary of Experience with White Leghorns

The total number of hens was 386, see table herewith. The average of the foundation flock was 106.88 eggs a hen. In the last year yards F and G, 116 hens, averaged 212.39 a hen, which is a percentage increase of 98.72.

In the main, the results with the Leghorns agree with those of the Plymouth Rocks. In the first generation of pedigreed stock there was a large increase. In the second year of pedigreed stock there was also a considerable increase. After that the increase was not particularly noteworthy. The variation in the increase in different years is not so marked. This agrees with the results secured with Barred Plymouth Rocks and White Leghorns.

On the whole the flock increases are much greater for the Oregons than for the Leghorns or Plymouth Rocks, if we calculate from the foundation stock. The first cross shows a considerable increase over both parent breeds. In the absence of selection, the only explanation for this increase is that there was an increase in the vigor, which was responsible for the increased production. The largest increase, however, came in subsequent generations of pedigreed stock, where the record shows high production of the ancestors. In this latter respect also, the results agree with those of the pure Leghorns and Barred Plymouth Rocks, the immediate parents apparently exercising a greater influence on the production of the flock than other ancestors. If we take the average of all the pullet flock and the average of the ancestors, it is clear that high production is inherited. But if certain individuals be selected from the table, it could be shown that there was both transmission and nontransmission of laying qualities. Certain individuals apparently inherited high production from high producers, and certain other individuals apparently inherited low production. But when the average of all the flock is taken, in connection with production of former generations of the strain, there can be no question that the factor of high production was inherited.

If there was lack of vigor in the parent stock and the first cross restored the vigor, then the production of the first generation of crosses, which was 135.49 eggs, should represent their actual inherited egg-laying capacity, and the increase from the first generation to the seventh of the crosses or "Oregons" represents the effect of selection of breeding stock, based on high production records. In that case, crossing alone did not produce high-record layers, the greatest increase coming from subsequent selection.

While crossing evidently increased the vigor and to that extent increased production, it is not assumed that the production of the first crosses was greater than may be secured from purebreds of good vigor; but in subsequent generations when trap-nest selection was practiced, a much greater increase was secured. It would appear that the opinion often expressed, that crossing usually gives good results in the first generation, but that in subsequent crosses there is reversion or a decrease, is not necessarily true. The increase in production of the pure White Leghorns and Barred Plymouth Rocks was secured not by maintaining merely breed purity, but by selection within the breed, or the use of breeding stock with high individual production records. Nor did we find that crossing the crosses brought about deterioration, but by similar selection of breeding stock we secured, on the contrary, a higher production than from the first generation of crosses. Inbreeding was avoided, because of the danger of defeating the purpose we had in view in crossing—that of increasing the vigor.

Summary of First Year's Production of White Leghorns (in average per hen)

Yard	Year	No. of birds	Pullets	Unidentified eggs added	Best two months	Dams	Dams' dams	Sires' dams
4 and 5	1908-09	50	106.14	106.88	32.54			
9	1909-10	21	104.67	104.67	30.71			
1	1910-11	10	164.60	164.60	41.30			
9	1912-13	14	207.85	208.93	47.64	223.85		229.00
3	1914-15	16	227.00	230.12	51.19	242.19	222.33	233.13
O	1914-15	42	212.86	216.72	49.24	236.97	221.68	237.25
I	1916-17	49	187.55	188.96	47.61	209.74	247.27	251.71
Q	1916-17	57	169.84	172.65	44.77	211.75	241.56	258.74
15	1916-17	11	215.00	225.27	51.63	227.64	245.09	242.18
F	1917-18	60	218.21	222.22	47.78	235.42	239.78	302.00
G	1917-18	56	199.63	201.85	48.57	234.50	247.95	236.81
Average		386	181.71	184.25	44.76	224.69	238.52	256.33

considerable, and does not conform to any statistical conception of breeding.

Summary of Experience with "Oregons"

There was a very marked increase from the unselected stock to the first generation of pedigreed stock, and the increase in subsequent generations is considerable, though

Summary of First Year's Production of Oregons (in average per hen)

Yard	Year	No. of birds	Pullets	Unidentified eggs added	Best two months	Dams	Dams' dams	Sires' dams
7	1909-10	33	129.94	131.06	32.09			
8	1909-10	30	137.96	140.36	33.80			
1	1910-11	37	147.16	147.16	36.43			
9	1912-13	20	219.45	220.80	48.40	207.45		224.42
E	1913-14	49	216.44	220.78	51.04	186.45	200.00	204.00
4	1914-15	10	246.00	250.20	52.40	308.00	201.00	228.00
5	1914-15	10	197.00	201.70	50.40	291.00	200.00	246.00
6	1914-15	14	211.14	217.27	50.86	211.43	283.50	286.51
J	1914-15	58	217.24	219.22	49.19	215.24	217.40	214.81
10	1915-16	13	234.39	237.77	50.31	274.00	222.50	297.46
7	1917-18	12	248.33	246.91	50.41	250.92	261.14	287.75
H	1917-18	63	224.38	228.57	52.80	287.79	269.44	269.62
Average		349	197.58	201.05	45.95	227.51	226.15	240.92

STUDIES IN EGG PRODUCTION AT MASSACHUSETTS EXPERIMENT STATION

Some Careful Experiments in Breeding Have Been Conducted at This Institution—Herewith Are Summarized Some of the Results Secured, Particularly with Reference to Effects of Inbreeding and Influence of Male on Fecundity

(NOTE:—For several years Dr. H. D. Goodale of the Massachusetts Experiment Station has been conducting investigations at that institution in inheritance of fecundity. Most of what has been published has appeared in scientific journals and is highly technical and not, therefore, adapted to reproduction here. However, the following extracts from Mass. Bul. 191, by Dr. Goodale, present in popular form some of the results of investigations in breeding at this institution.)

Inbreeding

THE poultryman often is in a quandary regarding inbreeding. On the one hand it is advocated, and on the other as strongly condemned. What, then, are the facts?

Inbreeding may be defined as the mating of relatives, and just as there are degrees of relationship so there are degrees of inbreeding. Line breeding involves inbreeding, so designed, however, as to keep its amount at a minimum.

In the work at this station close matings of various sorts have been made as well as unrelated matings. The results afford a practical answer to the question, shall I inbreed? The answer is found in a paraphrase of an old saying, which is applicable to all breeding: "Handsome is that handsome breeds"—that is, inbreeding is to be judged by its results. This is a special application of the well-known progeny test. Now experience shows that the results of some inbred matings are very good, while others are poor. But, contrariwise, sometimes the results of matings between unrelated birds are poor, while others are good. Nevertheless, matings between unrelated birds are almost universally approved, while inbreeding is often condemned.

In this bulletin no attempt will be made to answer most of the questions that arise concerning inbreeding, but evidence that inbreeding may be highly advantageous will be presented.

The Evidence

1. Male No. 8097 produced 10 daughters, by his sister, that averaged 155.7 eggs each. The father of this pair also came from a brother-sister mating. On the other hand, male No. 8097 by a half first cousin produced 14 pullets that averaged 156.7 eggs each. There is no relation here between the degree of kinship (or inbreeding) and egg production.

2. The offspring of male No. 8147 by 4 females furnish interesting comparisons. Eleven daughters by female No. 9420, which laid 221 eggs, averaged 190.1 eggs each. Female No. 9420 is distantly related to male No. 8147 five generations back through a single bird. By female No. 8652, laying 196 eggs, with a much inbred and tangled line of descent and closely related to himself, there were 9 daughters, with an average of 181.5 eggs. Moreover, the highest producing individual in the two families, viz., B2088, with a record of 237 eggs, was a daughter of female No. 8652. On the other hand, female No. 8418, laying 139 eggs, also closely related to male No. 8147, produced 11 pullets sired by him that averaged only 156 eggs each. Finally, there is the mating of female No. 8185 with male No. 8147. This female is related to male No. 8147 in exactly the same degree as female No. 9420 (though otherwise unrelated to No. 9420) and through the same great, great-grandparent. Unfortunately her 11 daughters were not all trap nested through the year, as they would have been if there had been any inkling of their importance. We are obliged to fall back on their winter records. Their average for the winter was 42.9 eggs each (which gives an estimated annual average of 142.9 eggs), while the daughters of female No. 9420 averaged 75.3 each; those of female No. 8652 averaged 62.4 each, and of female No. 8418, 50.5 each. These averages refer to the number of eggs laid before March 1 of the pullet year. Female No. 8185 was a good winter layer herself with a record of 85 eggs and a 365-day record of 185 eggs, but came from a mediocre family.

It may be of interest to note that the difference in average winter production of these groups of half sisters is closely related to average date of first egg. Thus, the average date of first egg of the daughters of female, No. 9420 was October 26, of female No. 8652, November 5, of female No. 8418, November 17, and of female No. 8185, December 12.

Thus, of 4 females mated with a single male, 2 were inbred and 2 were unrelated to the male. The offspring of one inbred female were good layers, while those of the other were relatively poor. The offspring of one unrelated female were good layers, and of the others relatively poor. The results of inbreeding, therefore, must be judged by the quality of the offspring. If it is good, utilize those

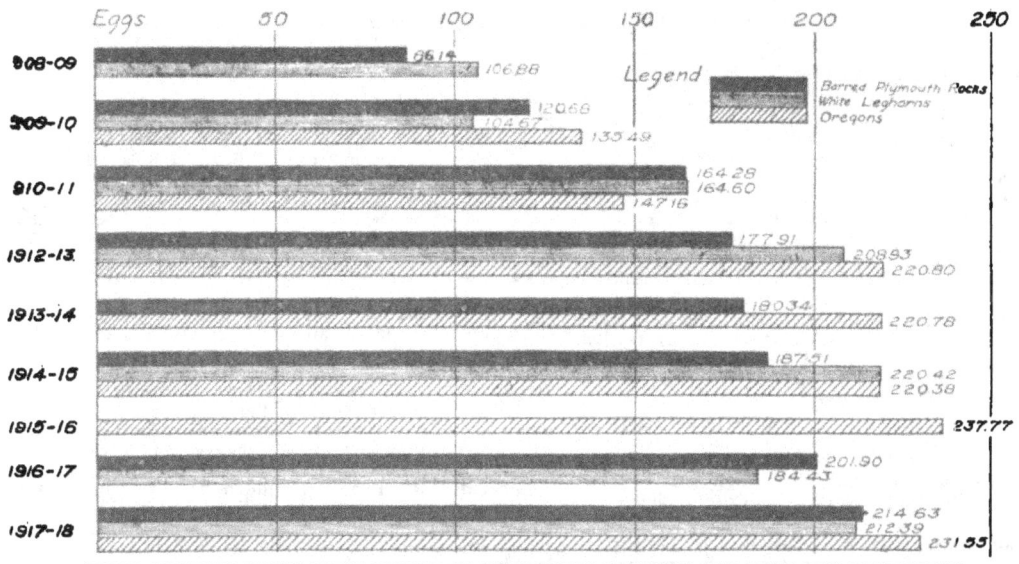

CHART SHOWING CUMULATIVE EFFECT OF SYSTEMATIC BREEDING FOR HIGH EGG PRODUCTION

This chart shows in graphic manner the increase in egg production at the Oregon Experiment Station brought about through the persistent breeding from high-production females mated to males from high-producing birds. Professor Dryden's new breed "Oregons" was not represented in the first year but regularly thereafter. Reproduced from Bulletin 180 of Oregon Experiment Station.

particular inbred matings. If not good, try other inbred matings.

A few instances may be added to show that the results of matings between unrelated birds may be very inferior to matings between closely related birds such as those just described, unless special attention is paid to egg production. Thus, female No. 6982 had 10 daughters, by an unrelated male, that averaged 101 eggs each, exactly their mother's record. The same male by female No. 5832, also unrelated, and laying 160 eggs, produced 7

A PEN OF HIGH-RECORD OREGONS AT INTERNATIONAL LAYING CONTEST

This breed developed by Professor Dryden has produced many high-record birds. The record of the above pen was 2,352, an average of 235.2 each.

daughters that averaged 151.7 eggs each. Another male, by an unrelated hen that laid 234 eggs, had 9 daughters that averaged 139.1 eggs each.

In these experiments, good, strong, healthy stock has been used. The families from which the males were to be chosen were selected on the basis of their sister's performance, and the strongest, most virile male in each family selected for breeding.

There is one thing more to be said on the subject. In our experience the very best results have come from outmatings, while the very poorest have come from close matings. It is clear, then, that very great care must be used when inbreeding, lest disaster overtake the breeder unawares. Very careful and accurate pedigrees and other records must be kept. Further, the provisional conclusion appears justified that the very best results are most likely to be obtained by crossing two distinct lines, each of which is inbred and which is doing well. Very likely the best way to renew the commercial egg flock is through the crossing of strong, high-producing, inbred lines, which will, of course, be maintained intact by inbreeding, and making the cross anew each season.

Is the Influence of the Male or of the Female the More Important

The view that high fecundity does not descend from mother to daughter but does descend from mother to son, or from father to both sons and daughters is now generally accepted. This, then, leads to the belief that the use of the sons of high layers insures high production in the progeny sired by such sons. The male is regarded as all important, the female of importance only as a producer of good males. This situation has arisen apparently from the attempt to describe certain modes of inheritance in everyday language. The scientific foundation, i. e., sex-linked inheritance of fecundity, on which the view mentioned is based, does not warrant the popular interpretation which it has received. However this may be, evidence is now available which indicates that high fecundity is not sex linked in some breeds, at any rate.

In speaking of fecundity the use of the terms high and low are not very precise for they are relative; we may, however, use them with this understanding of their limitation. In this bulletin winter production only is considered, because it is a fairly good index of a hen's inborn capacity to lay.

The important question to be answered is: Is it possible for high egg production to descend from mother to daughter? An experiment was made in which a male from a low line was mated with several high producers belonging to a high line and at the same time to several low producers. Of course, careful individual pedigrees were kept. The offspring of the high producers averaged 49.2 winter eggs against an average of the mothers and their sisters of 52.5 eggs. Nearly all were high producers. On the other hand, the offspring of the poor layers averaged only 11.6 winter eggs. In this experiment high production clearly descended from mothers to daughters.

In another experiment a male was used that came from a high-producing mother, but on the father's side production was poor. Some of his mates were good producers, some were poor. A few daughters were good layers, but most of them, regardless of whether their mother was a high or low producer, were mediocre to poor. In this experiment the influence of the male was more pronounced.

In other experiments males derived from high lines have been bred to low producers. For example, male B137, a high-line male, was bred to two low birds of low lines. The average winter production of the daughters was high, viz., 54.5 eggs. On the other hand, a case occurred where the production of the offspring of a male belonging to a high line, bred to a high-producing female, viz., No. 8185, of a mediocre line, was relatively poor compared with that obtained from the offspring of females belonging to high lines.

In still another experiment a male belonging to a low line was mated with a female belonging to another low line. Most of the offspring were high producers.

These experiments show that we are dealing with a situation that is complicated in many ways. It appears, however, to be perfectly clear that both male and female play a part in determining the egg production of their daughters. Whether one is more important than the other depends upon the particular individuals that are mated. In breeding for high production the influence of either must be judged by the production of the offspring. The aim of the breeder should be to produce a line that will give high average production and that will reproduce itself generation after generation. To this end the contribution of both father and mother must be made. Any male or any female, or a particular combination of a certain male with one or more females that gives high production consistently may well be used as breeders as long as they live, or till something better has been secured.

A STATISTICAL STUDY OF EGG PRODUCTION IN WYANDOTTES AT STORRS CONTEST, 1911-1921

A Summary of Contents of a Recent Storrs, Conn., Bulletin by Dr. L. C. Dunn, Geneticist

(NOTE:—Dr. Dunn of the Storrs Experiment Station has recently undertaken a complete study of the records of the Storrs Contests from January, 1911, to January, 1920, inclusive. These studies are to be published in a series of four bulletins, devoted respectively to Wyandottes, Plymouth Rocks, Rhode Island Reds and Leghorns. Through the kindness of Dr. Dunn, the authors have been permitted to examine the manuscript of the first bulletin, devoted to Wyandottes, and to make extracts from it for use herein. Of particular interest to many will be the discussion of laying periods and the influence of increased winter production upon the egg yield at other seasons. This bulletin divides the laying year into four periods: winter, spring, summer and autumn and discusses the performance of Wyandottes in these respective periods.

WINTER Period. Noting at present only the winter months, we cannot fail to remark the very great differences between the classes in the percentage of the total yearly production which occurs at this period. The rate of production is likewise different, being steadily upward in the low producers while in the high producers the large increase in December over November production is followed by a very slow increase in January and February. The increase of February over January is in fact probably not significant. March shows a sudden increase in both classes marking the opening of the spring period. The uniform rate of production of the high producers is in marked contrast with the rising rate of the low producers. No let-up in this rate occurs to indicate the looked-for pause between the winter and spring production. If we follow the percentage production in the winter months down through the various fecundity classes we find that the changes from the "low" type of winter production to the "high" type are continuous and gradual. High winter production is not attained by a discontinuous change such as we might expect if it were due to the addition intact of a separate and distinct cycle to the laying year. As far as the evidence from these Wyandottes is concerned, therefore, we must conclude that evidence of a distinct winter cycle is lacking, when all the birds are considered together or when the group is broken up into smaller classes differing in annual fecundity.

Spring Period. The usual spring cycle of production undoubtedly exhibited by these Wyandottes. This cycle is probably present in all breeds of poultry and species of wild birds, since upon fecundity, at this, the mating time, depends the reproduction and hence the survival of the species. Its inception is attended by an increase in mean production of March over February; a decrease in variability and the fact that practically every bird in the flock is laying in March. It is, however, defined better by the general character of egg production during the whole period than by conditions surrounding its beginning or its ending, since we have seen that by all the criteria applied, the change from winter production into spring production is a gradual one. The winter period of fecundity is a time of increasing egg production; the spring is a period of stationary production. It is a climax of which the winter production was a foreshadowing, and the subsequent production an aftermath. In the case of these Wyandottes, this reproductive climax is extraordinarily sustained, since production is maintained at the same level through three months, March, April and May.

Summer. The ending of the spring or normal reproductive cycle is as poorly defined as its beginning. Ordinarily, broodiness ensues after the heavy spring production and a marked reduction in laying activity is noticeable. Conventionally the last of May is taken to mark the close of the reproductive cycle and the opening of the summer period. In the Wyandottes, however, this cessation is delayed and egg production in June is very similar, in mean and variability, to egg production in the spring months. After June, egg production reaches a new stationary level lasting through July, August and September. June may then be considered, on the basis of fecundity, either as a part of the spring or summer cycle. The truth is undoubtedly that egg laying in this and in the summer months generally is not a result of a separate or induced physiological cycle, but merely a continuation of the one chief spring production period. The mechanism of reproduction, having been wound up, is only slowly running down. The chief factor in the running down in mean production after the spring cycle is the onset of broodiness and the cessation of laying on the part of broody birds. This has not had a serious effect on summer production in Wyandottes for a high mean production is maintained and the number of zero producers (in this case usually birds which have stopped laying) is absolutely small—2.6% of the flock in June, 4.1% in July, and 4.5% in August. The slight increase in mean production in August, which has been already noted, probably indicates only the maintenance of high egg production which is the striking feature of summer production, although it may be due to recovery from broodiness on the part of a few affected birds.

Autumn. If the characteristic of autumn production be the entrance of many of the birds into a period of reproductive quiescence accompanied by a molt, then in the Wyandottes this period is represented by but one month, October. September production differs only slightly from August in the mean, in variability and in numbers of birds which have ceased laying. September belongs more properly then with the summer months in that egg production is maintained and nearly all the birds are laying. The change from September to October is, however, sharp.

Mean egg production falls from 14.06 to 10.07; the coefficient of variability rises sharply from 50.10 to 75.09 and the percentage of the flock not laying rises from 6.7% to 20.7%. The onset of molting in so far as it can be gauged by egg production is, therefore, relatively late in Wyandottes. They are good summer and fall producers and their leading position among the breeds submitted to the contest probably owes much to their superiority in these periods.

The percentage of winter eggs increases steadily from 1911 to 1919, with only one pause in the increase in 1916. The absolute amount of this advance is considerable, a gross change in the nine years of 8.37%. The most remarkable aspect of this advance is that it is so persistent, seemingly unaffected by the low general production of 1913 and 1918 or by the long, hard winter of 1919-20. It signifies undoubtedly, that the efforts of the competing breeders to increase winter egg production in their fowls are meeting with some success.

Summary

The main facts established from a study of 903 Wyandotte records are:

1—Average egg production for the pullet year, 159.3 eggs.
2—Average winter egg production (Nov. 1 to Feb. 28), 35.8 eggs.
3—Average spring egg production (Mar. 1 to May 31), 53.8 eggs.
4—Average summer egg production (June 1 to Aug. 31), 45.8 eggs.

TYPE OF POULTRY HOUSE USED BY PROFESSOR DRYDEN FOR BREEDING FLOCKS
Many of Professor Dryden's breeding experiments were made in small portable houses like the above. Portable fences also are provided and the houses moved occasionally to new ground, thus to avoid any danger from soil contamination.

5—Average autumn egg production (Sept. 1 to Oct. 31), 24.1 eggs.
6—Highest individual year's record, 308 eggs.
7—Lowest individual year's record, 0 eggs.
8—Average annual egg production has increased by about 7 eggs from 1911 to 1920.
9—The percentage of annual production occurring in the winter months has increased on the average by about 8 per cent.
10—The relative intensities of spring, summer and autumn production have declined by a similar amount.
11—The percentage of high producers (birds laying 210 eggs and over) has increased slightly in the nine years.
12—The percentage of low producers (birds laying 104 eggs and under) has decreased slightly in the nine years.
13—The improvement of the Wyandotte breed as egg producers has been more rapid in England and the British Colonies and Dominions than in the United States.
14—The average egg production of Wyandottes in the nine contests is higher than that of any other breed except Leghorns, which are at present tied with the Wyandottes for first place.

www.ingramcontent.com/pod-product-compliance
Lightning Source LLC
Chambersburg PA
CBHW082328220526
45470CB00008B/2437